Semiconductor Silicon Crystal Technology

Leave the beaten track occasionally and dive into the woods. You will be certain to find something that you have never seen before.

Alexander Graham Bell

Semiconductor Silicon Crystal Technology

Fumio Shimura

Department of Materials Science and Engineering
North Carolina State University
Raleigh, North Carolina

Academic Press, Inc.
Harcourt Brace Jovanovich, Publishers
San Diego   New York   Berkeley   Boston
London   Sydney   Tokyo   Toronto

ACADEMIC PRESS, INC.
San Diego, Caliifornia 92101

United Kingdom Edition published by
ACADEMIC PRESS LIMITED
24-28 Oval Road, London NW1 7DX

Library of Congress Cataloging-in-Publication Data

Shimura, Fumio.
  Semiconductor silicon crystal technology.

  Includes bibliographical references and index.
  1. Semiconductors.   2. Silicon crystals.   I. Title.
TK7871.85.S523   1988   621.3815′2   88-6279
ISBN 0-12-640045-8   (alk. paper)

PRINTED IN THE UNITED STATES OF AMERICA
88 89 90 91   9 8 7 6 5 4 3 2 1

Contents

6. Crystal Characterization

215

7. Grown-In and Process-Induced Defects

279

8. Silicon Wafer Criteria for VLSI/ULSI Technology

378

Preface

Silicon, which has been and will be the dominant material in the semi-conductor industry, will carry us into the ultra-large-scale integration (ULSI) era. As the technology of integrated circuits (ICs) approaches ULSI, the performance of ICs is proving to be more sensitive to the characteristics of the starting material. My experience in working on semiconductor technology at an IC device manufacturer, at an electronic materials manufacturer, and at a university has convinced me that the understanding of basic science, silicon materials, IC device fabrication processes, and their interactions is indispensable for improving both the materials and processes, and ultimately for achieving high performance of ICs. Moreover, I strongly believe that engineers who have electronic device backgrounds should understand materials science more and those who have materials science backgrounds should understand device characteristics more. Although silicon is unarguably the most important material in the electronics and information era, no single source has covered the entire silicon crystal technology, from raw materials to silicon crystal modifications during the IC fabrication processes.

Excellent textbooks on specialized subjects such as semiconductor physics, crystal defects, and VLSI fabrication process engineering are available. There is no question that these textbooks contribute greatly to the understanding of the phenomena related to IC technology. However, the treatment in these books is often too narrow, too deep, or too general. On the other hand, several edited treatises related to silicon technology in which a group of authors each contributes a small portion of the whole are also available. These books are convenient for studying a specific subject in brief. Such edited books, however, typically suffer from a lack of unity and consistency of contents from chapter to chapter.

Semiconductor Silicon Crystal Technology will provide professionals involved in semiconductor technology and graduate students who study the above subject with a single comprehensive source of up-to-date knowledge of semiconductor silicon technology describing both theoretical and practical aspects. Although this book emphasizes silicon technology, there is no doubt that studying silicon, which has been extensively investigated and is highly established, provides a hint for the technology based on compound semiconductors.

At the end of each chapter, extensive references related to the subjects of interest are listed in order to make this book a useful guide that leads the reader to more detail in the fields of semiconductor technology.

In the course of the research and writing of this book, I have been encouraged, stimulated, and advised by many people. It would indeed be difficult to single out all of them for acknowledgment. I know, however, that my primary thanks must go to Dr. M. Uenohara of NEC Corporation and Dr. P. A. Tierney of Monsanto Electronic Materials Company, who gave me the opportunity to engage in research on semiconductor technology in Japan and the United States respectively. The various cultural differences in the two countries result in different ways of thinking and working.

Next, I must acknowledge Dr. H. R. Huff for his encouragement and stimulating discussions. I am also indebted to Dr. R. S. Hockett, Dr. J. P. DeLuca, Dr. R. A. Craven, Mr. C. J. Heinink, and Mr. M. Stinson for critical reading of the manuscript and valuable comments on the material covering their areas of specialization, and to Dr. G. A. Rozgonyi for his moral suppport. I am further indebted to Mr. T. Higuchi for his help in editing the index. Finally, I would like to express my sincere thanks to Ms. S. Hambee and Mr. R. Goodyear of the Graphic Service Department of Monsanto Company, who furnished hundreds of technical illustrations and photographs used in this book.

I alone, however, am responsible for the contents as presented here.

FUMIO SHIMURA

Chapter 1

Introduction

Solid-state electronics was launched with the "earth-shrinking" experimental discovery of voltage and power gain in a point-contact transistor by Bardeen and Brattain at Bell Laboratories on December 16, 1947.[1] The date of the invention of the transistor, however, has usually been taken as December 23, 1947, when the point-contact transistor was demonstrated to top executives of Bell Laboratories. The first public announcement of their discovery and the demonstration of their invention were not made until June 30, 1948.[2] The extensive investigation and development in solid-state electronics led to the invention of the "solid circuit," which was eventually judged by the courts to be the first semiconductor integrated circuit (IC), by Kilby at Texas Instruments in 1958.[3] Less than a year later, Noyce at Fairchild brought together the developments of the previous 10 years using the planar process and junction isolation: the prototype of today's ICs.[4] Since the creation of the first IC, the density and complexity of electronic circuits manufactured on a semiconductor chip have been increased from small-scale integration (SSI), to medium-scale integration (MSI), to large-scale integration (LSI), to very-large-scale integration (VLSI), and finally to ultra-large-scale integration (ULSI), which consists of 10^7 or more components per chip. The increasing component density has been achieved by shrinking the feature size, which is smaller than 1 μm for ULSI circuits.

Originally, germanium (Ge) was utilized as a semiconductor material for solid-state electronic devices.[5] However, the narrow bandgap (0.66 eV) limits the operation of germanium-based devices to temperatures of approximately 90°C due to the considerable leakage currents at higher temperatures. The wider bandgap of silicon (1.12 eV), on the other hand, results in electronic devices that are capable of operating up to around 200°C.[6] A more serious problem than the narrow bandgap is that germanium does not readily provide a stable passivation layer on the surface. For example, germanium dioxide (GeO_2) is water-soluble and dissociates at approximately 800°C.[7] Silicon, in contrast to germanium, readily accommodates itself to surface passivation by forming silicon dioxide (SiO_2), which provides a high degree of protection to the underlying devices. The stable SiO_2 layer results in a decisive advantage for silicon over germanium as the starting semiconductor material for electronic device fabrication. This advantage has been used to establish significant basic technologies, including the processes for diffusion doping and defining intricate patterns. In addition, other advantages of silicon from the environmental point of view are that silicon is entirely nontoxic, and that silica (SiO_2), the raw material of silicon, comprises approximately 60 % of the minerals in the earth's crust. This implies that the raw material of silicon can be steadily supplied to the IC industry. Moreover, electronic-grade silicon can be obtained at less than one-tenth the cost of germanium.[7] Consequently, silicon has almost completely replaced germanium in the semiconductor industry.

However, silicon is not an optimum choice in every respect. For example, compound semiconductors such as gallium arsenide (GaAs)[8] are superior to silicon in terms of electron mobility, resulting in devices with reduced parasitics and improved frequency response. The most serious disadvantage of silicon might be that silicon cannot be applied to optoelectronic devices because of its indirect bandgap. Some electronic and physical properties of germanium, silicon, and gallium arsenide are summarized in Table 1.1.[9] However, it is certain that silicon will continue to be the dominant material in the semiconductor industry as a whole. In particular, single-crystalline silicon grown by the Czochralski method will steadily carry us into the ULSI era.

A number of IC circuit chips are simultaneously fabricated on a silicon wafer through so-called *batch processing*. For example, Fig. 1.1 shows 1-Mbit dynamic random-access memory (DRAM) chips fabricated on a 150-mm-diameter Czochralski silicon wafer. In order to attain more IC chips per wafer, silicon wafers of an ever larger diameter have been demanded by IC manufacturers. Figure 1.2 shows the change in the maximum diameter of float-zone and Czochralski silicon wafers used in the silicon industry during the 25 years from 1960 to 1985.[10] The difference in the diameter of Czochralski silicon crystals is strikingly visualized in Fig. 1.3. The increased component density and device complexity, as well as the increased wafer diameter, are driving today's silicon wafers to ever more stringent specifications, since ICs and their fabrication processes are proving more sensitive to starting-material characteristics as IC technology approaches VLSI-ULSI.[11] Therefore, considering the interrelationships among silicon material characteristics, IC fabrication and circuit performance are crucial to the successful fabrication of VLSI/ULSI circuits.

Table 1.1　Physicochemical Properties of Three Principal Semiconductors[a]

Properties	Ge	Si	GaAs
Atoms/cm^3	4.42×10^{22}	5.0×10^{22}	4.42×10^{22}
Atomic weight	72.60	28.09	144.63
Breakdown field (V/cm)	$\sim 10^5$	$\sim 3 \times 10^5$	$\sim 4 \times 10^5$
Crystal structure	Diamond	Diamond	Zincblend
Density (g/cm^3)	5.3267	2.328	5.32
Dielectric constant	16.0	11.9	13.1
Effective density of states in conduction band, N_c (cm^{-3})	1.04×10^{19}	2.8×10^{19}	4.7×10^{17}
Effective density of states in valence band, N_v (cm^{-3})	6.0×10^{18}	1.04×10^{19}	7.0×10^{18}
Effective mass, m*/m$_0$			
Electron	$m_l^* = 1.64$ $m_t^* = 0.082$	$m_l^* = 0.98$ $m_t^* = 0.19$	0.067
Hole	$m_{lh}^* = 0.044$ $m_{hh}^* = 0.28$	$m_{lh}^* = 0.16$ $m_{hh}^* = 0.49$	$m_{lh}^* = 0.082$ $m_{hh}^* = 0.45$
Electron affinity (V)	4.0	4.05	4.07
Energy gap at 300 K (eV)	0.66	1.12	1.424
Intrinsic carrier concentration (cm^{-3})	2.4×10^{13}	1.45×10^{10}	1.79×10^6
Intrinsic Debye length (μm)	0.68	24	2250
Intrinsic resistivity (Ω cm)	47	2.3×10^{10}	1.79×10^6
Lattice constant	5.64613	5.43095	5.6533
Linear thermal expansion coefficient (°C^{-1})	5.8×10^{-6}	2.6×10^{-6}	6.86×10^{-6}
Melting point (°C)	937	1420	1238
Minority carrier lifetime (sec)	10^{-3}	2.5×10^{-3}	$\sim 10^{-8}$
Mobility (drift) (cm^2/V sec)			
Electron	3900	1500	8500
Hole	1900	450	400
Optical-phonon energy (eV)	0.037	0.063	0.035
Specific heat (J/g °C)	0.31	0.7	0.35
Thermal conductivity at 300 K (W/cm °C)	0.6	1.5	0.46

[a] After Sze.[9]

The development of silicon-based VLSI/ULSI technology has been achieved by a broad and interdisciplinary study of fields including physics, chemistry, metallurgy, statistics, and engineering. Future developments should be fueled by both previously obtained scientific knowledge and innovations in various scientific and engineering areas. The major theme in today's worldwide VLSI/ULSI development is to keep the silicon juggernaut rolling, and the minor theme might be to find a viable supplement or

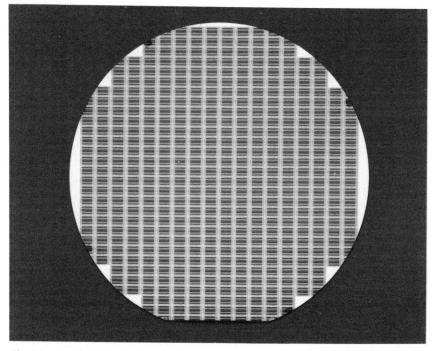

Fig. 1.1. DRAM chips (1 Mbit) fabricated on 150-mm-diameter Czochralski silicon wafer. (Courtesy of NEC Corporation.)

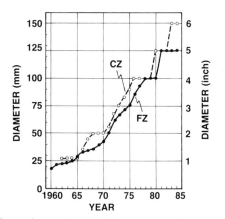

Fig. 1.2. Change in the maximum diameter of float-zone and Czochralski silicon wafers used in the silicon industry from 1960 to 1985. (After Abe.)

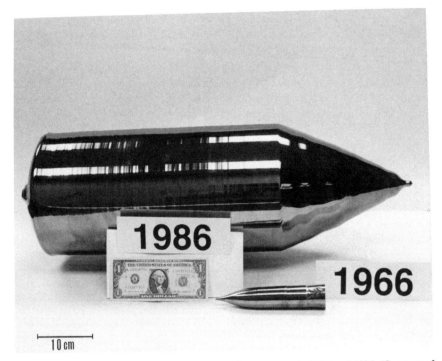

Fig. 1.3. Commercial Czochralski silicon crystal ingots grown in 1966 and 1986. (Courtesy of R. A. Frederick and H-D. Chiou, Monsanto Electronic Materials Company.)

alternative to the silicon-based technology.[12] In any case, understanding silicon material science and engineering is certainly essential.

This book consists of seven chapters following this introductory chapter.

Chapter 2 describes atomic structure and chemical bonds, which are basic to understanding of the electronic and chemical phenomena that occur in materials.

Chapter 3 describes basic crystallography and provides the fundamental aspects of crystalline solids with a regular atomic lattice configuation that leads to various physical characteristics of semiconductors.

Chapter 4 considers basic semiconductor physics necessary to understand the fundamental electronic phenomena that occur in semiconductor devices. Without the considerations based on semiconductor physics, one may not be able to solve the problems generated during device fabrication processes or improve the quality of silicon material in order to ensure high performance of VLSI/ULSI.

Chapter 5 outlines silicon crystal growth and wafer preparation processes from the raw material to polished wafers, and finally to silicon epitaxy. New and currently developed technology, particularly regarding crystal growth, will also be discussed in view of future innovation of silicon materials, which may lead to improved VLSI/ULSI performance.

Chapter 6 describes crystal characterization from the viewpoints of electrical, chemical, and physical characterization of silicon. Various diagnostic techniques that have been powerfully used in the silicon industry are reviewed in conjunction with practical application to the characterization of silicon.

Chapter 7 describes grown-in and process-induced defects in silicon crystals, characterized with the methodology discussed in the preceding chapter. The effects of these defects on electrical properties of silicon and electronic devices are discussed. Consequently, emphasis is placed on gettering techniques that control the generation of detrimental surface defects.

Chapter 8 outlines silicon wafer criteria for VLSI/ULSI technology, which may ensure the high performance of VLSI/ULSI circuits. Finally, it is emphasized that close working relationships will be required between the silicon wafer manufacturer and the IC manufacturer in order to effectively design and fabricate advanced IC products.

It should be noted that the material presented in this book is intended to serve as a foundation for silicon material science; however, some concepts may be obsolete tomorrow, since the field of electronics in general and semiconductor devices in particular has been changing so dramatically and rapidly. It is also true that there are many related phenomena that are by no means fully understood at this point. Therefore, the importance of understanding the fundamentals of materials science, which consists of physics, chemistry, and metallurgy, is emphasized. Finally, readers are strongly urged to study the extensive references listed at the end of each chapter for more detailed information.

References

1. W. Shockley, The path to the conception of the junction transistor. *IEEE Trans. Electron Devices* **ED-23**, 597–620 (1976).
2. G. K. Teal, W. R. Runyan, K. E. Bean, and H. R. Huff, Semiconductor materials. *In* "Materials and Processing" (J. F. Young and R. S. Shane, eds.), 3rd ed., Part A, pp. 219–312. Dekker, New York, 1985.
3. J. S. Kilby, Invention of the integrated circuit. *IEEE Trans. Electron Devices* **ED-23**, 648–654 (1976).
4. R. N. Noyce, Microelectronics. *Sci. Am.* **237**, 63–69 (1977).
5. G. K. Teal, Single crystals of germanium and silicon—Basic to the transistor and integrated circuit. *IEEE Trans. Electron Devices* **ED-23**, 621–639 (1976).
6. A. Bar-Lev, "Semiconductors and Electronic Devices," 2nd ed. Prentice-Hall, Englewood Cliffs, New Jersey, 1984.

7. S. Wolf and R. N. Tauber, "Silicon Processing for the VLSI Era," Vol. 1. Lattice Press, Sunset Beach, California, 1986.

8. H. J. Welker, Discovery and development of III–V compounds. *IEEE Trans. Electron Devices* **ED-23**, 664–674 (1976).

9. S. M. Sze, "Physics of Semiconductor Devices," 2nd ed. Wiley, New York, 1981.

10. T. Abe, Crystal fabrication. *In* "VLSI Electronics Microstructure Science" (N. G. Einspruch and H. R. Huff, eds.), Vol. 12, pp. 3–61. Academic Press, New York, 1985.

11. H. R. Huff, Chemical impurities and structural imperfections in semiconductor silicon. *Solid State Technol.* Feb., pp. 89–95; Apr., pp. 211–222 (1983).

12. J. A. Armstrong, The science of VLSI. *Phys. Today* Oct., pp. 24–25 (1986).

Chapter 2

Atomic Structure and Chemical Bonds

In order to understand the nature and formation of crystal structures, it is essential to have some understanding of atomic structure and atomic bonding mechanisms, which are the fundamentals of material science. In this chapter, some fundamental aspects of this subject that may help in understanding silicon crystal technology, the principal theme of this book, are presented. For a better understanding and creative research in the field of the semiconductor technology, it is strongly recommended that one learn as much as possible about modern atomic physics and chemical bonds with reference books such as Refs. 1–5 found at the end of this chapter.

2.1 Atomic Structure

2.1.1 Rutherford's Nuclear Atom Model

A modern atom theory originates in Rutherford's model proposed in 1911, which states that (1) practically all the mass of the atom is concentrated in a particle of some 10^{-12} cm diameter, that is, the central core or *nucleus*; (2) this nucleus is positively charged, and the number of units of charge, each numerically equal to the negative electron charge, on the nucleus is equal to the *atomic number* of the atom; and (3) *electrons* are diffusely spread through a region of atomic dimensions, 10^{-8} cm, around the nucleus. The number of electrons is equal to the atomic number.

A picture of this model, for C, is given by Fig. 2.1. The atom contains a nucleus and six electrons rotating around the nucleus.

Rutherford's nuclear atom model greatly developed the atomic concept of matter. However, it had to be amended shortly afterward by Niels Bohr in 1913 and in particular by the new *quantum mechanics* formulated by

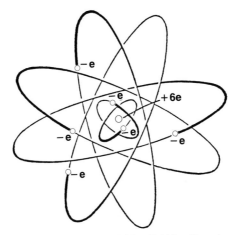

Fig. 2.1. Rutherfords's nuclear atom model for C. (After Humphreys and Beringer.[1])

Schrödinger in 1926. Rutherford's model did not say, for example, why the atom was stable, why it did not collapse due to the Coulomb attraction between the negatively charged electrons and positively charged nucleus. Electrons are charged particles, and a moving charged particle will continually radiate and in turn lose its energy according to classical physics. The orbiting electron should thus be doomed to lose its speed and spiral into the nucleus.

2.1.2 Bohr Theory

To solve the problems remaining in Rutherford's model, Bohr made two basic postulates:

1. The electron, while in a particular orbit, has *quantized* characteristic energy levels $E_1, E_2, \ldots, E_n$, which cannot change while it is in that orbit (Fig. 2.2).

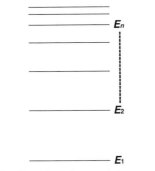

Fig. 2.2. Quantized characteristic energy levels.

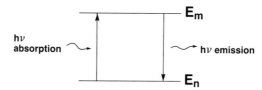

Fig. 2.3. Electromagnetic radiation from transition of an electron.

2. When an electron makes a transition from an orbit of energy E_m to another of E_n, electromagnetic radiation is emitted (Fig. 2.3). The frequency v of the radiation is given by

$$hv = E_m - E_n \qquad (2.1)$$

where h is Planck's constant.

By considering these postulates, Bohr introduced an important concept for the structure of hydrogen atom.

3. A hydrogen atom is a system in which one electron is rotating in circular orbits with the nucleus at the center.

4. The centripetal acceleration of the electron is due to electrostatic attraction by the nucleus, that is, a proton.

Calling the mass of the electron m, its velocity v, its orbit radius r, and its charge e, (3) and (4) are put in the form of the following equation:

$$\frac{mv}{r} = \frac{e^2}{r^2} \qquad (2.2)$$

5. Bohr also postulated that 2π times the angular momentum of the electron must be an integral number times h. In equation form,

$$2\pi M_a = nh \qquad (2.3)$$

where M_a is angular momentum and is given by

$$M_a = mvr \qquad (2.4)$$

From Eqs. (2.2), (2.3), and (2.4),

$$4\pi^2 m e^2 r = n^2 h^2 \qquad (2.5)$$

and finally,

$$r = \frac{n^2 h^2}{4\pi^2 m e^2} \qquad (2.6)$$

The energy E_n of the electron in its orbit r_n is

$$E_n = \tfrac{1}{2}mv^2 - \frac{e^2}{r_n} \qquad (2.7)$$

But from Eq. (2.2),

$$mv^2 = e^2/r_n \qquad (2.8)$$

Therefore,

$$E_n = -\frac{e^2}{2r_n} \qquad (2.9)$$

Putting Eq. (2.6) into Eq. (2.9),

$$E_n = \frac{-2\pi^2 me^4}{n^2 h^2} \qquad (2.10)$$

where n is called the *principal quantum number*. Each value of n (i.e., 1, 2, 3, ...) defines an energy for the electron in its orbit r_n. It can be seen that E_n will have a *discrete value* depending on n. Equation (2.6) shows that the nucleus and the electron are separated by an infinite distance as n approaches infinity, and then E_n in this limit is zero. As the electron gets closer to the nucleus (i.e., n gets smaller), E_n becomes more and more negative. The most negative energy represents the most stable system with respect to the infinitely separated electron and nucleus.

Accordingly, Eq. (2.1) is given with n_m and n_n associated with E_m and E_n, respectively:

$$hv = E_m - E_n$$
$$= \frac{2\pi^2 me^4}{h^2}\left[\left(\frac{1}{n_n}\right)^2 - \left(\frac{1}{n_m}\right)^2\right] \qquad (2.11)$$

The series transitions such as the Lyman, Balmar, Paschen, Brackett, and Pfund series in terms of the quantized energy levels are provided by Eq. (2.11).

2.1.3   Quantum Mechanics

The basis for our present understanding of the structure of atoms lies in the mathematically sophisticated theory known as quantum mechanics or wave mechanics, which explains the structure of atoms more successfully than the Bohr theory does. The following introduces a brief explanation of an atomic structure from a viewpoint of quantum mechanics, which may help to understand various subjects described in this book.

The Schrödinger wave equation describes the behavior of a single electron in a hydrogen atom as follows:

$$-\frac{h^2}{8\pi^2 \mu}\nabla^2\psi + V_0\psi = E\psi \qquad (2.12)$$

This equation is simply a symbolical way of stating that the total energy of a hydrogen atom, E, is the sum of the potential energy V_0 and its kinetic energy given by the first term of Eq. (2.12). Here ψ is the wave function, h is Planck's constant, and μ is reduced mass ($\mu = mM/m + M$, where m and M are the mass of the electron and the nucleus, respectively). The square of the absolute value of the wave function $|\psi|^2$, which must be solved, represents the *electron density* or the *probability* of finding the electron in some small volume of space dv near the nucleus. According to this interpretation, the electron may now be regarded as a hazy diffuse cloud rather than a small discrete particle. To distinguish from the old well-defined *Bohr orbits*, these electron clouds are called *orbitals* and are defined by the wave function ψ.

As a consequence of solving the Schrödinger equation, each of the orbital ψ terms is associated with three characteristic interrelated *quantum numbers* designated n, l, and m_l. The *principal quantum number* n determines the size of the orbital and also governs the allowed energy levels in the atom; n assumes the values of any integer, but not zero. The *azimuthal quantum number* l determines the shape of the orbital, and for any given value of n assumes all integral values from zero to a maximum of $n - 1$. The *magnetic quantum number* m_l has no effect on the size or shape of the orbitals but is related to the orientation of the orbital in space; m_l assumes $2l + 1$ different possible values for given l (i.e., $-l$, $-l + 1, \ldots, 0, \ldots, l - 1, l$). The orbitals are named according to their values of n, and the principal quantum number n appears in the name as an integer in front of the l value, which is designated by the letter $s, p, d, f, \ldots$, according to whether l is $0, 1, 2, 3, \ldots$, respectively. Solving the Schrödinger wave equation (2.12), s and p orbitals can be schematically represented by Fig. 2.4 and Fig. 2.5, respectively. It should be noted that these figures give a concept of the distribution of an electron occupying these orbitals. The s orbitals are spherically symmetrical, and as n gets larger the bounding sphere gets larger (Fig. 2.4). Since the s orbitals are spherically symmetrical, a bond can form in one direction as well as in any other. In contrast to s orbitals, the three p orbitals have boundary surfaces that

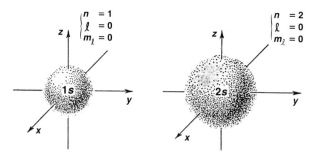

Fig. 2.4. The 1s and 2s orbitals.

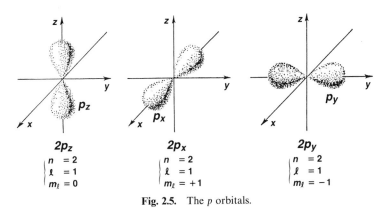

Fig. 2.5. The *p* orbitals.

resemble distorted dumbbells, are directed along the three Cartesian coordinates, and will tend to form bonds in these directions. The p_x, p_y, and p_z orbitals, shown in Fig. 2.5, are named because their lobes of maximum electron density lie along the x, y, and z axes in space, respectively.

Note here the existence of a fourth quantum number. One important feature of electrons is not directly explained by wave mechanics. This is the occurrence of *doublets* in a strong magnetic field. The electron has an intrinsic *spin*, which is quantized in such a fashion that the angular momentum of the electron may be only $\pm h/4\pi$. This spin or angular momentum of the electron causes a magnetic moment, as does a circulating electrical current. The magnetic moment may be oriented parallel or antiparallel to an external field and thus causes two energy states. In effect, the electron spin gives rise to a fourth quantum number s, the *spin quantum number*, which may have a value of $+\frac{1}{2}$ or $-\frac{1}{2}$.

2.1.4 Pauli Exclusion Principle

The rule regarding the type of orbitals occupied by the various electrons in a complex atom is clearly defined by the *Pauli exclusion principle*, which states that no two electrons in an atom may have the same four quantum numbers. As the number of electrons in an atom increases, added electrons fill orbitals of higher energy states characterized. That is, electron configurations are characterized by the principal quantum number $(1, 2, 3, \ldots)$ and the azimuthal quantum number $(s, p, d, f, \ldots)$ together with the number of electrons that can be accommodated at each energy level in accordance with the Pauli exclusion principle: that is, up to 2 electrons for s orbitals, 6 for p orbitals, 10 for d orbitals, and 14 for f orbitals.

Table 2.1 lists the electron configuration of each element. A schematic configuration of electrons, with notation ↑ or ↓ for the spin quantum number $+\frac{1}{2}$ or $-\frac{1}{2}$, in K, L, and M orbitals is given by Fig. 2.6.

Table 2.1   Electron Configuration of the Elements

		K	L		M			N				O				P			Q
												Energy level							
Element	Symbol	$1s$	$2s$	$2p$	$3s$	$3p$	$3d$	$4s$	$4p$	$4d$	$4f$	$5s$	$5p$	$5d$	$5f$	$6s$	$6p$	$6d$	$7s$
1 Hydrogen	H	1																	
2 Helium	He	2																	
Helium core		2																	
3 Lithium	Li	2	1																
4 Beryllium	Be	2	2																
5 Boron	B	2	2	1															
6 Carbon	C	2	2	2															
7 Nitrogen	N	2	2	3															
8 Oxygen	O	2	2	4															
9 Fluorine	F	2	2	5															
10 Neon	Ne	2	2	6															
Neon core		2	2	6															
11 Sodium	Na	2	2	6	1														
12 Magnesium	Mg	2	2	6	2														
13 Aluminum	Al	2	2	6	2	1													
14 Silicon	Si	2	2	6	2	2													
15 Phosphorus	P	2	2	6	2	3													
16 Sulfur	S	2	2	6	2	4													
17 Chlorine	Cl	2	2	6	2	5													
18 Argon	Ar	2	2	6	2	6													
Argon core		2	2	6	2	6													
19 Potassium	K	2	2	6	2	6		1											
20 Calcium	Ca	2	2	6	2	6		2											
21 Scandium	Sc	2	2	6	2	6	1	2											
22 Titanium	Ti	2	2	6	2	6	2	2											
23 Vanadium	V	2	2	6	2	6	3	2											
24 Chromium	Cr	2	2	6	2	6	5	1											
25 Manganese	Mn	2	2	6	2	6	5	2											
26 Iron	Fe	2	2	6	2	6	6	2											
27 Cobalt	Co	2	2	6	2	6	7	2											
28 Nickel	Ni	2	2	6	2	6	8	2											
Nickel core		2	2	6	2	6	10												
29 Copper	Cu	2	2	6	2	6	10	1											
30 Zinc	Zn	2	2	6	2	6	10	2											
31 Gallium	Ga	2	2	6	2	6	10	2	1										
32 Germanium	Ge	2	2	6	2	6	10	2	2										
33 Arsenic	As	2	2	6	2	6	10	2	3										
34 Selenium	Se	2	2	6	2	6	10	2	4										
35 Bromine	Br	2	2	6	2	6	10	2	5										
36 Krypton	Kr	2	2	6	2	6	10	2	6										
Krypton core		2	2	6	2	6	10	2	6										
37 Rubidium	Rb	2	2	6	2	6	10	2	6			1							
38 Strontium	Sr	2	2	6	2	6	10	2	6			2							

No.	Element	Symbol	1s	2s	2p	3s	3p	3d	4s	4p	4d	4f	5s	5p	5d	6s	6p
39	Yttrium	Y	2	2	6	2	6	10	2	6	1		2				
40	Zirconium	Zr	2	2	6	2	6	10	2	6	2		2				
41	Niobium	Nb	2	2	6	2	6	10	2	6	4		1				
42	Molybdenum	Mo	2	2	6	2	6	10	2	6	5		1				
43	Technetium	Tc	2	2	6	2	6	10	2	6	6		1				
44	Ruthenium	Ru	2	2	6	2	6	10	2	6	7		1				
45	Rhodium	Rh	2	2	6	2	6	10	2	6	8		1				
46	Palladium	Pd	2	2	6	2	6	10	2	6	10						
	Palladium core		2	2	6	2	6	10	2	6	10						
47	Silver	Ag	2	2	6	2	6	10	2	6	10		1				
48	Cadmium	Cd	2	2	6	2	6	10	2	6	10		2				
49	Indium	In	2	2	6	2	6	10	2	6	10		2	1			
50	Tin	Sn	2	2	6	2	6	10	2	6	10		2	2			
51	Antimony	Sb	2	2	6	2	6	10	2	6	10		2	3			
52	Tellurium	Te	2	2	6	2	6	10	2	6	10		2	4			
53	Iodine	I	2	2	6	2	6	10	2	6	10		2	5			
54	Xenon	Xe	2	2	6	2	6	10	2	6	10		2	6			
	Xenon core		2	2	6	2	6	10	2	6	10		2	6			
55	Cesium	Cs	2	2	6	2	6	10	2	6	10		2	6		1	
56	Barium	Ba	2	2	6	2	6	10	2	6	10		2	6		2	
57	Lanthanum	La	2	2	6	2	6	10	2	6	10		2	6	1	2	
58	Cerium	Ce	2	2	6	2	6	10	2	6	10	2	2	6		2	
59	Praseodymium	Pr	2	2	6	2	6	10	2	6	10	3	2	6		2	
60	Neodymium	Nd	2	2	6	2	6	10	2	6	10	4	2	6		2	
61	Prometium	Pm	2	2	6	2	6	10	2	6	10	5	2	6		2	
62	Samarium	Sm	2	2	6	2	6	10	2	6	10	6	2	6		2	
63	Europium	Eu	2	2	6	2	6	10	2	6	10	7	2	6		2	
64	Gadolinium	Gd	2	2	6	2	6	10	2	6	10	7	2	6	1	2	
65	Terbium	Tb	2	2	6	2	6	10	2	6	10	9	2	6		2	
66	Dysprosium	Dy	2	2	6	2	6	10	2	6	10	10	2	6		2	
67	Holmium	Ho	2	2	6	2	6	10	2	6	10	11	2	6		2	
68	Erbium	Er	2	2	6	2	6	10	2	6	10	12	2	6		2	
69	Thulium	Tm	2	2	6	2	6	10	2	6	10	13	2	6		2	
70	Ytterbium	Yb	2	2	6	2	6	10	2	6	10	14	2	6		2	
71	Lutetium	Lu	2	2	6	2	6	10	2	6	10	14	2	6	1	2	
72	Hafnium	Hf	2	2	6	2	6	10	2	6	10	14	2	6	2	2	
73	Tantalum	Ta	2	2	6	2	6	10	2	6	10	14	2	6	3	2	
74	Tungsten	W	2	2	6	2	6	10	2	6	10	14	2	6	4	2	
75	Rhenium	Re	2	2	6	2	6	10	2	6	10	14	2	6	5	2	
76	Osmium	Os	2	2	6	2	6	10	2	6	10	14	2	6	6	2	
77	Iridium	Ir	2	2	6	2	6	10	2	6	10	14	2	6	7	2	
78	Platinum	Pt	2	2	6	2	6	10	2	6	10	14	2	6	9	1	
	Platinum core		2	2	6	2	6	10	2	6	10	14	2	6	10		
79	Gold	Au	2	2	6	2	6	10	2	6	10	14	2	6	10	1	
80	Mercury	Hg	2	2	6	2	6	10	2	6	10	14	2	6	10	2	
81	Thallium	Tl	2	2	6	2	6	10	2	6	10	14	2	6	10	2	1
82	Lead	Pb	2	2	6	2	6	10	2	6	10	14	2	6	10	2	2
83	Bismuth	Bi	2	2	6	2	6	10	2	6	10	14	2	6	10	2	3
84	Polonium	Po	2	2	6	2	6	10	2	6	10	14	2	6	10	2	4
85	Astatine	At	2	2	6	2	6	10	2	6	10	14	2	6	10	2	5

(*continues*)

Table 2.1   (*Continued*)

		Energy level																		
		K	L		M			N				O				P			Q
Element	Symbol	1s	2s	2p	3s	3p	3d	4s	4p	4d	4f	5s	5p	5d	5f	6s	6p	6d	7s
86 Radon	Rn	2	2	6	2	6	10	2	6	10	14	2	6	10		2	6		
Radon core		2	2	6	2	6	10	2	6	10	14	2	6	10		2	6		
87 Francium	Fr	2	2	6	2	6	10	2	6	10	14	2	6	10		2	6		1
88 Radium	Ra	2	2	6	2	6	10	2	6	10	14	2	6	10		2	6		2
89 Actinium	Ac	2	2	6	2	6	10	2	6	10	14	2	6	10		2	6	1	2
90 Thorium	Th	2	2	6	2	6	10	2	6	10	14	2	6	10		2	6	2	2
91 Protactinium	Pa	2	2	6	2	6	10	2	6	10	14	2	6	10	2	2	6	1	2
92 Uranium	U	2	2	6	2	6	10	2	6	10	14	2	6	10	3	2	6	1	2
93 Neptunium	Np	2	2	6	2	6	10	2	6	10	14	2	6	10	4	2	6	1	2
94 Plutonium	Pu	2	2	6	2	6	10	2	6	10	14	2	6	10	5	2	6	1	2
95 Americium	Am	2	2	6	2	6	10	2	6	10	14	2	6	10	6	2	6	1	2
96 Curium	Cm	2	2	6	2	6	10	2	6	10	14	2	6	10	7	2	6	1	2
97 Berkelium	Bk	2	2	6	2	6	10	2	6	10	14	2	6	10	9	2	6		2
98 Californium	Cf	2	2	6	2	6	10	2	6	10	14	2	6	10	10	2	6		2
99 Einsteinium	Es	2	2	6	2	6	10	2	6	10	14	2	6	10	11	2	6		2
100 Fermium	Fm	2	2	6	2	6	10	2	6	10	14	2	6	10	12	2	6		2
101 Mendelevium	Md	2	2	6	2	6	10	2	6	10	14	2	6	10	13	2	6		2
102 Nobelium	No	2	2	6	2	6	10	2	6	10	14	2	6	10	14	2	6		2
103 Lawrencium	Lw	2	2	6	2	6	10	2	6	10	14	2	6	10	14	2	6	1	2

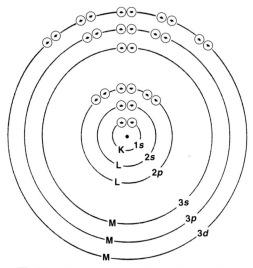

Fig. 2.6.   Electron configuration and energy levels.

2.2 Chemical Bond

2.2.1 Solid and Chemical Bond

Materials generally occur in three states of aggregation: the solid state, the liquid state, and the gaseous state. In solids and liquids the distance between neighboring atoms is of the order of a few Ångstroms (1 Å $= 10^{-8}$ cm); they contain 10^{22}–10^{23} atoms/cm^3. This may be compared with an average distance of approximately 30 Å for neighboring molecules (i.e., $\sim 10^{19}$ molecules/cm^3) in a gas at room temperature under 1 atm pressure. The nature of having approximately constant volume and shape distinguishes solids from liquids. This nature is because the configuration of the elements (i.e., atoms, ions, and molecules) of solids is almost fixed. The bond that joins or unites atoms or ions, resulting in molecules and materials, is called *the chemical bond*. Here, three general extreme types of chemical bonds are considered:

1. electrostatic bonds,
2. covalent bonds, and
3. metallic bonds.

Although each type of bond has well-defined properties, this classification is not rigorous since bonds of intermediate type (i.e., resonating bonds) exist in realistic chemical bonds.

Covalent bonds are particularly common in semiconductor materials such as carbon (C), silicon (Si), germanium (Ge), and gallium arsenide (GaAs), which make up the interests of this book. It is important to understand the nature of covalent bonds to comprehend the behavior of *donor-* and *acceptor-type dopants* in semiconductor crystals, as introduced in a later chapter of this book.

The consideration of the formation of a stable hydrogen molecule, H_2, may help to understand the nature of covalent bonds. As shown in Table 2.1, a hydrogen atom is a system that has a *nucleus* composed of one *proton* and one *electron* on the 1s orbital. Consider two isolated hydrogen atoms, each with its electron in its ground-state 1s orbital and spin antiparallel to each other, as they approach one another. In Fig. 2.7, the gound-state wave function of hydrogen is schematically shown by contours. As the atoms get closer and closer together, the 1s clouds containing the electrons begin to overlap. Each electron is attracted to the approaching nucleus, and overlap increases. The two atomic orbitals merge into one bigger cloud and the electrons will stay in the position where they are attracted most to both nuclei. That is, the merger halts when the repulsive forces between the positively charged nuclei have determined the position of closest approach. At this point the system of two

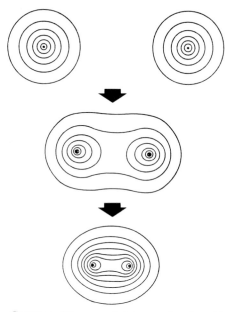

Fig. 2.7.   Contours of the ground-state wave function of hydrogen.

nuclei and two electrons has attained a stability much greater than that of the two isolated atoms. This is the way the *shared-electron-pair bond* or *covalent bond* forms. As already discussed, the Pauli exclusion principle does not allow two hydrogen atoms each with its parallel spin to merge their two atomic orbitals into one.

Figure 2.8 graphically shows that the total energy of the system consists of two hydrogen atoms, or more generally a diatomic molecule, as a function of the internuclear distance r. At large values of r, the energy of the system approximates that of two isolated atoms. This can be called the zero energy state, so that any more stable state of the system may be described by a negative energy. As r decreases, the energy decreases—that is, the stability increases—largely because of the dual nuclear attractive forces acting on the electrons. At the point of r_{AB}, the equilibrium internuclear distance, the energy minimizes, and in turn the stability maximizes. As two atoms get much closer, the energy increases because of the repulsive force between the two nuclei. In Fig. 2.8, r_{AB} and E_B define the *bond length*, 0.74 Å for H_2, and the *bond energy*, 4.75 eV for H_2, respectively. On the other hand, for two hydrogen atoms with parallel spins, the energy simply increases as r decreases, as shown by a dotted line in Fig. 2.8.

It is convenient to use the Lewis electronic structure formula,[5] which explains the shared-electron-pair bond of hydrogen atoms simply as follows:

$$H\cdot + \cdot H \rightarrow H:H \tag{2.13}$$

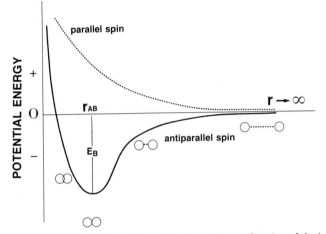

Fig. 2.8. Potential energy diagram for a diatomic molecule as a function of the internuclear distance *r*.

In the Lewis electronic formula the symbol of the element represents the *kernel* of the atom, consisting of the nucleus and the inner electrons, while the electrons in the valence shell, the *valence electrons*, are shown by dots. Dots between kernels represent shared electrons that result in the shared-electron-pair bond, the covalent bond. This idea of electron pairing gives a quantitatively satisfactory explanation of the covalent bonds.

2.2.2 Structure of the Silicon Atom

The structure of the silicon atom and its bonding is now examined. As shown in Table 2.1, the isolated silicon atom has 14 electrons, that is, $1s^2 2s^2 2p^6 3s^2 3p^2$. Since the neon core electrons (i.e., closed-shell electrons) do not directly contribute to the covalent bond, only 3*s* and 3*p* electrons are considered (Fig. 2.9). As mentioned, the Pauli exclusion principle limits *s*

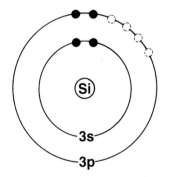

Fig. 2.9. Electron configuration of a silicon atom.

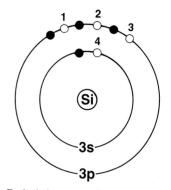

Fig. 2.10. Excited electron configuration of a silicon atom.

orbitals to two electrons and *p* orbitals to six. To complete the covalent bonding with four electrons (denoted with 1–4) of other silicon atoms (Fig. 2.10), one of the 3*s* electrons must be transferred to the 3*p* orbital, which results in *sp*³ *orbital hybridization.*

Since the 3*s* orbital is quite near the 3*p* in energy, one of the 3*s* electrons may be promoted to the 3*p* level, thus obtaining the *excited* configuration $3s^1 3p_x^1 3p_y^1 3p_z^1$ (see Fig. 2.5). These four unpaired electrons can be associated with four bonds. As illustrated by Fig. 2.11,[6] the 3*s* and three 3*p* orbitals may combine or smear together to result in four equivalent hybrid orbitals called *sp*³ or *tetrahedral hybrids.* The shape of the silicon molecule, and also of

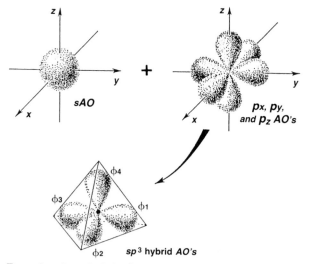

Fig. 2.11. Formation of tetrahedral *sp*³ hybrids. AO, atomic orbital. (After Companion.[6])

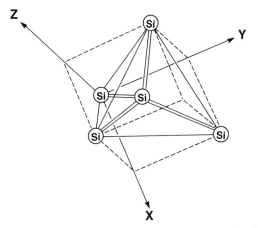

Fig. 2.12. Tetrahedral configuration of silicon molecule.

molecules of atoms with identical electron configuration in energy (e.g., C and Ge), is consequently like that of a tetrahedron (see Fig. 2.12). This tetrahedral covalent bond is a basis for the geometrical configuration not only of silicon but also of silicon oxide.

References

1. R. F. Humphreys and R. Beringer, "First Principles of Atomic Physics." Harper, New York, 1950.
2. L. I. Schiff, "Quantum Mechanics," 2nd ed. McGraw-Hill, New York, 1955.
3. L. Pauling, "The Nature of the Chemical Bond," 3rd ed. Cornell Univ. Press, Ithaca, New York, 1960.
4. C. Kittel, "Introduction to Solid State Physics," 4th ed. Wiley, New York, 1971.
5. R. C. Evance, "An Introduction to Crystal Chemistry." Cambridge Univ. Press, London and New York, 1948.
6. A. L. Companion, "Chemical Bonding." McGraw-Hill, New York, 1964.

Chapter 3

Basic Crystallography

Classic crystallography was established during the seventeenth to nineteenth centuries. The aims were to classify natural crystal morphologies based on observation and to study their macroscopic physical properties. However, it was the discovery of X-ray diffraction by the atoms of solids by Laue, Friedrich, and Knipping in 1912, proving for the first time the regular and periodic arrangement of the atoms in a crystal structure, that enabled the investigation of the atomic structure of materials. The aims of crystallography as developed by von Laue *et al.* and later by W. H. Bragg and W. L. Bragg have been to investigate the microscopic structure of materials and their dynamic behavior due to external and internal stimulation. Modern crystallography has contributed to wide academic areas, such as physics, chemistry, and biology. In these days, particularly, modern crystallography as "technological crystallography" has greatly contributed to marvelous progress of the electronics technology that is based mainly on semiconductor crystals.

In this chapter, "crystal" is defined first and then the structure of crystals determined by X-ray analyses is introduced. A basic idea of crystal lattice defects, which greatly affect the performance of electronic devices, is also discussed. Finally, the structure of the silicon crystal is presented. The knowledge of crystal structures and crystal defects surely helps in understanding the subjects that will be dealt with in the following chapters.

3.1 Solid-State Structure

3.1.1 Crystalline and Noncrystalline Materials

In Section 2.2.1, the three states of aggregation were described: the solid state, the liquid state, and the gaseous state. The solid state may be classified into

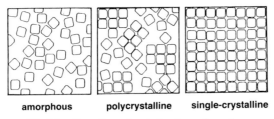

amorphous   **polycrystalline**   **single-crystalline**

Fig. 3.1.   Two-dimensional atomic configuration.

two groups from the viewpoint of atomic configuration; *crystalline* materials and *noncrystalline* materials or *amorphous* materials. A crystalline material is usually called simply a *crystal*. A crystal is defined as "a solid composed of atoms arranged in a three-dimensionally periodic pattern." On the other hand, amorphous materials do not have the periodicity of atomic structure. A majority of natural and manufactured solid materials are crystalline, since the energy of an ordered atomic arrangement is lower than that of an irregularly arranged atoms. However, amorphous materials may be formed when the atoms are not given an opportunity to be arranged properly for some reason, such as inhibiting their mobility.

When the periodicity of the atomic arrangement extends throughout a certain piece of material, the material is called *single crystal*. A crystalline material in which the periodicity of structure is interrupted at boundaries is called a *polycrystalline* material or simply a *polycrystal*. A polycrystal may be defined as a solid formed by many small single crystals, namely, *grains*, with different orientation. Figure 3.1 illustrates the difference in the two-dimensional atomic configuration of amorphous, polycrystalline, and single-crystalline materials. Although single crystals are the principal materials used as substrates for electronic device fabrication, there are many important electronic device parts to which polycrystalline and amorphous materials are applied.

3.1.2   Crystal Structure

Perfect Crystal   As defined in the preceding section, a single crystal is a solid in which atoms are arranged in a three-dimensionally periodic pattern; however, there may exist some imperfections, which are called *lattice defects* or *crystal defects* and are described in Section 3.4 as "realistic" crystals. To simplify the discussion, the structure of crystals that have no imperfection is first examined. A crystal with no imperfection is called a *perfect* or *ideal* crystal.

Lattice   In considering crystal structures, it is convenient to ignore the size of actual atoms composing the crystal and to think of a set of imaginary points that have a fixed relation in space to the atoms of the crystal. A

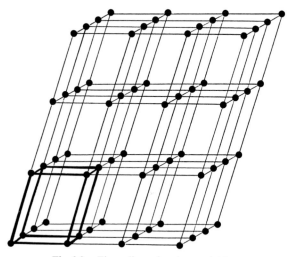

Fig. 3.2.   Three-dimensional space lattice.

three-dimensional framework composed of these points may be regarded as a *space lattice* on which the actual crystal is built. A space lattice is illustrated in Fig. 3.2. Note that the space lattice shown is formed by three-dimensional *translations* of the *cell* drawn with bold lines. Since all the cells of the space lattice in Fig. 3.2 are identical, any one can be chosen as a *unit cell*. The size and shape of the unit cell is defined by the three vectors **a**, **b**, and **c** drawn from one corner, at the origin of the cell (Fig. 3.3). These three vectors, called *primitive vectors*, also define the *crystallographic axes* of the cell. Primitive vectors are described in terms of their length (a, b, c) and the angles between them (α, β, γ), that is, the *lattice constants* or *lattice parameters* of the unit cell. Note that the primitive vectors define not only the unit cell but also the whole space lattice through the translations provided by these vectors. Thus by giving special values to the lattice constants (i.e., the axial lengths and angles), various kinds of space lattices can be defined. Eventually all crystal lattice structures can be classified into the seven *crystal systems* listed in Table 3.1. Semiconductor crystals such as Si, Ge, and GaAs belong to the *cubic system*, which is fortunately the simplest system.

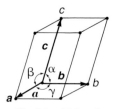

Fig. 3.3.   Unit cell.

Table 3.1 Crystal Systems and Bravais Lattices

System	Axial lengths and angles	Bravais lattice	Lattice symbol
Cubic	Three equal axes at right angles $a = b = c, \alpha = \beta = \gamma = 90°$	Simple Body-centered Face-centered	P I F
Tetragonal	Three axes at right angles, two equal $a = b \neq c, \alpha = \beta = \gamma = 90°$	Simple Body-centered	P I
Orthorhombic	Three unequal axes at right angles $a \neq b \neq c, \alpha = \beta = \gamma = 90°$	Simple Body-centered Base-centered Face-centered	P I C F
Rhombohedral (or trigonal)	Three equal axes, equally inclined $a = b = c, \alpha = \beta = \gamma \neq 90°$	Simple	R
Hexagonal	Two equal coplanar axes at 120°, third axis at right angles $a = b \neq c, \alpha = \beta = 90°, \gamma = 120°$	Simple	P
Monoclinic	Three unequal axes, one pair not at right angles, $a \neq b \neq c, \alpha = \gamma = 90° \neq \beta$	Simple Base-centered	P C
Triclinic	Three unequal axes, unequally inclined and none at right angles $a \neq b \neq c, \alpha \neq \beta \neq \gamma \neq 90°$	Simple	P

As schematically shown by a two-dimensional point lattice (Fig. 3.4), however, it is possible to choose some different unit cells, of which each point has identical surroundings, for a given lattice. This situation is of course the same for a three-dimensional space lattice. Bravais demonstrated in 1848 that there are 14 possible space lattices and no more. These space lattices are called *Bravais lattices*, after Bravais. The 14 Bravais lattices are also described in Table 3.1 and illustrated in Fig. 3.5.

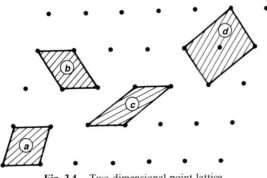

Fig. 3.4. Two-dimensional point lattice.

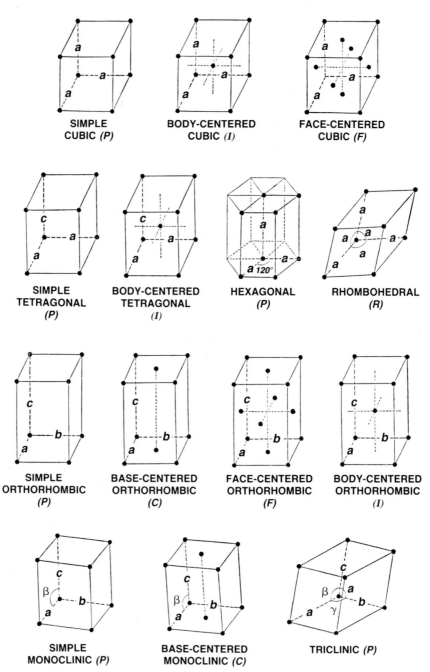

Fig. 3.5. Bravais lattices.

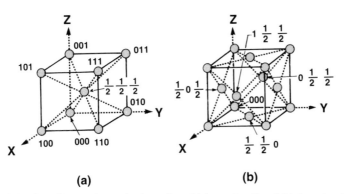

Fig. 3.6. Coordinates of some lattice points: (a) bcc unit cell and (b) fcc unit cell.

The number of lattice points N, the *coordination number*, which is defined as atoms per cell if there is one atom per lattice point, is given by

$$N = N_i + N_f/2 + N_c/8 \tag{3.1}$$

where N_i is the number of interior points that belong to the cell, N_f the number of points on faces that are shared by two cells, and N_c the number of points on corners that are shared by eight cells.

Lattice Coordinates   The position of any lattice point in a cell is given in terms of its *coordinates*. For example, the coordinates of some lattice points of a body-centered cubic (bcc) and a face-centered cubic (fcc) cell are given in Fig. 3.6a and b, respectively. If the vector from the origin of the unit cell to the given point has components xa, yb, and zc, where x, y, and z are fractions, the coordinates of the point are $x\,y\,z$. Thus, the coordinates of the origin are 0 0 0, those of the body-centered point are $\frac{1}{2}\frac{1}{2}\frac{1}{2}$, and those of the points on the corners are 0 1 1, 1 0 1, 1 1 0, etc. Face-centered points have coordinates such as $1\frac{1}{2}\frac{1}{2}, \frac{1}{2}0\frac{1}{2}, \frac{1}{2}\frac{1}{2}0$, etc. (Fig. 3.6b). Note that lattice coordinates are given by a simple enumeration $x\,y\,z$, but are *not* enclosed in parentheses or brackets or braces. These symbols are used to represent crystal directions and planes, which are described next.

Crystal Orientations and Planes   In considering crystal lattice structures, the *atomic net planes* in which atoms are arranged in a periodic pattern can be defined. Since each atomic net plane has its characteristic atomic arrangement, the properties such as mechanical and physical and electronic ones can be *anisotropic* in the crystal. It is very important to understand the difference of these crystal plane properties, particularly the electronic ones, in the fabrication of electronic devices. Actually, as described in detail in Section 3.3, silicon crystal substrates with different orientations have been separately applied to manufacturing certain devices because of the anisotropy. The following is a description of how to represent crystal or lattice directions and planes.

Imagine a line that passes the origin of the unit cell and any point having coordinates $u\,v\,w$. The orientation of this line is defined as $[uvw]$, which is called the *indices* of the orientation of the line. Note that these direction indices are given with square brackets [   ], and these numbers are not necessarily integral since the line with $[uvw]$ also passes through the points having coordinates such as $2u\,2v\,2w$ and $3u\,3v\,3w$. They are also the indices of any line parallel to the given line since the lattice is infinite and the origin can be taken chosen at any point. Thus, for instance, $[\frac{1}{2}\frac{1}{2}1]$, $[112]$, and $[224]$ all represent the same orientation; however, $[112]$, which consists of the smallest integers, is preferred. The orientation of a point having negative coordinates (e.g., $-u\,v\,w$ or $-u\,-v\,w$) is represented with the orientation indices written with a bar over the number: for instance, $[\bar{u}vw]$ or $[\bar{u}\bar{v}w]$, respectively, for the point just given. Some important orientation indices in the cubic system are illustrated in Fig. 3.7. In the cubic system, to which silicon crystals belong, directions with indices such as $[uvw]$, $[wuv]$, $[\bar{u}wv]$, and $[\bar{u}\bar{v}w]$ are all equivalent to each other. These equivalent orientations are called *orientations of a form* and are represented as $\langle uvw\rangle$ *en bloc*. Note that the orientation of a form is given with angular brackets $\langle$ $\rangle$. For instance, $\langle100\rangle$ is the general term for six orientations of $[100]$, $[\bar{1}00]$, $[010]$, $[0\bar{1}0]$, $[001]$, and $[00\bar{1}]$.

The orientation indices of $[uvw]$ are also the components of the vector. The intersectional angle between orientations $\mathbf{A}[u_1v_1w_1]$ and $\mathbf{B}[u_2v_2w_2]$, where

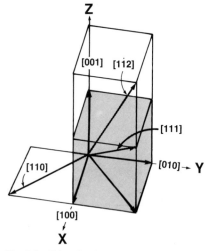

Fig. 3.7.   Direction indices in the cubic system.

$u_1, v_1, \ldots, v_2, w_2$ are the components of these vectors, is given by the following equation as the scalar product of the two vectors:

$$\mathbf{A} \cdot \mathbf{B} = |\mathbf{A}| \cdot |\mathbf{B}| \cos \theta$$

$$= u_1 u_2 + v_1 v_2 + w_1 w_2 \tag{3.2}$$

Thus,

$$\cos \theta = \frac{\mathbf{A} \cdot \mathbf{B}}{|\mathbf{A}| \cdot |\mathbf{B}|}$$

$$= \frac{u_1 u_2 + v_1 v_2 + w_1 w_2}{\sqrt{(u_1^2 + v_1^2 + w_1^2)} \sqrt{(u_2^2 + v_2^2 + w_2^2)}} \tag{3.3}$$

That is, when

$$u_1 u_2 + v_1 v_2 + w_1 w_2 = 0 \tag{3.4}$$

these two vectors fall at right angles each other, and when

$$u_1 u_2 + v_1 v_2 + w_1 w_2 = 1 \tag{3.5}$$

these are parallel to each other.

The orientation of planes in a lattice may also be represented symbolically according to the indices for the plane. Here consider the lattice plane that intercepts pa, qb, and rc of the crystal axes X, Y, and Z, respectively (Fig. 3.8), where a, b, and c are axial lengths. This plane is represented simply with p, q, and r, since a, b, and c are the lattice parameters, which are constant for a certain crystal system. However, a difficulty arises when the given plane is parallel to a certain crystal axis, because such a plane does not intercept that axis. In other words, its "intercept" can only be described as "infinity." To avoid this problem, the reciprocal of the fraction intercept is used, that is, $1/p$,

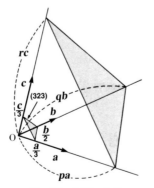

Fig. 3.8.   Miller indices.

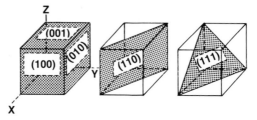

Fig. 3.9.   Miller indices of some planes in the cubic system.

$1/q$, and $1/r$ instead of p, q, and r. These indices h, k, and l are called *Miller indices*, and are integers that satisfy the following ratio:

$$1/p : 1/q : 1/r = h : k : l \qquad (3.6)$$

That is, the plane that intercepts pa, qb, and rc is defined with the Miller indices (hkl); note that the indices are written in parentheses.

For example, the Miller indices of the plane that makes fractional intercepts of $a/3$, $b/2$, and $c/3$ are (323), as shown in Fig. 3.8. From Eq. (3.6), it is clear that if a plane is parallel to a certain axis, the corresponding Miller index is zero since its fractional intercept on that axis is taken as infinity. If a plane intercepts a negative axis, the corresponding index is negative and is written with a bar over it. The Miller indices of some important planes in the cubic system are illustrated in Fig. 3.9. In the cubic system, planes with indices such as (hkl), (khl), $(\bar{h}\bar{k}l)$, and (lhk), are all equivalent each other. These equivalent planes are called *planes of a form* and are represented as $\{hkl\}$ *en bloc*. Note that the plane of a form is given with braces. For instance, $\{100\}$ is the general term for six planes of (100), $(\bar{1}00)$, (010), $(0\bar{1}0)$, (001), and $(00\bar{1})$ in the cubic system. Also, it is convenient to remember that a direction $[hkl]$ is always perpendicular to a plane (hkl) in the cubic system; however, $\langle hkl \rangle$ is not necessarily perpendicular to $\{hkl\}$ in noncubic systems.

Interplanar Distance and Angles   The planes $(nh\ nk\ nl)$ are parallel to the plane (hkl). An example for the relation among (100) and $(n00)$ planes is illustrated in Fig. 3.10. It is obvious that the plane spacing d of the plane $(nh\ nk\ nl)$, $d_{nh\ nk\ nl}$, is $1/n$-th of the spacing of the (hkl) plane, d_{hkl}; that is, $d_{nh\ nk\ nl} = (1/n)d_{hkl}$. The interplanar spacing d_{hkl} is a function of the plane indices (hkl) and the lattice constants (i.e., a, b, c, α, β, and γ) and is given by the equation

$$d_{hkl} = \sqrt{\frac{V^2}{S_{11}h^2 + S_{22}k^2 + S_{33}l^2 + 2S_{12}hk + 2S_{23}kl + 2S_{13}hl}} \qquad (3.7)$$

where V is the volume of the unit cell,

$$V = abc(1 - \cos^2\alpha - \cos^2\beta - \cos^2\gamma + 2\cos\alpha\cos\beta\cos\gamma) \qquad (3.8)$$

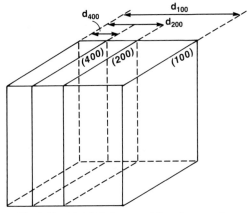

Fig. 3.10.   Parallel planes and the plane spacing.

and where

$$S_{11} = b^2c^2 \sin^2 \alpha \qquad (3.9)$$

$$S_{22} = a^2c^2 \sin^2 \beta \qquad (3.10)$$

$$S_{33} = a^2b^2 \sin^2 \gamma \qquad (3.11)$$

$$S_{12} = abc^2(\cos \alpha \cos \beta - \cos \gamma) \qquad (3.12)$$

$$S_{23} = a^2bc(\cos \beta \cos \gamma - \cos \alpha) \qquad (3.13)$$

$$S_{13} = ab^2c(\cos \gamma \cos \alpha - \cos \beta) \qquad (3.14)$$

For the cubic system, in which $a = b = c$ and $\alpha = \beta = \gamma = \pi/2$, d_{hkl} is given by a simple form:

$$d_{hkl} = \frac{a}{\sqrt{(h^2 + k^2 + l^2)}} \qquad (3.15)$$

The interplanar angle ϕ between the planes $(h_1k_1l_1)$ and $(h_2k_2l_2)$ is given by the following general equation:

$$\cos \phi = \frac{d_{h_1k_1l_1}d_{h_2k_2l_2}}{V^2} [S_{11}h_1h_2 + S_{22}k_1k_2 + S_{33}l_1l_2$$

$$+ S_{23}(k_1l_2 + k_2l_1) + S_{13}(l_1h_2 + l_2h_1)$$

$$+ S_{12}(h_1k_2 + h_2k_1)] \qquad (3.16)$$

For the cubic system, this equation is simplified in the same manner as above:

$$\cos \phi = \frac{h_1h_2 + k_1k_2 + l_1l_2}{\sqrt{(h_1^2 + k_1^2 + l_1^2)} \sqrt{(h_2^2 + k_2^2 + l_2^2)}} \qquad (3.17)$$

Table 3.2 Interplanar Angles between Planes $\{h_1 k_1 l_1\}$ and $\{h_2 k_2 l_2\}$ in Silicon (Cubic System)

$\{h_1 k_1 l_1\}$	$\{h_2 k_2 l_2\}$				
	100	110	111	211	511
100	2 of 90.00°	4 of 45.00°	4 of 54.74°	4 of 35.26°	4 of 15.80°
		2 of 90.00°		8 of 65.90°	8 of 78.90°
110	4 of 45.00°	4 of 60.00°	2 of 35.26°	4 of 30.00°	4 of 35.26°
	2 of 90.00°	1 of 90.00°	2 of 90.00°	2 of 54.74°	4 of 57.02°
				6 of 73.22°	2 of 74.20°
				2 of 90.00°	2 of 90.00°
111	3 of 54.74°	3 of 35.26°	3 of 70.53°	3 of 19.47°	3 of 38.95°
		3 of 90.00°		6 of 61.87°	6 of 56.25°
				3 of 90.00°	3 of 70.53°
211	1 of 35.26°	2 of 30.00°	1 of 19.47°	2 of 33.56°	1 of 19.47°
	2 of 65.90°	1 of 54.74°	2 of 61.87°	2 of 48.19°	2 of 38.22°
		2 of 73.22°	1 of 90.00°	2 of 60.00°	3 of 51.05°
		1 of 90.00°		1 of 70.53°	2 of 61.88°
				4 of 80.40°	2 of 80.97°
511	1 of 15.80°	2 of 35.26°	1 of 38.95°	1 of 19.47°	2 of 22.18°
	2 of 78.90°	2 of 57.02°	2 of 56.25°	2 of 38.22°	1 of 31.58°
		1 of 74.20°	1 of 70.53°	3 of 51.05°	2 of 65.95°
		1 of 90.00°		2 of 61.88°	2 of 70.53°
				2 of 71.68°	4 of 87.88°
				2 of 80.97°	

Note that Eq. (3.17) is equivalent to Eq. (3.2), since [*hkl*] is always perpendicular to (*hkl*) as noted in cubic systems. Table 3.2 lists the interplanar angles between important planes in the silicon semiconductor technology. The easiest way to visualize the relative positions of various planes is by means of a three-dimensional model such as the illustration shown in Fig. 3.11, which comprises the complete set of $\{100\}$, $\{110\}$, and $\{111\}$ planes of

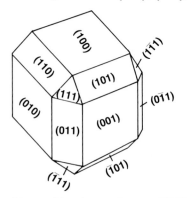

Fig. 3.11. Cubic system model comprising the complete set of $\{100\}$, $\{110\}$, and $\{111\}$ planes.

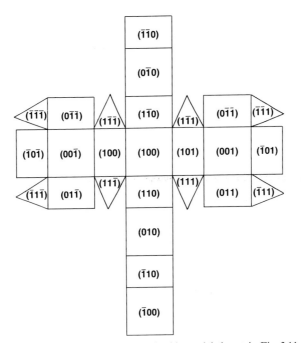

Fig. 3.12. Development pattern of cubic model shown in Fig. 3.11.

the cubic system. In the figure, notations of crystal planes and orientations are also shown. Figure 3.12 gives a development pattern of those planes for construction of the model. By pasting small lips to the insides of the faces, a three-dimensional model, such as one shown by Fig. 3.11, can be easily made.

3.2 X-Ray and Electron Diffraction

In this chapter a crystal was defined as "a solid composed of atoms arranged in a three-dimensionally periodic pattern" and the structure of crystals was described. As noted at the outset of this chapter, X-ray diffraction study by von Laue, W. H. Bragg, and W. L. Bragg and others first proved the periodicity of the arrangement of atoms within a crystal. Since then, X-ray diffraction study and electron diffraction study have greatly expanded our knowledge of the microstructure and dynamic behavior of materials. The nature of diffraction has been widely applied to characterization of semiconductor crystals, as will be introduced in Section 6.3. In order to help understand those characterization techniques, the fundamentals of X-ray and electron diffraction by crystal lattices will be introduced in this section.

3.2.1 Properties of X-Rays

X-Rays, discovered in 1895 by Röntgen and so named because their nature
was unknown at that time, are electromagnetic radiation. Therefore X-rays,
like other electromagnetic waves, reflect, diffract, refract, and transmit.
X-Rays are also dealt with as photons. The energy E of an X-ray photon is
related to its wavelength λ by

$$E = h\nu = hc/\lambda \qquad (3.18)$$

where c is the velocity of light.

Figure 3.13 gives the complete electromagnetic spectrum.[1] The boundaries
between regions are broad; no sharp upper or lower limits can be assigned for
each category. X-Rays occupy the region between gamma ($\lambda \approx 10^{-2}$ Å) and
ultraviolet ($\lambda \approx 10^2$ Å) rays. The X-rays used in crystal diffraction studies
have wavelengths lying approximately in the range of 0.5–2.5 Å, which is
roughly equivalent to crystal lattice spacings. Soft X-rays, whose wavelength

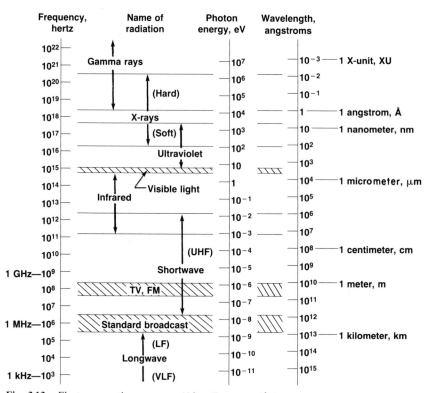

Fig. 3.13. Electromagnetic spectrum. (After Enge *et al.*[1] Reproduced by permission from
Addison-Wesley Publishing Co., Inc.)

is ~ 3 Å or longer, have recently attracted application as the radiation source in X-ray lithography for submicrometer patterning of VLSI/ULSI devices.[2] In addition, X-rays

1. are effective in blacking a photographic film in much the same way as visible light is;
2. excite fluorescent compound materials, such as zinc sulfide containing a trace of nickel, resulting in visible light emission;
3. ionize atoms whether they are in a gas or a solid;
4. have a refractive index n that is very close to 1, such that the deviation δ, 10^{-5}–10^{-6} for most materials, is given by

$$\delta = 1 - n = Ne^2\lambda^2/2\pi mc^2 \tag{3.19}$$

where N is the number of electrons in unit volume of the material; and
5. have a high transmissivity.

The principal means used to detect X-rays (i.e., photographic films, fluorescent screens, and counters) depend on properties 1–3. Because of property 4, X-rays are not similar to light beams, which converge or diverge, due to its high refractive index for materials. In other words, there is no "lens" for X-rays. That is, an X-ray image can neither be focused nor enlarged by a "lens." Property 5 allows X-rays to be utilized as a tool for the characterization or identification of materials.

3.2.2 Production of X-Rays

X-Rays are produced when any electrically charged particles, usually electrons, with sufficiently high kinetic energy are rapidly decelerated. Figure 3.14 illustrates the production of X-rays. A high voltage, such as some tens of kilovolts, accelerates thermoelectrons emitted by a filament and causes them

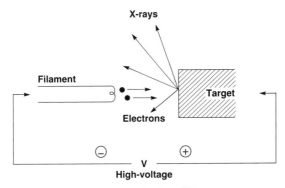

Fig. 3.14. Production of X-rays.

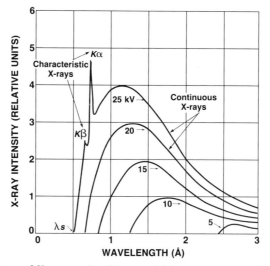

Fig. 3.15. Spectrum of X-rays produced by a molybdenum target as a function of applied voltage. (After Cullity.[3] Reproduced by permission from Addison-Wesley Publishing Co., Inc.)

to strike a target or anode with very high velocity. X-Rays are produced at the point of impact and radiate in all directions. Figure 3.15 schematically shows the spectrum of X-rays produced by a molybdenum (Mo) target as a function of applied voltage.[3] The X-rays analyzed consist of a mixture of different wavelengths with variation of intensity depending on the applied voltage. X-Rays may be categorized into *continuous or white X-rays*, which have continuous wavelengths, and *characteristic X-rays*, which have a sharp intensity peak at certain wavelengths.

The continuous X-rays are due to the rapid deceleration or bremsstrahlung (German for "breaking radiation") of the electrons striking the target. Such electrons, or decelerated charges, emit energy according to Eq. (3.18). Since every electron is decelerated differently (i.e., some give up all their energy at once by one impact, while others lose successively), resulting spectra may have continuous wavelengths longer than a certain wavelength called the shortest wavelength λ_{min}. The shortest wavelength depends on applied accelerating voltage V and is given by

$$\lambda_{min} = hc/eV \tag{3.20}$$

Setting the physical constants[4] $e = 1.602 \times 10^{-19}$ C, $h = 6.626 \times 10^{-34}$ J sec and $c = 2.998 \times 10^{8}$ m sec^{-1} into Eq. (3.19), the relation in laboratory units is obtained by

$$\lambda_{min} \quad (\text{Å}) = 12.4/V \quad (\text{kV}) \tag{3.21}$$

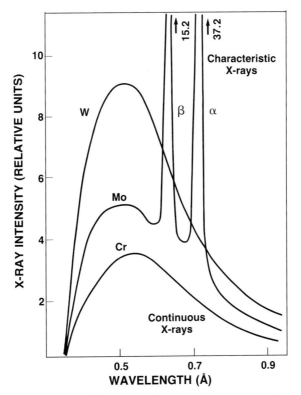

Fig. 3.16.   X-Ray spectrum of W, Mo, and Cr targets at certain generation conditions.

The intensity of continuous X-rays, I_w, depends both on the atomic number Z of the target element and on the X-ray tube current i:

$$I_w \propto iV^2Z \qquad (3.22)$$

Figure 3.16 shows the X-ray spectrum of tungsten (W), molybdenum (Mo), and chromium (Cr) targets at certain values of V and i. Note that the relative intensity depends on Z, but λ_{min} does not depend on the element, as Eq. (3.21) shows.

The characteristic X-rays are due to electron transition (see Fig. 2.3). Figure 3.17 schematically explains the mechanism of characteristic X-ray production. Consider an atom as consisting of a central nucleus surrounded by electrons lying in various shells, where the designations K, L, M, ... correspond to the principal quantum number $n = 1, 2, 3, ...$ (refer to Fig. 2.6 and Table 2.1). If one electron with a high velocity (i.e., sufficient kinetic energy) hits a K-shell electron, the K-shell electron is knocked out of the shell, leaving a vacancy in the K shell. One of the outer-shell electrons

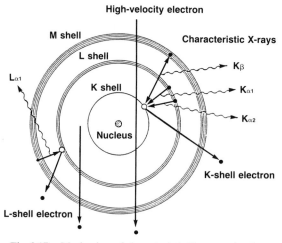

Fig. 3.17. Mechanism of characteristic X-ray production.

immediately falls into the vacancy. In the process, emission of corresponding energy, the characteristic K-radiation, occurs. The K-shell vacancy may be filled by an electron from any one of the outer shells producing K series of characteristic X-rays—for example, Kα (Kα$_1$, Kα$_2$) and Kβ resulting from the filling of a K-shell vacancy by an electron from the L (L$_{III}$, L$_{II}$) and M shells, respectively. Although it is possible to fill a K-shell vacancy with an electron from either the L or M shells, it is more probable that a K-shell vacancy will be filled by an L-shell electron than by an M-shell electron; in fact, as shown in Figs. 3.15 and 3.16, the intensity of Kα X-rays is stronger, by about two times, than that of Kβ X-rays. Characteristic X-rays of L series are produced in a similar way, as shown in Fig. 3.17; an L-shell electron is knocked out of the L-shell and the vacancy is filled by an electron from an outer shell. Because of their stronger intensity, characteristic X-rays of the K series, particularly Kα, are usually used for X-ray diffraction studies. The wavelength of Kα X-rays, λ_K, is given by following equation [also see Eq. (2.1)]:

$$E_n - E_k = hc/\lambda_k \tag{3.23}$$

where E_n is the energy of a bound electron in the L, M, N, ... shell, and E_k is of one in the K shell. Note that the wavelength λ of characteristic X-rays does not depend on the applied voltage, but depends on the atomic number Z of the target element by *Moseley's law*:

$$1/\sqrt{\lambda} = C(Z - s) \tag{3.24}$$

where C and s are constants depending on the spectra. Notice in Fig. 3.15 that the characteristic X-rays can be seen in the curve for 25 kV applied voltage

Table 3.3   Excitation Voltage and Wavelength of K-Characteristic X-Rays

Target	Excitation voltage (kV)	Wavelength (Å)		
		$K\alpha_2$	$K\alpha_1$	$K\beta$
Cr	5.98	2.294	2.290	2.085
Fe	7.10	1.940	1.936	1.757
Co	7.71	1.793	1.789	1.621
Cu	8.86	1.544	1.541	1.392
Mo	20.0	0.7135	0.7093	0.6323
Ag	25.5	0.5638	0.5594	0.4858
W	69.3	0.2188	0.2090	0.1844

but not for lower applied voltages. This is because the characteristic X-rays are not produced by lower applied voltages than the critical *excitation voltage*. For example, the critical excitation voltage V_k for K characteristic X-rays is given by

$$V_k = E_k/e \qquad (3.25)$$

Table 3.3 lists the excitation voltage and wavelength of K characteristic X-rays for several targets that are commonly used in X-ray diffraction studies. The intensity I_c of characteristic X-rays is experimentally given by

$$I_0 \propto i(V - V_0)^n \qquad (3.26)$$

where V_0 is the excitation voltage and n is a constant that depends on V; $n \approx 2$ at $V \leq 3V_0$ and $n \approx 1$ at $V > 3V_0$.

3.2.3   Absorption and Scattering of X-Rays

Several effects such as absorption, scattering, and electron displacement occur when X-rays pass through a substance as shown in Fig. 3.18. When X-rays encounter any substance, they are partially absorbed and partially

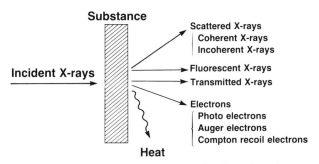

Fig. 3.18.   Effects produced by X-rays passing through a substance.

transmitted. The fractional decrease in the intensity, which has the initial intensity of I_0, upon passing through any homogeneous substance, is proportional to the distance traversed t, and the resulting intensity I is given by

$$I = I_0 \exp(-\mu t) \tag{3.27}$$

where μ is called the *linear absorption coefficient* and depends on the density ρ of the substance as well as on the wavelength of the X-rays. Therefore, the quantity μ/ρ, called the *mass absorption coefficient*, is a constant for the substance and is independent of its physical state (i.e., solid, liquid, or gas). Equation (3.27) may then be rewritten in a more general form:

$$I = I_0 \exp\left(-\frac{\mu}{\rho} \cdot \rho t\right) \tag{3.28}$$

The mass absorption coefficient of material containing more than one element is obtained simply by the weighted average of the mass absorption coefficients of its constituent elements. It does not depend on whether the material is a mechanical mixture, a solution, or a chemical compound, or even on whether it is in the solid, liquid, or gaseous state. For example, $w_1, w_2, \ldots$ are the weight fractions of elements $1, 2, \ldots$ in the material and $(\mu/\rho)_1$, $(\mu/\rho)_2, \ldots$ are their mass absorption coefficients, with the mass absorption coefficient of the material $(\mu/\rho)_c$ given by

$$(\mu/\rho)_c = w_1(\mu/\rho)_1 + w_2(\mu/\rho)_2 + \cdots \tag{3.29}$$

As illustrated in Fig. 3.18, when incident X-rays strike an atom of a substance, two different types of secondary X-rays are produced: scattered X-rays and fluorescent X-rays. Scattered X-rays consist of *coherent X-rays* due to scattering by tightly bound electrons and *incoherent X-rays* due to scattering by more loosely bound electrons, which scatter part of the incident beam. The former is called *elastic scattering* or *Thomson scattering*, and the latter *inelastic scattering* or *Compton scattering*; both kinds occur simultaneously and in all directions. Since coherently scattered waves move in phase with the incident waves and undergo reinforcement in certain directions, Thomson scattering can be used in crystal diffraction studies. On the other hand, incoherent scattering due to the Compton effect results in a change in wavelength between incident and scattered waves. This then gives rise to an incoherent background.

If the incident X-rays are of high enough energy (i.e., of short enough wavelength), the emission of photoelectrons and characteristic fluorescent radiation occurs. The fluorescent X-rays are emitted during recovery from the ionized state of an atom ionized by the incident X-rays. The Compton recoil electrons are the loosely bound electrons knocked out of the atom by X-ray quanta, while Auger electrons are those ejected from an atom by characteristic X-rays produced within the atom. Since the fluorescent X-rays and Auger electrons are characteristic of the element, they are used for the chemical analysis of substances.

3.2.4 X-Ray Diffraction

When X-rays impinge on a crystal, the electrons in each atom scatter X-rays in all directions as already described. In Fig. 3.19, consider the incoming X-rays OA and O'B inclined at the angle α_0 to an atomic net plane, and the scattered rays AP and BP' forming the angle α_n with the plane. Here α_n is arbitrary. The difference between the total path lengths OAP and O'BP' is given by

$$\Delta p_1 = \text{OAP} - \text{O'BP'}$$
$$= \text{AD} - \text{CB}$$
$$= a(\cos \alpha_0 - \cos \alpha_n) \tag{3.30}$$

When

$$\Delta p_1 = 0 \tag{3.31}$$

that is,

$$\alpha_0 = \alpha_n \tag{3.32}$$

then these scattered rays are in phase with each other; that is, the waves of the individual rays arriving at PP' form a common wave front. Equation (3.32) is the condition for in-phase scattering by one atomic net plane in a crystal.

Next consider diffraction of X-rays in crystal planes (*hkl*) shown in Fig. 3.20. Notice the incoming rays OA and O'B', and the scattered rays AP and BP'. Both the incoming and scattered rays form the angle with (*hkl*) planes according to Eq. (3.32). The difference between total path lengths of OAP and O'BP' is given by

$$\Delta p_n = \text{OAP} - \text{O'BP'}$$
$$= \text{CA} + \text{AD}$$
$$= 2d \sin \theta \tag{3.33}$$

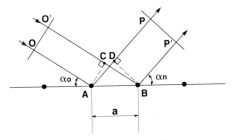

Fig. 3.19. X-Ray scattering.

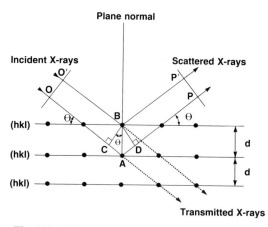

Fig. 3.20. Diffraction of X-rays in crystal planes (*hkl*).

In order that the scattered rays AP and BP' be in phase, or the two planes scatter in phase, the path length difference Δp_n must equal an integral number of wavelengths, $n\lambda$. Thus the condition for in-phase diffraction by a set of parallel crystal planes is

$$n\lambda = 2d \sin \theta \qquad (3.34)$$

Note that the diffraction process is similar to reflection of X-rays by a set of parallel planes (*hkl*). It is common therefore to interchange the words diffraction and reflection of X-rays. Moreover, note that the parallel planes (*hkl*) are not necessarily parallel to the surface of the sample being considered.

Equation (3.34) is known as the *Bragg equation*, and it shows that in-phase diffraction occurs only at certain values of θ_B, corresponding to specific values of n and d. The angles θ_B and 2θ are referred to as the *Bragg angle* and the *diffraction angle*, respectively. The integer n is called the *order of reflection*. The nth reflection of (*hkl*) planes with spacing d is usually considered to be the first ($n = 1$) reflection of planes of (*nh nk nl*) with spacing d/n. Therefore, Eq. (3.34) may be rewritten as

$$\lambda = 2 \frac{d}{n} \sin \theta$$

$$= 2d_{nh\,nk\,nl} \sin \theta \qquad (3.35)$$

One can also derive the following equation from Eq. (3.34):

$$\sin \theta = n\lambda/2d \le 1 \qquad (3.36)$$

That is, the wavelength λ of X-rays must be less than $2d$ to be diffracted by crystal planes.

3.2.5 Electron Diffraction

Fast electron beams are obtained by a filament operating in much the same way as shown by Fig. 3.14. As briefly discussed in Section 2.1, electrons are charged particles and the paths are described by their associated waves. The theory of wave mechanics indicates that this wavelength λ is given by the ratio of Planck's constant h to the momentum M_p of the particle:

$$\lambda = h/M_p \tag{3.37}$$

The momentum M_p is given by

$$M_p = mv \tag{3.38}$$

The kinetic energy of the electrons at the applied voltage V is given by

$$mv^2/2 = eV \tag{3.39}$$

Therefore

$$mv = \sqrt{2meV} \tag{3.40}$$

and then from Eqs. (3.36), (3.37), and (3.39),

$$\lambda = \sqrt{h/2meV}$$
$$= \sqrt{150/V} \tag{3.41}$$

where λ is in angstroms and V is in volts.

Equation (3.41) requires a small correction at higher applied voltages, higher than several tens of kilovolts, due to the variation of electron mass with velocity according to the theory of relativity. The relativistic correction for electron mass is given by

$$m = m_0/[1 - (v/c)^2] \tag{3.42}$$

where m_0 is the rest mass of electron. Equation (3.41) is then rewritten as

$$\lambda = \sqrt{150/V}/\sqrt{1 + 0.9788 \times 10^{-6}} \tag{3.43}$$

According to Eq. (3.43), Fig. 3.21 shows the relative velocity (v/c), as a fraction of the velocity of the light c, and the wavelength λ (Å), of the accelerated electron as a function of applied voltages. Since practical applied voltages in diffraction studies lie in the range of less than several tens of kilovolts, only Eq. (3.41) may be needed. Table 3.4 lists the λ (Å) and v (cm/sec) values of electrons in the range of practical applied voltages. Note that the electron wavelength is considerably shorter than that of X-rays. Moreover, note that the wavelength of the primary electron beam, unlike X-rays, is continuously variable with applied voltages.

During passage of an electron wave through a substance, scattering of the electrons occurs at the atoms in the substance in all directions. Moreover,

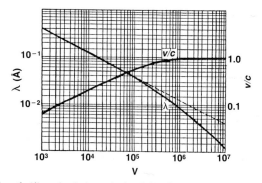

Fig. 3.21. Wavelength (λ) and relative velocity (v/c) of accelerated electrons as a function of applied voltage.

electrons, similar to X-rays, are diffracted by crystals; here the Bragg equation also applies. Table 3.5 compares[5] the similarities and differences between X-ray and electron diffraction. The fundamental characteristic differences between them are in (1) wavelength, (2) interaction with substances, and (3) reflection effects. Those differences are next described in more detail.

First, the wavelength of electrons is much shorter than that of X-rays (compare Tables 3.3 and 3.4). This shorter-wavelength primary beam allows the obtaining of diffraction from higher-order crystal planes [see Eqs. (3.35) and (3.36)]. Because of the short wavelength, and in turn the small diffraction angle, the function $\sin \theta$ in the Bragg equation (3.34) can be replaced with θ. That is,

$$\sin \theta \approx \theta \qquad (3.44)$$

This simplification has an important consequence when analyzing diffraction patterns using the *reciprocal lattice*.[6]

Table 3.4 Wavelength and Velocity of Electrons as a Function of Applied Accelerating Voltage

Applied voltage (kV)	Wavelength (Å)	Velocity ($\times 10^8$ m/sec)
1	0.3876	0.1874
10	0.1220	0.5848
50	0.0536	1.2377
100	0.0370	1.6441
200	0.0251	2.0855
500	0.0142	2.5878
1000	0.0087	2.8223
2000	0.0050	2.9363

Table 3.5 Comparison of X-Ray Diffraction (X) and Electron Diffraction (E)[a]

Similarities
1. Nature of superposition of waves
   Bragg's law
   Structure factor
   Extinction laws
2. Types of diffraction patterns
   Laue
   Debye-Scherrer
   Texture pattern

Differences
1. Nature of scattering at individual atom
   E: Scattering by atom nucleus
   X: Scattering by shell electrons
2. Wavelength of radiation
   E: So small that diffraction angle is only 0–2°, therefore sin θ and the diffraction pattern is approximately a plane section through reciprocal lattice
   X: All diffraction angles up to 180° occur, therefore the locus of diffraction spots is the Ewald sphere
3. Intensity of diffraction spots
   Because stronger interaction with the atom nucleus occurs with E, the intensity is 10^6–10^7 times that observed with X.
4. Penetration of radiation
   As a consequence of (3); E, order of magnitude 5 μm or less, X, order of magnitude 100 μm or more.
5. Affected sample volume
   As a consequence of (4) and beam cross section; E, order of magnitude 10^{-9} mm^3; X, order of magnitude 5 mm^3.

---

[a] After von Heimendahl.[5]

Second, electrons interact with substances much more than X-rays do, which means that electrons are scattered much more intensely than X-rays, so that even a very thin layer of material gives a strong diffraction. However, due to this stronger interaction, electrons penetrate much less into material than X-rays and are very easily absorbed, even by air. Electron diffraction is therefore well suited to the study of thin surface layers. In addition, because of this strong interaction with substances, the *dynamical theory* as well as the *kinematical theory* is required to understand the electron diffraction phenomena precisely.[6]

Third, it is important to note that electrons, unlike X-rays, can be deflected and focused by electrostatic or electromagnetic lenses, in complete analogy to the focusing of light waves by glass lenses. This nature of electron beams has been widely applied to electron microscopy since 1932, when the idea of using electron beams for producing enlarged images was first carried out by two independent research groups[5]: Knoll and Ruska, who produced a magnetic-type electron microscope, and Bruche and Johannson, who produced an electrostatic type. When electrons, with charge e and velocity v, approach the lens nearly parallel to the optical axis, the *Lorentz force* **F** exerted by the magnetic field **B** on an electron is given by

$$\mathbf{F} = e[\mathbf{V} \times \mathbf{B}] \tag{3.45}$$

The force acts normal to the direction of the magnetic field and normal to the velocity of electrons. Therefore, the force **F** accelerates the electrons out of the axial direction and into a helical path. The electrons passing through the field are thus deflected toward the optical axis. Accordingly, the effect of the electron lens on the electron beam is comparable to the reflection of a glass lens acting on light rays.

3.3   Properties of Silicon Crystal

3.3.1   Structure of Crystal Lattice

As described briefly in Section 3.1, silicon belongs to the cubic system and has the *diamond structure*.[7] Figure 3.22 shows the structure of the unit cell of silicon crystal. The diamond structure consists of two interpenetrating fcc subcubes with the origin of one displaced $\frac{1}{4}\frac{1}{4}\frac{1}{4}$ from the other (i.e., one atom of the second subcube occupies the site at one-fourth of the distance along a

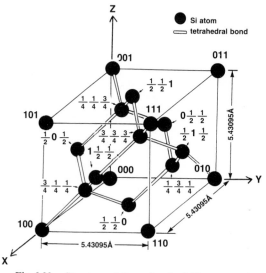

Fig. 3.22. Structure of the unit cell of silicon crystal.

major diagonal of the first subcube). According to Eq. (3.1), the coordination number for this structure is eight. Therefore, the atomic density n_a of the silicon crystal is given by

$$n_a = 8/(5.43095 \times 10^{-8})^3 \quad \text{(atoms/cm}^3\text{)}$$
$$\approx 5 \times 10^{22} \quad \text{(atoms/cm}^3\text{)} \tag{3.46}$$

Each silicon atom has four valence electrons, which provide tetrahedral covalent bonding with the nearest neighbors. The distance between two neighboring atoms is $(\sqrt{3}/4)a$, that is, 2.35167 Å. The tetrahedral covalent radius of the silicon atom is thus 1.17584 Å, assuming a "hard-sphere" model for atoms. A simple calculation using the hard-sphere model shows 34% of the silicon lattice is occupied by atoms. By way of comparison, the packing of a close-packed structure (i.e., Cu, Fe, etc.) is approximately 74%; thus the silicon lattice has a relatively loosely packed structure, which allows considerable incorporation of impurities.

3.3.2 Properties of Crystal Planes

General Remarks Mechanical, physical, chemical, and electronic properties of crystalline materials strongly depend on the crystal orientation.[8] In fact, many electronic device fabrication processes are sensitive to the crystallographic planes of semiconductor substrates used. These characteristic properties are attributable principally to the planar configuration of atoms and spacing of atomic net planes. Consider the properties of silicon {100}, {110}, and {111} planes, which are the most important in the semiconductor technology. The orientation dependence of the properties is perhaps most graphically described by Fig. 3.23, in which the planar atomic configuration

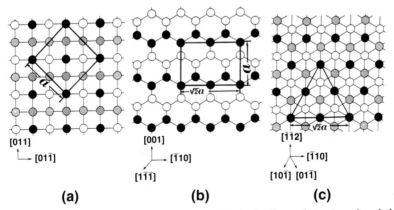

Fig. 3.23. Planar atomic configuration and channeling in the diamond structure viewed along (a) [100], (b) [110], and (c) [111].

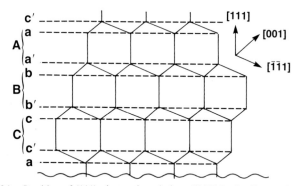

Fig. 3.24. Stacking of (111) planes viewed along [110] in the diamond structure

and channeling in a silicon crystal viewed along [100], [110], and [111] are shown; different shadowing for atoms refers to atoms in different atomic layers. In each figure, (100), (110), and (111) planes in the unit cell are also shown with solid lines. Note that the (100) and (110) unit planes consist of the atoms illustrated with solid circles in Fig. 3.23a and b, respectively, whereas the (111) plane is made up of the atoms illustrated with solid and open circles in Fig. 3.23c, since a {111} plane is defined by two sublayers as shown in Fig. 3.24 viewed along [110]. From this view it is understandable that the ion implantation profiles in crystalline targets with different orientations are very much distinguishable from each other. Moreover, the implant range distributions in single-crystal targets are different from those in amorphous targets, primarily because of the possibility that the implanted ions can channel along open directions, which results in several times more penetration into the lattice.[9]

Table 3.6[10–12] lists several properties of silicon planes oriented to ⟨100⟩, ⟨110⟩, and ⟨111⟩. The anisotropic properties of crystalline silicon are

Table 3.6 Properties of Silicon Crystal Planes

Properties	Orientation		
	100	110	111
Spacing (Å)	5.43	3.84	3.13
Young's modulus (dyn/cm^2)[10]	1.3	1.7	1.9
Surface energy (J/m^2)[11]	2.13	1.51	1.23
Atomic density (10^{14}/cm^2)	6.78	9.59	15.66
Available bond density^a (10^{14}/cm^2)[12]	6.78	9.59	11.76

^{a} The density of available bonds for the reaction with H_2O molecules was calculated with consideration of steric hindrance, although each surface silicon atom, in general, has two bonds capable of reaction.[12]

principally explained by the surface free energy, which is a function of the number of free bonds per unit area of the plane. The following is a discussion of silicon orientation-dependent properties, which are very important to silicon device fabrication; some of these will be discussed in more detail in later chapters.

Dissolution and Growth Rate   The orientation effect on the dissolution rate or the etch rate, in a more practical sense, is attributed to the density of surface free bonds; that is, the relative etch rate increases with the number of surface free bonds.[13,14] For example, the etch rates for $\langle 100 \rangle$, $\langle 110 \rangle$, and $\langle 111 \rangle$ oriented silicon are approximately $50 : 30 : 3 \ \mu m/hr$, respectively, in the case of the water–ethylenediamine–pyrocatechol solution,[15] and $11.1 : 7.0 : 0.7 \ \mu m/hr$, respectively, in the case of the water–copper nitrate–ammonium fluoride solution.[14] This anisotropic etching for (100) silicon wafers through a patterned SiO_2 mask creates precise V-grooves, with the edges being {111} planes at an angle of $54.74°$ from the (100) surface. As illustrated in Fig. 3.25, the depth d_g, of the groove is determined by the pattern width W on the (100) surface, since etching will stop automatically at the point where the {111} planes intersect each other and the (100) bottom surface no longer exist. Therefore, d_g is given by

$$d_g = (W/2) \tan 54.74° \tag{3.47}$$

This preferential etching technique has been applied to dielectric isolation in the integrated circuit technology.[16–18] However, note that pyramidal etch hillocks bounded by four convergent {111} planes or their approximation can be formed on a {100} silicon slice surface by improper etching or by surface dirt.[19]

Anisotropic etching rates have been found also in vapor-phase etching, which, in addition to the chemical cleaning of the substrates, normally precedes epitaxial silicon deposition (see Section 5.6 for details). *In situ*

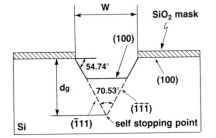

Fig. 3.25.   Anisotropic etching for a (100) silicon wafer through a patterned SiO_2 mask.

vapor-phase etch rates of $\langle 100 \rangle$-, $\langle 110 \rangle$-, and $\langle 111 \rangle$-oriented silicon sub-strates were found to be 3.4 : 3.0 : 1.5 μm/min, respectively, in the case of 5 % HCl in H_2 carrier gas at 1200°C.[8]

The orientation-dependent crystal growth rate is explained similarly to the dissolution rates and also depends on the crystal orientation. In general, the crystal orientations that grow most slowly are those perpendicular to the plane with the closest packing; that is, the more close-packed planes have the lower specific surface energy (see Table 3.6) and hence would be the planes most likely to develop as interfaces.[20] Therefore, the preferred growth habit of silicon is octahedral, that is, bounded by eight $\{111\}$ family planes.[21] As expected from the surface energy and the areal atomic density listed in Table 3.6, the growth rate of silicon crystals formed on a free melt surface (grown under the condition without any constraints) shows the maximum in a $\langle 100 \rangle$ orientation and the minimum in a $\langle 111 \rangle$ orientation.[22] Note, however, that $\langle 111 \rangle$ silicon crystals have been grown practically as fast as $\langle 100 \rangle$ crystals by state-of-the-art silicon crystal growth processing, which is computer pro-grammed.

Historically $\langle 111 \rangle$-oriented silicon crystals, because of their higher crystal-lographic quality, have been used for semiconductor device applications such as alloy junction devices and bipolar transistors; however, $\langle 100 \rangle$ oriented crystals would be more preferable for most of the silicon electronic device fabrication, not only because of their faster growth rate but for also other factors that will be discussed later in this section. In fact, the consumption rate of $\langle 100 \rangle$ silicon, particularly of large-diameter crystals, has rapidly increased in the semiconductor technology. Although $\langle 110 \rangle$ silicon has inherent attractive features for some semiconductor device applications such as the vidicon-type target design,[17] the $\langle 110 \rangle$ crystal growth is much less favorable than $\langle 111 \rangle$ or $\langle 100 \rangle$ due to the $\{111\}$ twin or dislocation plane (see next section for the details), which is perpendicular to the $\{110\}$ growth plane.[8,10] When the $\langle 110 \rangle$ crystal dislocates during crystal growth, the dislocation plane parallels the $\langle 110 \rangle$ growth axis or inclines at 54.74° away from it since dislocations are constrained to lie on $\{111\}$ planes. Consequent-ly, any dislocations formed continue to grow throughout the length of the crystal. On the other hand, there is no dislocation plane perpendicular to a $\{111\}$ or $\{100\}$ growth plane. Therefore, the removal of the dislocations on inclined planes, which are at 70.53° and 54.74° in the $\langle 111 \rangle$ and $\langle 100 \rangle$ growth, respectively, can be accomplished by forming a thin neck according to the Dash technique,[23,24]

Mechanical Properties   The $\{111\}$ planes have the highest density of atoms and the highest elastic modulus, so that the forces between adjacent $\{111\}$ planes are much weaker than those of other planes. The $\{110\}$ planes have the

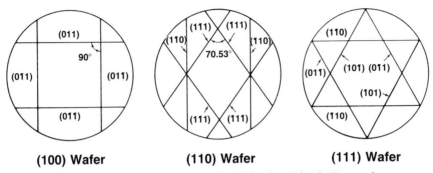

(100) Wafer   **(110) Wafer**   **(111) Wafer**

Fig. 3.26. Cleavage planes normal to {100}, {110}, and {111} silicon wafers.

second most number of atoms and second highest elastic modulus. As a result, silicon crystals cleave most easily and most often between {111} planes. The {110} planes are the next most preferred cleavage planes and are used to advantage in silicon device processes such as chip dicing. Figure 3.26 compares the cleavage planes normal to {100}, {110}, and {111} silicon wafers. It is quite understandable that {100} wafers are most favored from the view point of device chip dicing.

Wafer edge chiping, which may produce scratches on the wafer surface or weaken the wafer and may lead to breakage during subsequent processing, is occasionally induced by improper wafer shaping processes such as slicing, grinding, and lapping (see Sections 5.5 and 7.2 for details). Through experience, it has been found that the {100} and {110} wafers chip more easily than {111} wafers with similar shaping processes. It has also generally been noticed that {100} and {110} wafers are more difficult to polish to a mirror-like surface both mechanically and chemically. This may be attributable to less tightly packed atomic structures and faster etch rate of {100} and {110} planes.

Thermal Oxidation Rate   The formation of the surface oxide of silicon on a silicon wafer is one of the most important processes in silicon electronic device fabrication. The oxide growth is explained by the widely accepted linear-parabolic model,[25] by which the oxide growth rate is governed by linear and parabolic rate constants. It has been experimentally observed that the linear rate constant strongly depends on the crystal orientation,[8] although the parabolic rate constant is relatively independent of the crystal orientation. The growth rate for $\langle 111 \rangle$ silicon has been widely found to be considerably greater than that for $\langle 100 \rangle$[26–28]; however, some arguments about the comparison between the growth rates for $\langle 111 \rangle$ and $\langle 110 \rangle$ silicon have been made.[12,29–33]

The dependence of the oxidation rate on the crystal orientation has been explained by the available surface bond density[12,34] (see Table 3.6). The number of Si–Si bonds capable of reacting with H_2O molecule was calculated for the surface-oriented $\langle 100 \rangle$, $\langle 110 \rangle$, and $\langle 111 \rangle$ with the consideration that reacting H_2O molecules are sufficiently large that other H_2O molecules are presumably shielded from reacting with adjacent Si–Si bonds. This geometric effect is called *steric hindrance*. Consequently, the higher growth rate of silicon oxide is attributed to more available bonds.

Oxide Charges and Traps   Various charges and traps[35] are associated with thermally oxidized silicon, since the Si–SiO_2 interface contains a transition region (between crystalline silicon and amorphous silica), both in terms of atomic position and stoichiometry. It has been observed that the oxide fixed charge density and the interface-trapped charge density strongly depend on the crystal orientation.[35–37] The lowest values of those are obtained for $\langle 100 \rangle$-oriented silicon and the highest are for $\langle 111 \rangle$ silicon. This result has been correlated with the number of available bonds per unit area of silicon surface.[12,35] As a result, the $\langle 100 \rangle$-oriented silicon wafers are preferable for integrated circuit technology, particularly for the fabrication of metal oxide semiconductor field-effect transistors (MOS FETs).[34,38]

Pattern Displacement during Epitaxy   Epitaxial silicon film technology has given rise to great innovations in bipolar transistor device fabrication. Metal oxide semiconductor (MOS) technology is also beginning to appreciate the advantages of using the epitaxial silicon structure.[39] In bipolar device fabrication processes, one or more diffusions to form buried layers are applied to the substrate prior to epitaxial silicon film growth. After the growth of an epitaxial layer, the features replicated in the epitaxial surface usually show some pattern displacement, that is, pattern shift and pattern distortion. This pattern displacement strongly depends on the orientation of the substrate[40] as well as on the epitaxial conditions. It has been recognized that silicon substrates misoriented by 3–5° from the $\langle 111 \rangle$ direction toward the nearest $\langle 110 \rangle$ direction show a minimum pattern distortion, while for $\langle 100 \rangle$ silicon the best results are obtained when the substrate is precisely oriented.[41,42]

Oxidation-Induced Stacking Faults   The generation of stacking faults in the surface region during oxidation of silicon wafers is very common, and these defects, which are called oxidation-induced stacking faults (OSF or OISF), are most detrimental to electronic device performance.[43–46] It has been observed that the growth rate of OSF depends on the orientation of the silicon substrate[47-49] as well as on oxidation conditions. The growth rate of OSF is found to be about the same for $\langle 100 \rangle$- and $\langle 110 \rangle$-oriented surfaces, but is faster by about a factor of three for $\langle 111 \rangle$-oriented surfaces.[47] The durability to the attack by HF acid has also been observed to depend on the orientation of silicon crystals.[50,51] The HF acid attacks and disorders more the $\langle 100 \rangle$-oriented surface, which results in more OSFs than the $\langle 111 \rangle$ surface.

Comprehensive Judgment   As discussed, many factors depend on the crystal orientation. The $\{100\}$, $\{110\}$, and $\{111\}$ wafers have their advantages and disadvantages from the viewpoint of their use in electronic device fabrication. Although $\langle 511 \rangle$ crystallographic orientation is not commonly recognized in silicon device technology, the $\{511\}$ wafers, because of their low surface state density close to the value for $\{100\}$ wafers[37] and the more uniform radial dopant distribution compared to $\{111\}$ or $\{100\}$ wafers,[52] are considered to be favorable for the fabrication of integrated circuits.[53] In addition, the crystal growth of $\langle 511 \rangle$-oriented silicon does not present particular difficulties, and the shaping of $\{511\}$ wafers is as easy as for $\{111\}$ and $\{100\}$ wafers.

In summary, Table 3.7 compares several factors for $\langle 100 \rangle$-, $\langle 511 \rangle$-, $\langle 110 \rangle$-, and $\langle 111 \rangle$-oriented silicon wafers by judging each factor with simplified ranking from the viewpoints of wafer production and application to device

Table 3.7   Practical Comparison of Silicon Wafers with Different Orientation

Factors	Wafer orientation[a]			
	$\langle 100 \rangle$	$\langle 511 \rangle$	$\langle 110 \rangle$	$\langle 111 \rangle$
Grown-crystal quality	○	○	●	◎
Growth rate	◎	○	◑	○
Crystal production cost	◎	●	●	◎
Mechanical strength	○	○	○	◎
Square-chip dicing	◎	◑	◑	◑
Radial dopant distribution	◎	◎	◑	○
Oxidation rate	○	○	◎	◎
OISF growth rate	◑	◑	◑	○
Surface state density	◎	◎	○	◑
Epi-pattern displacement	◎	◎	○	◑
Comprehensive judge	◎	◑	●	○

[a] ◎, Excellent; ○, good; ◑, acceptable; ●, poor.

fabrication. Finally, it is concluded that {100} wafers are most preferred for silicon device fabrication.

3.4 Crystal Defects

As will be discussed in later chapters, crystal defects have great influence, whether detrimental or beneficial, on device performance. Crystal defects or imperfections in crystals may include various types, such as structural defects and electronic defects. The discussion in this section will concentrate on structural defects in single-crystal silicon. It is appropriate to highlight crystal defects in silicon according to their geometry, although most crystal defects in silicon crystal processed in device fabrication do not exist independently but interact each other. Table 3.8 lists those defects categorized according to their geometry; most of the defects listed are shown in Fig. 3.27, which illustrates a two-dimensional simple cubic array for convenience. It should be noted that, as shown in Figs. 3.22–3.24, the silicon crystal lattice has the diamond structure, which differs from the lattice shown in Fig. 3.27.

In addition to those crystal defects listed in Table 3.8, there are dangling bonds at the crystal surface and the ends of crystal defects, such as dislocations and incoherent precipitates. These dangling bonds in the surface region are directly related to surface charges and traps as discussed in the previous section. Moreover, the dangling bonds caused by crystal defects play

Table 3.8   Crystal Defects in Silicon[a]

Geometry	Defects
Point	Intrinsic point defect
	Vacancy[a]
	Self-interstitial[b]
	Extrinsic point defect
	Substitutional impurity atom[c]
	Interstitial impurity atom[d]
Line	Dislocation
	Edge dislocation[e]
	Screw dislocation
	Dislocation loop (extrinsic type[f] and intrinsic type[g])
Plane	Stacking fault (extrinsic type and intrinsic type)
	Twin
	Grain boundary
Volume	Precipitate[h]
	Void (negative crystal)[i]

[a–i] Defects illustraed in Fig. 3.27.

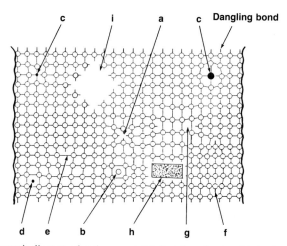

Fig. 3.27.   Schematic diagram of various crystal defects in a simple cubic lattice (modified from the figure in Ref. 54 by Zulehner and Huber). Refer to Table 3.8.

a very important role in intrinsic gettering (IG),[55] which will be discussed in Section 7.4.

3.4.1   Point Defects

Intrinsic Point Defects   In general, a relatively small fraction of the atoms leave their usual sites in the lattice to go into the spaces between the normal atom positions due to thermal lattice fluctuation. These displaced atoms are referred to as *self-interstitials*, while the lattice site vacated by the atom is referred to as a *vacancy*. Self-interstitials and vacancies are classified as *intrinsic point defects*. A self-interstitial and vacancy pair is called a *Frenkel defect* and is schematically shown in Fig. 3.28a. Vacancies can also be created in the lattice by the diffusion of self-interstitials to the crystal surface, where they deposit themselves on the surface to form new layers. The defects formed this way are referred to as *Schottky defects* and are illustrated in Fig. 3.28b. It

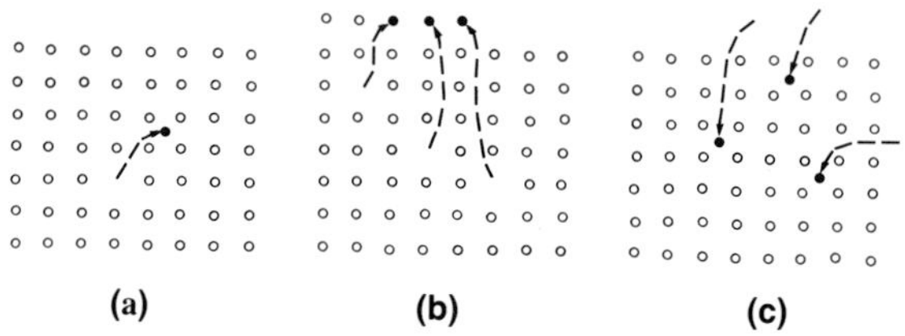

Fig. 3.28.   Intrinsic point defects: (a) Frenkel defect, (b) Schottky defects, and (c) extrinsically induced self-interstitials.

is evident that Frenkel defects consist of equal numbers of vacancies and self-interstitials in the crystal lattice, while Schottky defects leave only vacancies in the lattice. In addition to the formation of Frenkel defects, self-interstitials in the crystal lattice may be induced extrinsically as schematically illustrated in Fig. 3.28c. In fact, self-interstitials extrinsically generated at the interface of growing silicon oxide, either the surface SiO_2 layer or interior oxygen precipitates (SiO_x), play a very important role in silicon device processing; those self-interstitials form oxidation-induced stacking faults (OSF) and strongly affect the diffusion of impurities as well as oxygen precipitation in the silicon crystal. The behavior of extrinsically generated self-interstitials will be discussed in detail in Chapter 7. The intrinsic point defects may be associated or combined in several different ways according to the *law of mass action*.[56] The combination of a vacancy and a self-interstitial obviously leads to their mutual annihilation. Two vacancies may come together to form a vacancy pair or so-called *divacancy*. The arrangement of the bonds of those atoms surrounding the vacant lattice site can be imagined as "dangling" as shown in Fig. 3.27. However, as shown in Fig. 3.29a, it may be considered that the atoms labeled B, C, D, and E have an uncompleted covalent bond because of the vacancy.[57] The unshared electron of the atom E is shared with those of B and D. Similarly, the unshared electron of the atom B is shared with C and E, and so on. The bonds shared with three atoms instead of two as in the normal covalent bond are referred to as *defect bonds*.[57] The possible bond arrangements for agglomerates of multiple vacancies or a negative crystal (see part i in Fig. 3.27) are also diagrammatically shown in Fig. 3.29b. The presence of defects in the crystal lattice changes the internal energy as well as the entropy. Therefore, their equilibrium concentration is a function of the energy of formation and of temperature. Consequently, the equilibrium concentration of intrinsic point defects may be theoretically calculated by applying thermodynamics to purely geometric lattice points.[20]

For the case of vacancies,

$$X_v = \frac{n_v}{N_L} = \exp\left(\frac{\Delta S_v}{k}\right)\exp\left(\frac{-\Delta H_v}{kT}\right) \tag{3.48}$$

where X_v, n_v, N_L, ΔS_v, and ΔH_v are the fraction of total sites vacant, number of vacancies, number of normal lattice sites, vibrational entropy change associated with vacancy formation, and enthalpy of vacancy formation, respectively. Similarly, for the case of interstitials,

$$X_i = \frac{n_i}{N_i} = \exp\left(\frac{\Delta S_i}{k}\right)\exp\left(\frac{-\Delta H_i}{kT}\right) \tag{3.49}$$

where X_i, n_i, N_i, ΔS_i, and ΔH_i are the fraction of interstitial sites, number of interstitials, total number of interstitial sites, vibrational entropy change associated with interstitial formation, and enthalpy of interstitial formation,

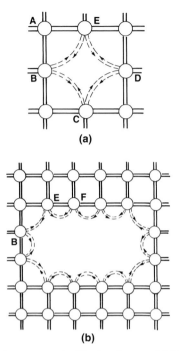

Fig. 3.29.   Two-dimensional arrangement of the bonds around (a) a single vacancy and (b) an agglomerate of multiple vacancies. (After Champion.[57])

respectively. Since the ΔS terms are not expected to be very important compared with the ΔH terms in Eqs. (3.48) and (3.49), the concentration of a particular type of defect in a crystal may be determined primarily through the magnitude of ΔH.[20] That is, a large positive value of ΔH will result in a small value of concentration.

However, it is important to understand that, as illustrated in Fig. 3.28, intrinsic point defects can be created in three different ways and the formation mechanisms may act independently of one another. In addition, the presence of various sources and sinks for point defects is inevitable in real crystals; in particular, the interaction of intrinsic point defects with impurities must be considered in semiconductor crystals used for electronic device fabrication. That is, intrinsic point defects react not only with each other but also with chemical impurities in the crystal. As a result, the magnitude of the terms ΔS and ΔH in practice may depend considerably on the crystal quality. It has been calculated that during the growth of a silicon crystal from the melt, a high density ($\sim 10^{18}/cm^3$) of vacancies is created at temperatures near the melting point (1420°C). The concentration of self-interstitials at these temperatures, however, is believed to be much less ($\sim 10^7/cm^3$).[56] Considering the interaction between these point defects, the concentrations of these intrinsic

point defects at room temperature may depend strongly on the cooling rate of the grown crystal. Therefore, the incorporation of intrinsic point defects is very much affected by the crystal growth conditions. It has been observed that the type and concentration of point defects that remain in the grown crystal depends on the ratio $G_p/\Delta T$, where G_p is the crystal growth rate and ΔT the temperature gradient at the solid–melt interface.[58] That is, the predominant defects are vacancies if $G_p/\Delta T > \xi$, while self-interstitials are predominant defects if $G_p/\Delta T < \xi$, where ξ is a certain constant. The question of how many intrinsic point defects are present in silicon crystals is still a hotly argued subject.[59–61] As a result, several different values of ΔS and ΔH for vacancies and self-interstitials in silicon have been separately reported.[4,54,59,61–64]

Impurities   The importance of silicon in the electronic device technology relies on semiconducting properties, which can be controlled by small additions of impurity elements, that is, *dopants*, such as Group III elements (e.g., boron) for *acceptors* and Group V elements (e.g., phosphorus) for *donors*. Although these dopant impurities play a key role in semiconductor device operation, they are referred to as *extrinsic point defects* from a crystallographic point of view. In addition to those impurities intentionally doped, other impurities such as oxygen, carbon, and transition metals innevitably are incorporated into silicon crystals during device fabrication as well as during crystal growth and wafer shaping processes (see Chapters 5 and 7). These are also called extrinsic point defects involving foreign atoms. The foreign atoms that occupy the silicon lattice sites are referred to as *substitutional impurities*, while those located at interstitial sites are *interstitial impurities*. Substitutional impurities, which generally are either larger or smaller than the host atom, expand or contract the lattice regularity depending on the size of the impurity. These extrinsic point defects are shown in Fig. 3.27.

As noted in Section 3.3.1, the diamond lattice is a relatively loosely packed structure. Figure 3.30 shows the preferred interstitial sites in the unit cell of the diamond lattice. There are five sites that are tetrahedrally surrounded by other atoms within the unit cell.[56] By applying Eq. (3.1), a total of eight interstitial sites per unit cell is obtained for the diamond structure. Note that the number of preferred interstitial sites is identical with that of the regular lattice sites [see Eq. (3.46)]. As a result, interstitial atoms can be easily accommodated in silicon crystals, although there is a slight restriction of their radius, which must be smaller than $0.885R$ (R is the radius of the host atom) in order to pass from one interstitial site to another.[65] This corresponds to a atomic radius of 1.04 Å or less for the silicon lattice.

There is a maximum specific concentration that the crystal lattice can

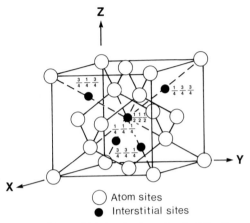

Fig. 3.30.   Preferred interstitial sites in the unit cell of the diamond lattice.

accept in a solid solution of itself and the impurity. This maximum concentration is referred to as the *solid solubility* and depends on the element as well as the temperature. *Hume-Rothery's rule*,[66] based on the empirical observation, in general regulates the solid solubility of the impurity (solute) in host (solvent) by the following three factors: (1) the atomic size, (2) the electrochemical effect, and (3) relative valency effect. According to the atomic size factor, a low solubility can be generally expected if the atomic radii of the solvent and solute atoms differ by more than 15%. For the diamond lattice crystals, to which silicon belongs, the tetrahedral covalent radius of the impurity relative to that of the host should be taken into account. Table 3.9

Table 3.9   The Tetrahedral Covalent Radius, the Ratio to That of Silicon, and the Site in Silicon Lattice[a]

Ib	IIb	IIIb	IVb	Vb	VIb
		$5B^S$	$6C^S$	$7N^I$	$8O^I$
		0.88	0.77	0.70	0.66
		0.75	0.66	0.60	0.56
		$13Al^S$	$14Si$	$15P^S$	$16S^-$
		1.26	1.17	1.10	1.04
		1.08	1.00	0.94	0.89
$29Cu^{I,S}$	$30Zn^-$	$31Ga^S$	$32Ge^S$	$33As^S$	$34Se^-$
1.35	1.31	1.26	1.22	1.18	1.14
1.15	1.12	1.07	1.04	1.01	0.97
$47Ag^-$	$48Cd^-$	$49In^S$	$50Sn^S$	$51Sb^S$	$52Te^-$
1.53	1.48	1.44	1.40	1.35	1.32
1.31	1.26	1.23	1.20	1.16	1.13

[a] Superscripts: S, substitutional site; I, interstitial sites; —, data not available.

lists some elements with the tetrahedral covalent radius and the ratio to that of silicon according to the periodic table. The site of each element in the silicon lattice is also shown with the superscript, where S refers to the substitutional site, I the interstitial sites, and—to data not available.[4] Moreover, the electrochemical and relative valency effects dominate the magnitude of the solubility of impurities in silicon.[56] As discussed in Section 2.2.2, the atoms of the silicon lattice are held together in a tetrahedral arrangement by four covalent bonds of the sp^3 type. Since those elements belonging to Groups II–V of the periodic table have their valence electrons in the s and p states (see Table 2.1), they may fit in with the sp^3 bonding of silicon by electron sharing or hybridizing. This type of bonding is easiest and strongest for Group IV elements; therefore, silicon and germanium are mutually soluble in all proportions.[67] The impurity elements belonging to Groups III and V, which are electrically active when located at substitutional sites, and have the highest solubilities in the diamond lattice.[56] The incorporation of the transition-metal impurities (e.g., Cr, Fe, Co, Ni, and Cu)[68] and Group Ib elements (e.g., Cu, Ag, and Au) in the diamond lattice causes relatively large electrostatic effects and lattice distortions; therefore, the solubilities of these impurities are very low compared with those of other impurities mentioned above.

Figure 3.31 shows the solid solubility of some impurities in silicon as a function of temperatures.[69,70] The solid solubility of these impurities is observed to increase with temperature and to reach a peak value, then to fall sharply as the temperature approaches the melting point of silicon. This behavior is commonly termed a *retrograde solid solubility.*

Impurities such as carbon and oxygen, which are generally incorporated into silicon crystals during the growth, play a very special role in generation of process-induced defects; therefore, these impurities will be discussed in detail in Chapters 5 and 7. To understand the behavior of an impurity in silicon, it is essential to consider the solubility, and the diffusion and segregation coefficients. It should be noted that the solubility of impurities as well as the concentration of intrinsic point defects depends considerably on the quality of the crystal. Thus the solubility data empirically obtained by early workers and those of today using state-of-the-art silicon crystals may differ from each other due to the difference in the density of grown-in defects.

3.4.2   Line Defects

General Remarks   Line defects or one-dimensional defects in crystals take the form of *dislocations*, which, as the terminology implies, are geometrical faults or disturbances in an otherwise perfect crystal lattice. When an external force, such as a tensional, compressional, or shearing force, is applied to a crystal, the crystal will be deformed either elastically or plastically, depending

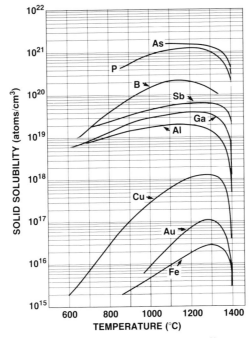

Fig. 3.31. Solid solubility of impurities in silicon. (After Trumbore[69] and Vick and Whittle.[70])

on both the strength of the force and the crystal. That is, in the case of elastic deformation, the crystal recovers to its original state when the force is removed; however, when the force exceeds the elastic limit or *yield strength*, the crystal is plastically deformed, resulting in dislocations.

In this section, a simple cubic lattice is considered first to understand the fundamentals of dislocations, and then dislocations in the diamond lattice, which is more complex, are discussed.

Edge Dislocations Figure 3.32 is a representation of a crystal with an applied shearing force along ABCD. Glide has occurred over the region AEFD of the *slip plane* ABCD. The boundary between the slipped and unslipped regions is indicated by the broken line EF normal to the slip direction shown by arrows. Here an *extra half-plane* EFGH is present in the otherwise regular lattice. This line EF of atomic misfit is referred to as the *dislocation line* and is defined as an *edge dislocation*. The dislocation line is commonly indicated with a symbol of ⊥. When this takes place from the right to the left face, the crystal section above the slip plane will be displaced by one repeat distance *b* relative to the section below the slip plane. The vector **b** is referred to as the *Burgers vector*. In the case that **b** is perpendicular to the

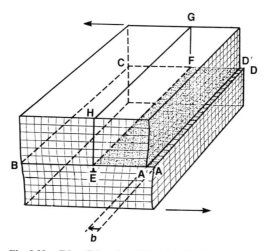

Fig. 3.32. Edge dislocation EF in the slip plane ABCD.

dislocation line, such a dislocation is called a *pure edge dislocation*. The edge dislocation illustrated in Fig. 3.32 represents a "positive edge dislocation." If, on the other hand, it were turned through 180° (i.e., the extra half-plane of atoms was inserted below the slip plane), the dislocation would be referred to as "negative" for convenience. By itself this has no significant meaning, but it becomes important when discussing the interaction of dislocations.

The edge dislocation could also be created theoretically by cutting the crystal along EFGH and inserting the extra half-plane EFGH into the two exposed faces. From this schematic arrangement, it is obvious that there must be considerable internal strain associated with the extra half-plane in the vicinity of the dislocation line EF. That is, the atoms in the section above the slip plane are compressed along the direction of the arrow, while those below are in tension. A detailed discussion of the energy associated with dislocations can be found in Refs. 71 and 72.

Edge dislocations are extremely mobile in their own slip plane under the application of a shearing force. Figure 3.33 illustrates the initiation of an edge dislocation, its movement along the slip plane AB, and disappearance at the right-hand face in the sequence. The mechanism for such a movement is called *slip*. The *slip system*, which consists of the slip direction and slip plane, is characteristic for a given crystal system in most cases. The preferred slip direction is almost invariably that of the shortest lattice vector—that is, the slip direction is determined almost entirely by the crystal structure. However, the slip plane differs from one crystal to another of the same structure, and may change even in a given crystal if the direction of the applied stress or the temperature is changed. Nevertheless, the general rule is that the operative

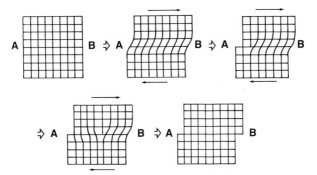

Fig. 3.33. A sequence of simplified diagrams showing the movement of a edge dislocation from its initiation through disappearance.

slip planes are those that are most densely packed with atoms.[71] For the diamond lattice, this leads to {111} slip planes and ⟨110⟩ slip directions, corresponding to the shortest Burgers vector $\frac{a}{2}$ ⟨110⟩.

Screw Dislocations Figure 3.34 shows the manner in which a regular crystal lattice is subjected to a shearing force in order to establish another type of dislocations. The crystal is imagined to be cut along the section AEFD parallel to its slip plane ABCD by applying a shearing force as indicated by arrows. The two halves are displaced one atomic distance AA' or DD', that is, the Burgers vector **b**, in the direction of slip. The dislocation line EF is parallel to its Burgers vector **b**; a dislocation that has its axis parallel to its Burgers vector is called a *screw dislocation*. With continued application of the

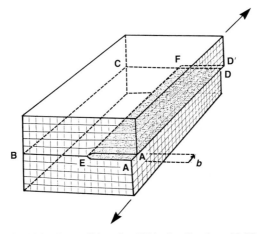

Fig. 3.34. Screw dislocation EF in the slip plane ABCD.

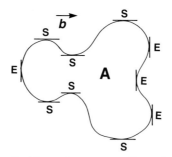

Fig. 3.35.   Dislocation loop of orbitrary shape surrounding region A that has slipped by the amount *b*.

external shear force, the screw dislocation line EF moves by advancing normal to itself until it finally disappears at the opposite crystal surface BC.

As "positive" and "negative" were defined for edge dislocations, a positive screw dislocation is defined as a right-handed screw and a negative one as a left-handed screw. The positive screw moves upward while the negative one moves downward. As in the case of edge dislocations, the same crystal displacement is produced by the motion of positive and negative dislocations in opposite directions.

Dislocation Loops   The dislocation lines discussed above terminate on an external surface of the crystal. If the dislocation lies entirely within the crystal, the dislocation line forms a closed loop, which is referred to as a *dislocation loop*. Figure 3.35 illustrates a hypothetical dislocation loop of arbitrary shape that bounds the slipped region A with the Burgers vector **b** in the crystal. The sections of the dislocation labeled E are pure edge-type and those labeled S are pure screw-type. Most of the dislocation line is of mixed type, partly edge and partly screw; that is, its Burgers vector makes an arbitrary angle with the dislocation line. Therefore the mixed dislocation may vary in character from being pure edge-type to pure screw-type.

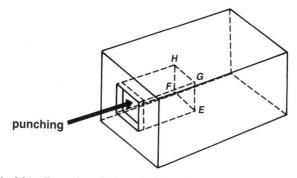

Fig. 3.36.   Formation of prismatic dislocation loop EFGH by punching.

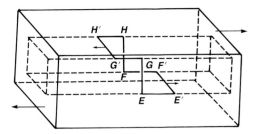

Fig. 3.37. Dislocation segment movement by applying shearing forces with opposite directions.

In Fig. 3.36, another type of dislocation loop EFGH, which might be produced by plastically indenting one face of the crystal with a punch, is illustrated. The system behaves as if the punch pushed cylinders or prisms of the section of the crystal as rigid units. The prisms appear to be polygonal in cross section, the planar surfaces being slip planes and the direction of the axis being slip directions. This driving is termed *prismatic punching*, and those dislocations generated by punching are referred to as *prismatic punched-out dislocation loops* or simply *prismatic dislocation loops*.[73] By applying shearing forces with opposite directions to the top and bottom faces of the crystal as shown in Fig. 3.37, the segment EF moves toward E'F' while GH moves toward G'H'. The segments EG and FH are not acted on directly by the shear force, and, if these remained fixed, new segments EE', FF', GG', and HH' would be formed. These segments lie along the slip direction and are therefore screw-type.[72]

Furthermore, multiple intrinsic point defects may exist in a crystal. They normally first exhibit a nonordered structure; however, as the clusters grow they condense to form plates or disks, which are more energetically favored. That is, when enough intrinsic point defects agglomerate, they may collapse into a disk along a particular lattice plane. Once the disk has a sufficiently large radius, the portions of the crystal on either face of the disk will join to form a general dislocation loop, instead of forming defect bonds that were shown in Fig. 3.29b. It may be evident that the aggregates of vacancies will be more stable as general dislocation loops having a high degree of registry than as spheroidal or needle-like aggregates.[74] Figure 3.38 illustrates the sequence of intrinsic (or vacancy) type dislocation loop formation. An extrinsic (or interstitial) type dislocation loop (Fig. 3.39) may also be formed in an identical sequence.

Climb of Dislocations   In addition to slip, *climb* is an alternate way in which a dislocation can move in a crystal. Since the edge of an extra half-plane consists of a row of atoms having incomplete lattice bonding, atoms can

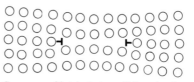

Fig. 3.38. Sequence of intrinsic-type dislocation loop formation.

easily be added or removed from these sites, referred to as climb. For an edge dislocation, the climb motion is illustrated in Fig. 3.40. If the extra half-plane of atoms absorbs self-interstitial atoms, substitutional atoms may also be involved in this process, as represented in Fig. 3.40a; the edge dislocation moves downward, that is, negative climb occurs. On the other hand, positive climb occurs when vacancies are captured by the dislocation as shown in Fig. 3.40b. The same situation happens when the atoms constituting the extra half-plane are removed from the dislocation. This climbing motion, therefore, requires the mass transport of atoms by a diffusion process leading the extra half-plane to extend by additional atoms or contract by absorption of vacancies. Dislocation loops such as shown in Figs. 3.38 and 3.39 can also change their size by a climbing motion. In either case, an edge dislocation moves perpendicular to its slip plane. If climb occurs at a part of the

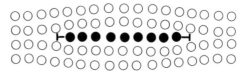

Fig. 3.39. Extrinsic-type dislocation loop.

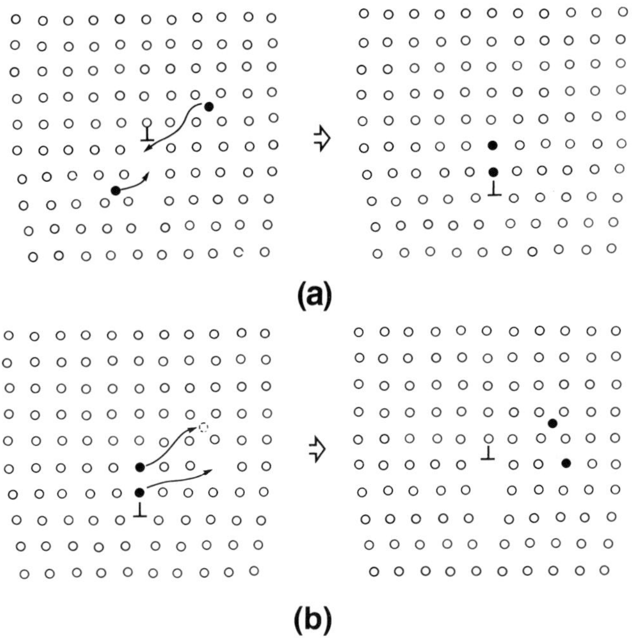

Fig. 3.40. Climb motion in an edge dislocation: (a) negative climb by absorbing self-interstitials and (b) positive climb by capturing vacancies.

dislocation line, *jogs* are formed, as illustrated in Fig. 3.41. A jog represents a small displacement of the dislocation line and consists of two steps perpendicular to the slip plane. In other words, the climb of the edge dislocation accompanies the nucleation and subsequent movement of jogs along its axis. The absorption or removal of the point defects at these jogs is presumably facilitated by their rapid diffusion along the core of the dislocation.[74]

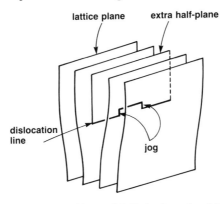

Fig. 3.41. Jogs formed by partial climb of an edge dislocation.

Interaction of Dislocations   When two edge dislocations having Burgers vectors of equal value but an opposite sign come together on neighboring slip planes, their respective extra half-planes combine to form a complete crystal lattice and result in the elimination of the dislocations as shown in Fig. 3.42a. If the dislocations move on the same slip plane, however, a row of interstitial atoms is formed in the process of annihilation of two half-planes, as in Fig. 3.42b. In addition, as shown in Fig. 3.42c, a row of vacancies will result when the slip planes are separated by a distance of two atomic spacings. Similarly, two screw dislocations having equal *b* values but opposite in sign and moving on the neighboring slip plane will also interact mutually. The result of interaction between dislocations in more complex situations depends on both their orientations and their Burgers vectors. The details have been discussed in refs. 71 and 72.

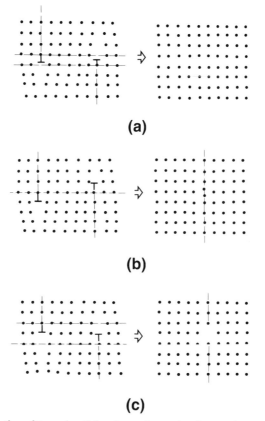

Fig. 3.42.   Interaction of two edge dislocations of opposite sign moving on (a) adjacent slip planes, (b) the same slip plane, and (c) slip planes separated by two atomic distances.

Dislocations in Diamond Lattice   The structure of the diamond lattice to which silicon belongs was shown in Fig. 3.22. Here, for the sake of showing the structures of dislocations more clearly, the diamond lattice is drawn with one of its $\langle 111 \rangle$ bonds vertical as in Fig. 3.43. A simplified two-dimensional projection of this orientation on the (110) plane was given in Fig. 3.24. The structure of the dislocations in the silicon lattice has been discussed in detail by Hornstra[75] and Tan.[76] Here, the geometry of the dislocation structures is described according to the publication by Hornstra. The shortest distance between two equivalent atoms in the diamond lattice is along a $\langle 110 \rangle$ direction and is equal to the shortest lattice vector that is allowed as the Burgers vector $\frac{a}{2} \langle 110 \rangle$, that is, half the diagonal of a cubic face. Taking into account the Burgers vector $\frac{a}{2} \langle 110 \rangle$ and the most energetically stable structure, three types of simple dislocations in the diamond lattice are considered: (1) a screw dislocation, (2) a so-called 60° dislocation, and (3) a pure edge dislocation. These three simple dislocations are illustrated in Fig. 3.44.

The character of the narrowest form of a screw dislocation (Fig. 3.44a) is seen by comparing the normal hexagon labeled 7–8–9–10–11–7 with the screw structure 13–2–3–14–15–16–17. In the first case atom 7 is the beginning and end of the loop; however, in the latter case there is a gap between 13 and 17 that is equal to the Burgers vector **b** of the dislocation whose axis is indicated by the arrow in the diagram. That is, the screw dislocation is characterized with a $\{111\}$ slip plane and a $\langle 110 \rangle$ direction.

The 60° dislocation, illustrated in Fig. 3.44b, lies on the (111) slip plane with the axis along the $[1\bar{1}0]$ direction and its Burgers vector is 60° to it in the

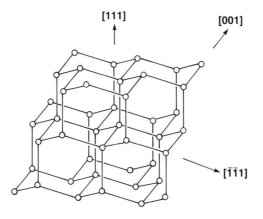

Fig. 3.43.   The diamond lattice with [111] vertical.

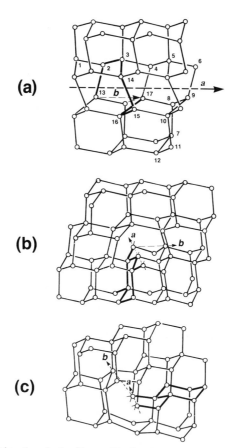

Fig. 3.44. Simple dislocations in the diamond lattice: (a) screw dislocation, (b) 60° dislocation, and (c) pure edge dislocation. (After Hornstra.[75] Reprinted with the permission of Pergamon Journals, Ltd.)

[$\bar{1}$01] direction. The extra half-plane of atoms shown as the (11$\bar{1}$) plane is indicated by heavy lines in the diagram. This plane ends in dangling bonds, which are assumed to play a significant role in both the electrical and the physical behavior of the dislocation, and which are drawn as short vertical lines at tetrahedral angles to the other bonds.

The axis of the pure edge dislocation, illustrated in Fig. 3.44c, lies along [$\bar{1}\bar{1}$2], the Burgers vector is $\frac{a}{2}$[110], and the slip plane is (001). However, this slip plane is not likely to occur in practical crystals with the diamond lattice, since, as discussed in the previous section, the {111} is the most dominant slip plane; indeed, only this slip plane has been found experimentally.[75] The extra half-plane is drawn with heavy lines in the diagram. The plane ends in atoms with two dangling bonds each.

Other dislocation types whose geometries are more complicated than the simple dislocation described above presumably exist in the diamond lattice.[75,76] These may occur when the direction of simple dislocations changes from one ⟨110⟩ direction to another. However, such a rearrangement can only take place by diffusion of vacancies or self-interstitials and hence should occur at elevated temperatures or under heavily strained circumstances.

3.4.3 Plane Defects

General Remarks As listed in Table 3.8, plane defects or two-dimensional defects include *stacking faults, twins,* and *grain boundaries.* Among these, stacking faults are the most important and most commonly induced defect in processed silicon crystals, and degrade the performance of integrated circuits if they exist in the device region. On the other hand, twins and grain boundaries are common defects in lower-grade silicon materials used for energy conversion devices such as solar cells, but not in silicon crystals for the VLSI application. For that reason, the discussion in this section will focus on the structure of stacking faults.

Stacking Faults and Partial Dislocation The close-packing plane, that is, {111} planes, in the fcc lattice is illustrated in Fig. 3.45, where the ordered sequence of atomic layers is indicated by ABC. The atoms in the plane of the figure are designated B and the positions of the atoms above and below this

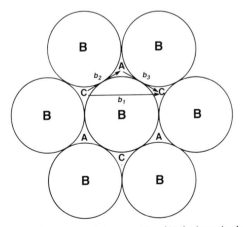

Fig. 3.45. Ordered sequence of close-packing {111} planes in the fcc lattice.

plane are marked A and C, respectively. The vector $\mathbf{b}_1 \left(= \frac{a}{2}[10\bar{1}] \right)$ defines one of the observed slip directions as discussed in Section 3.5.2. However, if atoms are regarded as hard balls rolling over the plane, it is easier to displace the (111) plane in the direction of slip $\mathbf{b}_1$ $(= [101])$ by the zigzag path $C \rightarrow A \rightarrow C$ following the vectors $\mathbf{b}_2 \left(= \frac{a}{6}[2\bar{1}\bar{1}] \right)$ and $\mathbf{b}_3 \left(= \frac{a}{6}[11\bar{2}] \right)$.[72] The dislocation dissociation equation can then be written as

$$\frac{a}{2}[10\bar{1}] = \frac{a}{6}[2\bar{1}\bar{1}] + \frac{a}{6}[11\bar{2}] \qquad (3.50)$$

If slip takes place from C to A positions, the order of stacking of (111) layers in the crystal is altered. That is, the normal sequence ...ABCABC... becomes ...ABCABCBCABC... This is termed a *stacking fault*. Thus the stacking fault and its associated shear can be produced by the motion of the dislocation with the vector $\mathbf{b} \left(= \frac{a}{6}\langle 211 \rangle \right)$. The types of dislocations, such as $\mathbf{b}_1$ in Fig. 3.45, that have been discussed in Section 3.5.2 are usually termed *perfect dislocations* because their Burgers vectors are equal to a unit lattice distance; in turn, the perfect structure of the lattice is maintained during the passage of such dislocations. On the other hand, a dislocation with a Burgers vector with a part of the lattice vector and such that a new atomic configuration is produced is termed *imperfect dislocation* or *partial disloca-tion*. Partial dislocations of the type $\frac{a}{6}\langle 211 \rangle$, which can move on a $\{111\}$ plane by slip motion, are called *Schockley partials*.

Another type of partial dislocation associated with stacking faults can be understood by considering the diagram in Fig. 3.46 in which the (111) planes are illustrated by lines. The stacking order ...ABCABC... is normal except in the center, where part of the additional A plane is inserted (a) or part of the C plane is missing (b). The former type is referred to as an *extrinsic-type* stacking fault (ESF), while the latter an *intrinsic-type* stacking fault (ISF). Note that an edge dislocation with a Burgers vector $\frac{a}{3}\langle 111 \rangle$ (i.e., the displacement is normal to the $\langle 111 \rangle$ direction) is formed at each end of this faulted region. Partial dislocations of the type $\frac{a}{3}\langle 111 \rangle$ are called *Frank partials*. Dislocations of this type can only move in the (111) glide plane by climb, that is, by the transport of atoms by diffusion to or from the edge of the incomplete atomic plane, and are called *sessile* since they can not move in the normal manner by slip in a $\langle 111 \rangle$ direction, in contrast to other type of dislocations, which are termed *glissile*.

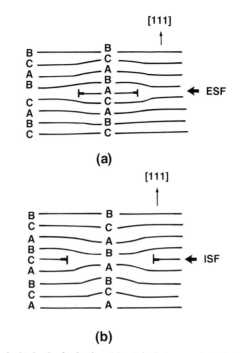

Fig. 3.46. Stacking faults in the fcc lattice: (a) extrinsic type and (b) intrinsic type.

For convenience, although stacking faults are associated with partial dislocations, the term of *dislocation*, which is not associated with a faulted plane, is distinguished from the term *stacking faults* in this book unless specifically noted.

Stacking Fault As shown in Fig. 3.24, the layer structure of the diamond lattice contains {111} planes in the sequence a, a′, b, b′, c, c′, ... In the projection normal to (111), the layers a and a′ are seen to project to the same type of position, as do b and b′, and c and c′. In analogy with the fcc lattice, stacking faults can be created by the insertion or the removal of *pairs of layers*, that is, aa′, bb′, etc., because this results in low-energy faults in which there is no change in the four nearest-neighbor covalent bonds in the lattice; all other faults disturb the nearest-neighbor bonding and are high-energy faults. Here, therefore, A denotes the layer for a and a′, B for b and b′, and C for c and c′, as indicated in Fig. 3.24. The packing can then be described by the sequence ABCABC.... Two types of stacking faults—an extrinsic type caused by inserting a (111) lattice plane and an intrinsic type caused by removing a (111) lattice plane—are illustrated in Fig. 3.47. Instead of the regular stacking ABCABC..., the irregular stacking ABACAB... and

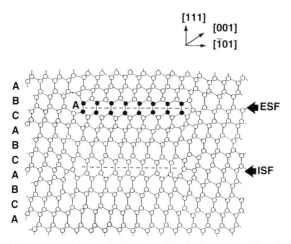

Fig. 3.47. Extrinsic-type stacking fault (ESF) and intrinsic-type stacking fault (ISF) in the diamond lattice.

ABC^{V}BC are seen at the extrinsic- and intrinsic-type stacking faults, respectively, in the diagram. The stacking faults observed in silicon crystals are almost exclusively the extrinsic type. Intrinsic-type stacking faults have only been found in epitaxial layers (see Section 5.6).[54]

Twins   A twin is formed when a portion of the crystal lattice is plastically deformed in a specific direction, and the deformed and undeformed parts remain in intimate contact over their bounding planes. Figure 3.48 shows the two-dimensional diagram of the structure of a twin where the open circles represent the positions of the atoms before twinning and the solid circle the positions after twinning.[77] The tip of an advancing twin corresponds to a dislocation pileup. The figure also illustrates how the atoms have moved into their twin positions by a shear mechanism in the direction parallel to the *twin boundary* or *twinning plane*. As shown in Fig. 3.48, when the atoms of this

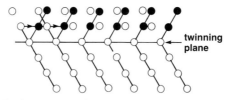

Fig. 3.48.   Diagram of twin structure, where open circles represent the atom positions before twinning and solid circles the positions after twinning. (After Hall.[77])

plane are common to both the twinned and untwinned parts, it is referred to as the *composition plane*. In the diamond lattice, however, the {111} twinning planes are arranged in a sequence of alternating long and short interspacings in which the twin boundary is not the composition plane.

In ribbon-growth silicon crystals, the most common twins are normally aligned parallel to the $\langle 211 \rangle$ growth direction and perpendicular to the {110} ribbon surface.[78,79] During the crystal growth of silicon ingots by the Czochralski technique, described in Section 5.2, {111} twins may also occur due to improper crystal growth conditions.

Grain Boundaries A grain boundary is the interface where two or more single crystals of different orientation join in such a manner that the material is continuous across the boundary. A grain boundary can be curved; however, in thermal equilibrium, it is planar in order to minimize the boundary area and hence the boundary energy. The *low-angle grain boundary* as illustrated in Fig. 3.49 consists of edge dislocations and is believed to be formed, during some stage in the crystal growth, by the migration of the

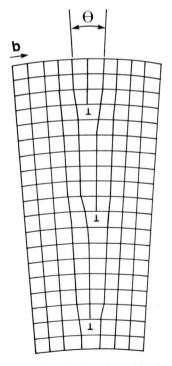

Fig. 3.49. Low-angle grain boundary with edge dislocations.

dislocations along their glide planes and their subsequent climb normal to these planes. In the case where the angle of tilt across the boundary is large (e.g., several degrees), this simple dislocation configuration loses its physical significance, and in turn the single crystal loses its identity and becomes polycrystalline.

3.4.4   Volume Defects

General Remarks   Volume defects in crystals include *voids* or *negative crystals* and agglomerates of impurities. Considering the fabrication of VLSI/ULSI devices, the most influential volume defects are unquestionably clusters of metallic impurities in the device active region and oxygen precipitates in the interior region of silicon substrates. As will be discussed later, metallic impurities in device active regions directly deteriorate electrical properties of the device, and oxygen precipitates generate secondary defects that strongly affect the characteristics of silicon substrates.

Voids   Voids (see Fig. 3.27) are defects mostly observed in oxide crystals, but rarely in silicon crystals, grown by the Czochralski method.[80–82] The occurrence of voids in grown crystals depends on the growth rate, the viscosity of the melt,[82] and the crystal rotation rate, which is related to centrifugal acceleration effects.[81] A void usually forms a polyhedron that consists of crystallographic faces; thus, such a void is called a *negative crystal*. The shapes of void polyhedra supply information regarding possible low-energy planes in the crystal. For silicon, as listed in Table 3.6, the surface energies diminish in the order of (100) > (110) > (111). Indeed, the preferred growth habit of silicon is an octahedron bounded by eight {111} planes.[21,83] Therefore, most negative crystals in silicon generally have especially prominent facets of this type.[84] However, the regular octahedron is often truncated by {100} and {311} planes.[83,84] When {311} facets are present, these are generally smaller than their {100} counterparts. However, since 24 {311} facets can occur in a closed structure compared with six individual {100} facets, {311} planes make a significant contribution to the surfaces of a negative crystal. Thus, both {100} and {311} planes should possess relatively low surface energies.[84] In fact, {111}, {100}, and {311} planes have been observed to appear and coexist during the vapor growth of Si and Ge.[85]

Precipitates   In general, the precipitation of impurities in a crystal occurs when the concentration of the impurity exceeds the solid solubility limit. The solid solubility of impurity A as a function of temperatures is schematically illustrated in Fig. 3.50, where RT refers to as room temperature and T_m the melting point of solvent crystal B (see also Fig. 3.31). Assume that the

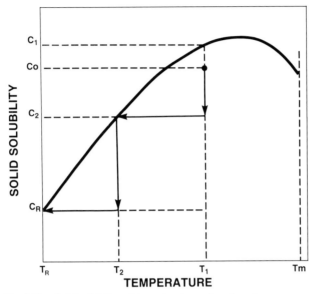

Fig. 3.50.   Solid solubility of impurity A as a function of temperature.

concentration of impurity A is $[C_0]$ at temperature T_1. At this point no precipitation of A occurs since $[C_0]$ is less than the solid solubility $[C_1]$ at T_1. However, when crystal B is cooled to T_2, the impurity A of $[C_0 - C_2]$, the *supersaturated* amount, will precipitate either as a single phase A or compound A–B. Similarly, when cooled down from T_2 to RT, A of $[C_2 - C_R]$ will precipitate. Precipitation in silicon crystals usually results in silicides for metal impurities and silicon oxide (SiO_x: $x \approx 2$) for oxygen.

Precipitation of impurities proceeds through three steps: (1) nucleation, (2) diffusion, and (3) growth. When a precipitate is formed, it starts very small and then increases in size. Initially it has a high surface-to-volume ratio, which tends to make it unstable because of its high surface energy. The initially formed particles are called *embryos*. In general, there is a region of temperature in which the embryo can be stabilized to form a stable particle, a *nucleus*, capable of further growth. The nucleation process may occur in two basic ways: (1) *homogeneous nucleation*, by which nuclei are formed homogeneously by a random composition fluctuation of the solute, and (2) *heterogeneous nucleation*, in which crystal defects, such as point defects and dislocations, serve as favorable low-energy sites for nucleation. Heterogeneous nucleation is by far the more commonly observed in nature. The rate of impurity precipitation depends on many factors, such as the temperature, the degree of the supersaturation, and the diffusivity of the impurity. Thus, the impurity with higher diffusivity precipitates faster in general.

References

1. H. A. Enge, M. R. Wehr, and J. A. Richards, "Introduction to Atomic Physics." Addison-Wesley, Reading, Massachusetts, 1972.
2. R. K. Watts, X-ray lithography. *In* "VLSI Handbook" (N. G. Einspruch, ed.), pp. 365–380. Academic Press, New York, 1985.
3. B. D. Cullity, "Elements of X-Ray Diffraction" 2nd ed. Addison-Wesley, Reading, Massachusetts, 1978.
4. F. Shimura and H. R. Huff, VLSI silicon material criteria. *In* "VLSI Handbook" (N. G. Einspruch, ed.), pp. 191–269. Academic Press, New York, 1985.
5. M. von Heimendahl, "Einfuhrung in die Elektronenmikroskopie." Friedrich Vieweg, 1970.
6. P. B. Hirsch, A. Howie, R. B. Nicholson, D. W. Pashley, and M. J. Whelan, "Electron Microscopy of Thin Crystals." Butterworth, London, 1965.
7. R. W. G. Wyckoff, "Crystal Structures." Wiley (Interscience), New York, 1960.
8. K. E. Bean and P. S. Gleim, The influence of crystal orientation on silicon semiconductor processing. *Proc. IEEE* **57**, 1469–1476 (1969).
9. J. M. Gibbons, Ion implantation in semiconductors. Part I. Range distribution theory and experiments. *Proc. IEEE* **56**, 295–319 (1968).
10. L. D. Dyer, Dislocation-free Czochralski growth of ⟨110⟩ silicon crystals. *J. Cryst. Growth* **47**, 533–540 (1979).
11. R. J. Jaccodine, Surface energy of germanium and silicon. *J. Electrochem. Soc.* **110**, 524–527 (1963).
12. J. R. Ligenza, Effect of crystal orientation on oxidation rates of silicon in high pressure steam. *J. Phys. Chem.* **65**, 2011–2014 (1961).
13. H. C. Gatos and M. C. Lavine, Chemical behavior of semiconductors: Etching characteristics. *Prog. Semicon.* **9**, 1–46 (1962).
14. W. K. Zwicker and S. K. Kurtz, Anisotropic etching of silicon using electrochemical displacement reactions. *In* "Semiconductor Silicon 1973" (H. R. Huff and R. R. Birgess, eds.), pp. 315–326. Electrochem. Soc., Princeton, New Jersey, 1973).
15. R. M. Finne and D. L. Klein, A water–amine-complexing agent system for etching silicon. *J. Electrochem. Soc.* **114**, 965–970 (1967).
16. T. J. Rodgers and J. D. Meindl, Epitaxial V-groove bipolar integrated circuit process. *IEEE Trans. Electron Devices* **ED-20**, 226–232 (1973).
17. K. E. Bean and J. R. Lawson, Application of silicon crystal orientation and anisotropic effects to the control of charge spreading in devices. *IEEE J. Solid-State Circuits* **SC-9**, 111–117 (1974).
18. K. E. Bean and W. R. Runyan, Dielectric isolation: Comprehensive, current and future. *J. Electrochem. Soc.* **124**, 5C–12C (1977).
19. F. Shimura, TEM observation of pyramidal hillocks formed on (001) silicon wafers during chemical etching. *J. Electrochem. Soc.* **127**, 910–913 (1980).
20. R. A. Swalin, "Thermodynamics of Solids," 2nd ed. Wiley, New York, 1972.
21. W. R. Runyan, "Silicon Semiconductor Technology." McGraw-Hill, New York, 1965.
22. T. F. Ciszek, Crystallographic growth forms of silicon on a free melt surface. *J. Electrochem. Soc.* **132**, 422–427 (1985).
23. W. C. Dash, Silicon crystals free of dislocations. *J. Appl. Phys.* **29**, 736–737 (1958).
24. W. C. Dash, Growth of silicon crystals free from dislocations. *J. Appl. Phys.* **30**, 459–474 (1959).
25. B. E. Deal and A. S. Grove, General relationship for thermal oxidation of silicon. *J. Appl. Phys.* **36**, 3770–3778 (1965).

26. D. Hess and B. E. Deal, Kinetics of the thermal oxidation of silicon in O_2/HCl mixtures. *J. Electrochem. Soc.* **124**, 735–739 (1977).

27. B. E. Deal, Thermal oxidation kinetics of silicon in pyrogenic H_2O and 5% HCl/H_2O mixtures. *J. Electrochem. Soc.* **125**, 576–579 (1978).

28. R. R. Razouk, L. N. Lie, and B. E. Deal, Kinetics of high pressure oxidation of silicon in pyrogenic steam. *J. Electrochem. Soc.* **128**, 2214–2220 (1981).

29. W. A. Pliskin, Separation of the linear and parabolic terms in the steam oxidation of silicon. *IBM J. Res. Dev.* **10**, 198–206 (1966).

30. E. A. Irene, The effects of trace amounts of water of the thermal oxidation of Si in oxygen. *J. Electrochem. Soc.* **121**, 1613–1616 (1974).

31. E. A. Irene, H. Z. Massoud, and E. Tierney, Silicon oxidation studies: Silicon orientation effects on thermal oxidation. *J. Electrochem. Soc.* **133**, 1253–1256 (1986).

32. H. Z. Massoud, J. D. Plummer, and E. A. Irene, Thermal oxidation of silicon in dry oxygen. *J. Electrochem. Soc.* **132**, 1745–1753 (1985).

33. H. Z. Massoud, J. D. Plummer, and E. A. Irene, Thermal oxidation of silicon in dry oxygen growth-rate enhancement in the thin regime. I. Experimental results. *J. Electrochem. Soc.* **132**, 2685–2700 (1985).

34. E. H. Nicollian and J. R. Breuws, "MOS (Metal Oxide Semiconductor) Physics and Technology." Wiley, New York, 1982.

35. B. E. Deal, M. Sklar, A. S. Grove, and E. H. Snow, Characteristic of the surface-state charge (Qss) of thermally oxidized silicon. *J. Electrochem. Soc.* **114**, 266–274 (1967).

36. P. Balk, P. J. Burkhardt, and L. V. Gregor, Orientation dependence of built-in surface charges on thermally oxidized silicon. *Proc. IEEE* **53**, 2133–2134 (1965).

37. E. Arnold, J. Ladell, and G. Abowitz, Crystallographic symmetry of surface state density in thermally oxidized silicon. *Appl. Phys. Lett.* **13**, 413–416 (1968).

38. S. M. Sze, "Physics of Semiconductor Devices," 2nd ed. Wiley, New York, 1981.

39. H. M. Liaw, J. Rose, and P. L. Fejes, Epitaxial silicon for bipolar integrated circuits. *Solid State Technol.* May, pp. 135–143 (1984).

40. S. P. Weeks, Pattern shift and pattern distortion during CVD epitaxy on (111) and (100) silicon. *Solid State Technol.* Nov., pp. 111–117 (1981).

41. C. M. Drum and C. A. Clark, Geometrical stability of shallow surface depressions during growth of (111) and (100) epitaxial silicon. *J. Electrochem. Soc.* **115**, 664–669 (1968).

42. C. M. Drum and C. A. Clark, Anisotropy of macrostep motion and pattern edge displacements during growth of epitaxial silicon near {100}. *J. Electrochem. Soc.* **117**, 1401–1405 (1970).

43. H. Shiraki, J. Matsui, T. Kawamura, M. Hanaoka, and T. Sakaki, Bright spots in the image of silicon vidicon. *Jpn. J. Appl. Phys.* **10**, 213–217 (1971).

44. G. H. Schwuttke, K. Brack, and E. W. Hearn, The influence of stacking faults on leakage currents of FET devices. *Microelectron. Reliab.* **10**, 467–470 (1971).

45. H. Strack, K. R. Mayer, and B. O. Kolbesen, The detrimental influence of stacking faults on the refresh time of MOS memories. *Solid-State Electron.* **22**, 135–140 (1979).

46. L. C. Parrillo, R. S. Payne, T. E. Seidel, M. Robinson, G. W. Reutlinger, D. E. Post, and R. L. Field, Jr., The reduction of emitter-collector shorts in a high-speed all-implanted bipolar technology. *IEEE Trans. Electron Devices* **ED-28**, 1508–1514 (1981).

47. A. W. Fisher and J. A. Amick, Defect structure on silicon surfaces after thermal oxidation. *J. Electrochem. Soc.* **113**, 1054–1060 (1966).

48. Y. Sugita, T. Kato, and M. Tamura, Effect of crystal orientation on the stacking fault formation in thermally oxidized silicon. *J. Appl. Phys.* **42**, 5847–5849 (1971).

49. S. M. Hu, Anomalous temperature effect of oxidation stacking faults in silicon. *Appl. Phys. Lett.* **27**, 165–167 (1975).

50. C. M. Drum and W. van Gelder, Stacking faults in (100) epitaxial silicon caused by HF and thermal oxidation and effects on P–N junctions. *J. Appl. Phys.* **43**, 4465–4468 (1972).
51. C. M. Hsieh and D. M. Maher, Nucleation and growth of stacking faults in epitaxial silicon during thermal oxidation. *J. Appl. Phys.* **44**, 1302–1306 (1973).
52. T. F. Ciszek, Characteristic of [115] dislocation-free float-zoned silicon crystals. *J. Electrochem. Soc.* **120**, 799–802 (1973).
53. C. H. J. Van Den Brekel, Growth rate anisotropy and morphology of autoepitaxial silicon films from SiCl$_4$. *J. Cryst. Growth* **23**, 259–266 (1974).
54. W. Zulehner and D. Huber, Czochralski-grown silicon. *In* "Crystals 8: Silicon-Chemical Etching" (J. Grabmaier, ed.), pp. 1–143. Springer-Verlag, Berlin and New York, 1982.
55. F. Shimura, H. Tsuya, and T. Kawamura, Thermally-induced defect behavior and effective intrinsic gettering sinks in silicon wafers. *J. Electrochem. Soc.* **128**, 1579–1583 (1981).
56. R. G. Rhodes, "Imperfections and Active Centers in Semiconductors" Macmillan, New York, 1964.
57. F. C. Champion, Some physical consequences of elementary defects in diamonds. *Proc. R. Soc. London, Ser. A* **234**, 541–556 (1956).
58. V. V. Voronkov, The mechanism of swirl defects formation in silicon. *J. Cryst. Growth* **59**, 625–643 (1982).
59. A. Seeger, H. Foll, and W. Frank, Self-interstitials, vacancies and their cluster in silicon and germanium. *Conf. Ser.—Inst. Phys.* **31**, 12–29 (1977).
60. U. Gösele, F. Morehead, H. Föll, W. Frank, and H. Strunk, The predominant intrinsic point defects in silicon: Vacancies or self-interstitials? *In* "Semiconductor Silicon 1981" (H. R. Huff, R. J. Kriegler, and Y. Takeishi, eds.), pp. 766–778. Electrochem. Soc., Princeton, New Jersey, 1981.
61. T. Y. Tan and U. Gösele, Point defects, diffusion processes, and swirl defect formation in silicon. *Appl. Phys. [Part] A* **A37**, 1–17 (1985).
62. G. D. Watkins, An EPR study of the lattice vacancy in silicon. *J. Phys. Soc. Jpn.* **18**, Suppl. II, 22–26 (1963).
63. M. Yoshida and K. Saito, Dissociative diffusion of nickel in silicon and self-diffusion of silicon. *Jpn. J. Appl. Phys.* **6**, 573–581 (1967).
64. J. A. Van Vechten and C. D. Thurmond, Comparison of theory with quenching experiments for the entropy and enthalpy of vacancy formation in Si and Ge. *Phys. Rev. B: Solid State* [3] **14**, 3551–3557 (1976).
65. J. N. Hobstetter, Equilibrium, diffusion and imperfections in semiconductors. *Prog. Met. Phys.* **7**, 1–63 (1958).
66. R. E. Smallman, "Modern Physical Metallurgy," 2nd ed. Butterworth, London, 1963.
67. M. Hansen, "Constitution of Binary Alloys," 2nd ed. McGraw-Hill, New York, 1958.
68. E. R. Weber, Transition metal impurities in silicon. *Appl. Phys. [Part] A* **A30**, 1–22 (1983).
69. F. A. Trumbore, Solid solubilities of impurity elements in germanium and silicon. *Bell Syst. Tech. J.* **39**, 205–233 (1960).
70. G. L. Vick and K. M. Whittle, Solid solubility and diffusion coefficients of boron in silicon, *J. Electrochem. Soc.* **116**, 1142–1144 (1969).
71. F. R. N. Nabarro, "Theory of Crystal Dislocations." Oxford Univ. Press, London and New York, 1967.
72. J. P. Hirth and J. Lothe, "Theory of Dislocations." McGraw-Hill, New York, 1968.
73. F. Seitz, Prismatic dislocations and prismatic punching in crystals. *Phys. Rev.* **79**, 723–724 (1950).
74. R. M. Thomson and R. W. Balluffi, Kinetic theory of dislocation climb. I. General models for edge and screw dislocations. *J. Appl. Phys.* **33**, 803–816 (1962).
75. J. Hornstra, Dislocations in the diamond structure. *J. Phys. Chem. Solids* **5**, 129–141 (1958).

76. T. Y. Tan, Atomic modeling of homogeneous nucleation of dislocations from condensation of point defects in silicon. *Philos. Mag. [Part] A* **44**, 101–125 (1981).
77. E. O. Hall, "Twinning and Diffusionless Transformations in Metals." Butterworth, London, 1954.
78. K. V. Ravi, The growth of EFG silicon ribbons. *J. Cryst. Growth* **39**, 1–16 (1977).
79. R. Gleichmann, B. Cunningham, and D. G. Ast, Process-induced defects in solar cell silicon. *J. Appl. Phys.* **58**, 223–229 (1985).
80. K. Nassau and A. M. Broyer, Calcium tungstate: Czochralski growth, perfection, and substitution. *J. Appl. Phys.* **33**, 3064–3073 (1962).
81. J. R. Carruthers and K. Nassau, Nonmixing cells due to crucible rotation during Czochralski crystal growth. *J. Appl. Phys.* **39**, 5205–5214 (1968).
82. F. Shimura and Y. Fujino, Crystal growth and fundamental properties of $LiNb_{1-y}Ta_yO_3$. *J. Cryst. Growth* **38**, 293–302 (1977).
83. F. Shimura, Octahedral precipitates in high temperature annealed CZ-silicon. *J. Cryst. Growth* **54**, 589–591 (1981).
84. A. G. Cullis, T. E. Seidel, and R. L. Meek, Comparative study of annealed neon-, argon-, and krypton-ion implantation damage in silicon. *J. Appl. Phys.* **49**, 5188–5198 (1978).
85. G. A. Wolff, Surface energy of germanium and silicon. *J. Electrochem. Soc.* **110**, 1293–1294 (1963).

Chapter 4

Basic Semiconductor Physics

Although this book aims to provide comprehensive and up-to-date knowledge of the semiconductor silicon crystal technology, the ultimate objective of this effort is not only to contribute to the production of state-of-the-art VLSI/ULSI devices but also to create future electronic devices on the basis of silicon materials. To this end, understanding semiconductor physics is unquestionably helpful to the further use of silicon materials.

In this chapter, the basic semiconductor physics considered to be essential to develop an understanding of semiconductor crystals is described. The subjects emphasized in this chapter include (1) electrical conductivity and (2) basic electronic-device operation physics. For detailed consideration of semiconductors and their device physics, the reader should consult the standard text books such as Refs. 1–5 listed at the end of this chapter.

4.1 Semiconductors

4.1.1 Properties of Semiconductors

Definition of Semiconductor   What is a semiconductor? What are its properties? Although the term "semiconductor" is well known, it may not be easy to answer these questions quickly. A semiconductor is often defined simply from the viewpoint of electrical *conductivity* σ. That is, semiconductors are materials with values of the electrical *resistivity* ρ $(= 1/\sigma)$ at room temperature generally in the range of $\sim 10^{-2}$ to $10^{9}\,\Omega$ cm, intermediate between *conductors* ($\sim 10^{-6}\,\Omega$ cm) and insulators ($\sim 10^{4}$ to $\sim 10^{22}\,\Omega$ cm). Indeed, the term "*semi*conductor" is from such an electrical conductive characteristic.

Electrical Properties   A semiconductor has been defined as a material that has intermediate conductivity. However, a pure and perfect crystal of most semiconductors behaves as an insulator at the temperature absolute zero, contrary to many metals and alloys whose electrical resistivity drops suddenly to zero at a sufficiently low temperature, often a temperature in the range of liquid helium (i.e., ~ 4 K). That is, semiconductors possess a *negative temperature coefficient* of resistance in some range of temperatures. The negative coefficient is to be connected with the excitation of electrons from the ground state. Accordingly, the characteristic semiconducting properties are usually brought about by thermal excitation and impurities. Most semiconductor device operation, as a matter of fact, depends on the conducting characteristics, which are very sensitive to the impurities added. It should be emphasized that this property of semiconductors can be a great advantage for control device operation purposes; on the other hand, undesirable impurities may create a very severe problem of maintaining the purity of the material.

Another important electrical property of semiconductors is *photoconductivity*. The conductivity of a semiconductor is increased by the absorption of photons of light depending on the wavelength, which may range from the ultraviolet to the infrared. This property has been applied to photoconductor devices.

Table 4.1 Properties of Important Semiconductors[a]

Semiconductor		Bandgap (eV) 300 K	0 K	Band[b]	Mobility at 300 K (cm²/V sec) Electron	Hole
Element	C	5.47	5.48	I	1800	1200
	Ge	0.66	0.74	I	3900	1900
	Si	1.12	1.17	I	1500	450
	Sn		0.082	D	1400	1200
IV–IV	a-SiC	2.996	3.03	I	400	50
III–V	GaSb	0.72	0.81	D	5000	850
	GaAs	1.42	1.52	D	8500	400
	GaP	2.26	2.34	I	110	75
	InSb	0.17	0.23	D	80000	1250
	InAs	0.36	0.42	D	33000	460
	InP	1.35	1.42	D	4600	150
II–VI	CdS	2.42	2.56	D	340	50
	CdSe	1.70	1.85	D	800	
	CdTe	1.56		D	1050	100
	ZnO	3.35	3.42	D	200	180
	ZnS	3.68	3.84	D	165	5

[a] After Sze.[4]
[b] I, Indirect; D, direct.

4.1.2   Semiconducting Materials

Although silicon (Si), germanium (Ge), and gallium arsenide (GaAs) are the most well-known semiconductors, there are many other semiconducting materials, such as listed in Table 4.1. Semiconductors are usually classified into two groups: (1) *elemental* or *monoatomic semiconductors,* and (2) *compound semiconductors,* which are further categorized into several subgroups according to their structure. Among those materials, as mentioned in Chapter 1, by far the most important in engineering use is silicon. The compound semiconductors, usually compounds of Groups III and V or II and VI of the periodic table, are becoming more important in use for specific purposes such as optoelectronic devices, which cannot be realized with silicon.

4.2   Electrical Conductivity

4.2.1   Electrical Conduction Phenomena

Mobility and Conductivity   When an electric field is applied to materials, an electric current, which is represented in terms of the number of charged carriers present and their drift velocity, is observed. The electric current density J_e is defined as the charge transported through a unit area in a unit time and is given by

$$J_e = n_{cp} z e v_d \tag{4.1}$$

where n_{cp} is the number of charged particles per unit volume, ze the total charge per particle (i.e., z the valence and e the absolute value of the electric charge), and v_d the drift velocity. The electrical conductivity σ is defined by

$$\sigma = J_e/\xi \tag{4.2}$$

where ξ is the electric field strength. Consequently,

$$\sigma = (n_{cp} z e) v_d / \xi \tag{4.3}$$

The drift velocity is directly proportional to the electric field strength, and this ratio is defined as the *drift mobility*:

$$\mu_D = v_d / \xi \tag{4.4}$$

The conductivity, then, is the product of the concentration, the total charge per particle, and the drift mobility of charged carriers:

$$\sigma = (n_{cp} z e) \mu_D \tag{4.5}$$

The particle that transports the charge resulting in the electric current is called a *carrier.*

Considering the charged carriers as initially having a random movement with an average drift velocity v of zero, one finds the equation of motion resulting from the application of a steady average external force F is given by

$$m^* (dv_d/dt + v_d/t_r) = F = ze\xi \tag{4.6}$$

where m^* is the effective mass of the carrier and t_r a characteristic *relaxation time*[†] governing the time required to reach equilibrium between the charged carriers and the crystal lattice. The term dv_d/dt describes inertial effects and must be included when v_d is time-dependent. When a steady state has been reached (i.e., $dv_d/dt = 0$),

$$v_d = zet_r\xi/m^* \tag{4.7}$$

and by comparison with Eq. (4.4),

$$\mu_D = zet_r/m^* \tag{4.8}$$

Furthermore, from Eqs. (4.3) and (4.7),

$$\sigma = n_{cp}z^2e^2t_r/m^* \tag{4.9}$$

That is, the electrical conductivity is proportional to the charge density (zen_{cp}) and the acceleration of charge in a given field, which is proportional to ze/m^* and t_r corresponding to the time that these forces act on the charge between collisions and random motion.

Fermi–Dirac Distribution Function In an ideal crystal, in addition to all atoms being on the regular sites, the electrons should be in the lowest energy configuration. The behavior of electrons, in general, is limited by the Pauli exclusion principle, which have already been mentioned in Section 2.1.5. The rule states that no more than one electron can be in any state. For transitions to take place, there must be an empty state to which the electron may go. Statistical mechanics extends the Pauli principle to describe electron occupancy as a function of energy. This distribution is called the *Fermi–Dirac distribution function* $f(E)$:

$$f(E) = 1/\{1 + \exp[(E - E_F)/kT]\} \tag{4.10}$$

where E_F is referred to as the *Fermi level* or *Fermi energy*, which can be determined from the charge neutrality condition. The distribution function $f(E)$ gives the probability that the state of energy E is filled. Figure 4.1 diagrammatically shows $f(E)$ at different temperatures T_1, T_2, T_3 $(T_1 > T_2 >$

[†] The symbol τ is often used to denote both the relaxation time and lifetime, which is discussed later. To avoid confusion, t_r is used to denote the relaxation time and τ is reserved for the lifetime in this text.

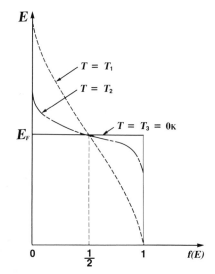

Fig. 4.1. Fermi–Dirac distribution function at different temperatures.

$T_3 = 0$ K). At higher temperatures thermal excitation gives an equilibrium distribution in some higher energy states.

Energy-Band Structure   The energy-band structure of a solid is theoretically obtained by solving the Schrödinger equation of an approximate single-electron problem[7] [see Eq. (2.12)]. The energy-band structures for a conductor (or a metal), a semiconductor, and an insulator are diagrammatically illustrated in Fig. 4.2. There is a forbidden energy range where no states can exist. Energy states are permitted above and below this region. The upper region is called the *conduction band*, while the lower region is the *valence band* or *filled band*.

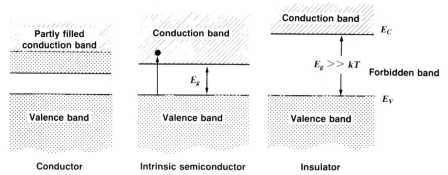

Fig. 4.2.   Energy-band structures for a conductor, a semiconductor, and an insulator.

Bandgap The bottom of the conduction band is designated E_C and the top of the valence band E_V. The difference between the energy of the lowest conduction band and that of the highest valence band, $E_C - E_V$, is defined as the *bandgap* or *energy gap*, E_g, which is one of the most important parameters in the physical properties of a semiconductor. One of the main reasons for silicon replacing germanium for utilization in semiconductor devices is that the bandgap of silicon is greater than that of germanium, resulting in silicon electronic devices capable of operation at higher temperatures than their germanium counterparts. A practical maximum working temperature for Ge devices is around 90°C, and for Si devices it is around 200°C.[8] Table 4.1 also lists the bandgap of several semiconductors.[4] Those values listed are for high-purity materials. The bandgap of impurity-doped materials becomes smaller. In the table, the type of bandgap, that is, *direct bandgap* (D) or *indirect bandgap* (I), is also indicated. In direct-bandgap semiconductors such as GaAs, an electron transfers from the top of the valence band to the bottom of the conduction band or vice versa and changes only its energy. On the other hand, in indirect-bandgap semiconductors such as Ge and Si, both energy and momentum must be simultaneously changed. The type of bandgap is very important from the viewpoint of utilization of semiconductor materials for different purposes of electronic devices. For example, many important optoelectronic devices, such as light-emitting diodes, lasers, and infrared photodetectors, are not possible with indirect-bandgap semiconductors.

In addition, the bandgaps of most semiconductors decrease with increasing temperature. The temperature dependence of the bandgap is given by[9]

$$E_g(T) = E_g(0) - \alpha T^2/(T + \beta) \tag{4.11}$$

where $E_g(T)$ is the bandgap at T, $E_g(0)$ the bandgap at 0 K, and α and β are the constants. Figure 4.3 shows variation of bandgap as a function of temperature for the three most important semiconductors, Si, Ge, and GaAs.[10]

Electrical Conductivity Electrical current flow results from the net movement of charged carriers. Completely filled or completely unfilled bands have no ability to conduct electricity. In conductors, there is no energy barrier that electrons need to overcome to be promoted to higher energy states; that is, there is always a finite concentration of electrons in the conduction band. In contrast, in semiconductors and insulators, a completely filled valence band at 0 K is separated from a completely unfilled conduction band by the bandgap; this is the forbidden energy bandgap. Nevertheless, in pure or *intrinsic semiconductors* the bandgap is not large compared to the thermal energy at any nonzero absolute temperatures, so that a few electrons are

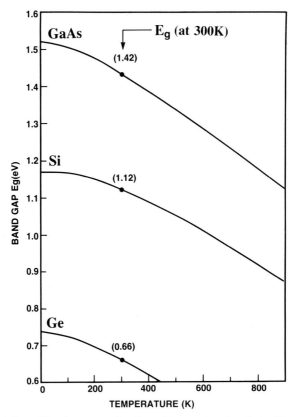

Fig. 4.3. Variation of bandgaps as a function of temperature for GaAs, Si, and Ge. (After Thurmond.[10] Reprinted with the permission of The Electrochemical Society, Inc.)

thermally excited into the conduction band, leaving empty electron positions, that is, *holes* in the valence band. The electron density n_e near the bottom of the conduction band is given by

$$n_e = N_C \exp[-(E_C - E_F)/kT] \qquad (4.12)$$

where N_C is the effective density of states in the conduction band. The hole density n_h near the top of the valence band is given by

$$n_h = N_V \exp[-(E_F - E_V)/kT] \qquad (4.13)$$

where N_V is the effective density of states in the valence band.[4]

In insulators, however, the bandgap is so large that thermal excitation is insufficient to excite many electrons into the conduction band; therefore, the conduction band is almost completely empty of electrons and the next lower band of energy, the valence band, is almost completely full with no vacant sites at normal temperatures of interest.

4.2.2 Intrinsic Semiconductors

Intrinsic Conductivity   In intrinsic semiconductors, as has been mentioned, some small fraction of electrons gains sufficient energy by thermal excitation except at low temperatures to be excited into the conduction band from the valence band. These electrons leave an equal number of holes in the valence band, that is,

$$n_e = n_h = n_i \tag{4.14}$$

where n_i is the intrinsic carrier density. Figure 4.4 shows the temperture dependence of n_i for Ge, Si, and GaAs.[10] As expected from Eqs. (4.12) and (4.13), the larger bandgap results in the smaller intrinsic carrier density. At room temperature (300 K), the n_i for Ge, Si, and GaAs is obtained as 2.4×10^{13}, 1.45×10^{10}, and 1.79×10^6 cm^{-3}, respectively.[4]

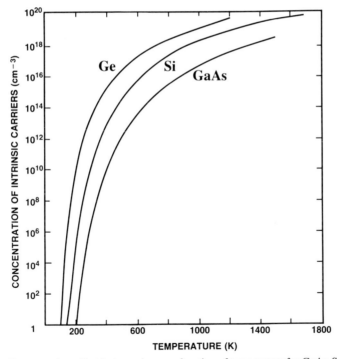

Fig. 4.4. Concentration of intrinsic carriers as a function of temperature for GaAs, Si, and Ge. (After Thurmond.[10] Reprinted with the permission of The Electrochemical Society, Inc.)

The electrons excited into the conduction band are able to move through the crystal giving rise to an electrical conductivity, that is, *intrinsic conductivity,* proportional to the concentration of conduction electrons n_e, the electron charge e, and the electron mobility μ_e, as described in Eq. (4.5). At the same time, holes in the valence band also contribute to an intrinsic conductivity proportional to their concentration n_h, absolute value of the electronic charge e, and mobility μ_h. These electrons and holes are referred to as *intrinsic carriers.* The electron mobility μ_e and the hole mobility μ_h are given by comparing with Eq. (4.8):

$$\mu_e = et_{r/e}/m_e \tag{4.15}$$

and

$$\mu_h = et_{r/h}/m_h \tag{4.16}$$

where $t_{r/e}$ and $t_{r/h}$ are the relaxation time for an electron and a hole, respectively. Consequently, for intrinsic semiconductors, the total conductivity is given by

$$\sigma = e(n_e \mu_e + n_h \mu_h) \tag{4.17}$$

In most semiconductors, as shown in Table 4.1, the electron mobility is higher than the hole mobility. For example, the electron mobility is higher by a factor of about three in Si; however, sometimes, as in GaAs and InSb, it is higher by more than one order of magnitude. A semiconductor material of a high carrier mobility is preferable for the devices that require high-speed switching. For example, since GaAs has an electron mobility five times higher than that of Si, it has been expected to meet the high switching speed requirements of those circuits as compared to the Si IC technology.[11]

The presence of acoustic phonons[12] and ionized impurities[13] results in carrier scattering events, which significantly limit the mobility. As a result, the mobility decreases with the increased impurity concentration.

Fermi Level in Intrinsic Semiconductor   According to the charge neutrality condition, $n_e = n_h$ [see Eqs. (4.12) and (4.13)], the Fermi level in an intrinsic semiconductor E_{Fi} is given by

$$E_{Fi} = \frac{E_C + E_V}{2} + \frac{kT}{2} \ln \frac{N_V}{N_C} \tag{4.18}$$

Equation (4.18) anticipates that E_{Fi} is approximately midway between E_C and E_V. Figure 4.5 represents the situation schematically for an intrinsic semiconductor by showing (a) the simplified band diagram, (b) the density of states $N(E)$, (c) the Fermi–Dirac distribution function $f(E)$, and (d) the carrier concentrations n_e and n_h.

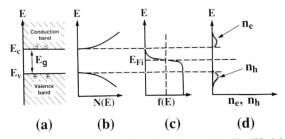

Fig. 4.5. Schematic situation in an intrinsic semiconductor: (a) simplified band diagram, (b) density of states $N(E)$, (c) Fermi–Dirac distribution function, and (d) carrier concentrations. (After Sze.[4] Adapted by courtesy of John Wiley & Sons, Inc.)

4.2.3   Extrinsic Semiconductors

Donors and Acceptors   At room temperature, the intrinsic carrier density is too small from the viewpoint of electronic device operation. For the purpose of the semiconductor device fabrication, *extrinsic* or *impurity conductivity* must be induced into the material by adding certain types of impurities. Such impurities drastically affect the electrical properties of a semiconductor. The effect results from the fact that the impurity atoms introduce new localized energy levels for electrons intermediate between the valence band and the conduction band.

As discussed in Section 3.4.1, the diamond lattice consists of each atom forming four covalent bonds, one with each of its four nearest neighbors, corresponding to the chemical valence four. If a pentavalent impurity atom (i.e., the Group V elements such as phosphorus, arsenic, and antimony; see Table 3.9) is now substituted into the diamond lattice in place of a host atom, there will be one valence electron from the impurity atom left over after the four covalent bonds are formed with the nearest neighbors. The situation, as illustrated for Si in Fig. 4.6a, is that the structure has an excess electron from the impurity atom, which has lost one electron, resulting in an excess positive charge. An impurity atom that results in an excess electron is referred to as a *donor*. Donor impurities introduce new energy levels called *donor levels*, as shown in Fig. 4.7a, which are filled by electrons close to the energy level of the conduction band. Therefore, electrons may be easily excited from impurity atoms into the conduction band. The electrons excited into the conduction band are able to contribute to the extrinsic conductivity. Semiconductors in which electrons are the dominant carriers are called *n-type semiconductors*.

If a trivalent impurity atom (i.e., the Group III elements such as boron and aluminum; see Table 3.9), is substituted in the diamond lattice, a hole will be bound to the impurity, just as an electron is bound to a pentavalent impurity, shown in Fig. 4.6b. That is, the lattice has an excess free hole from the impurity atom, which has taken up an electron from the host atom, resulting

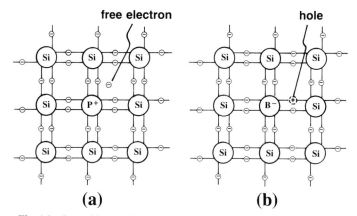

Fig. 4.6. Impurities in silicon lattice: (a) donor (P) and (b) acceptor (B).

in an excess negative charge. An impurity atom that brings about an excess hole is called an *acceptor*. Acceptors also introduce new energy levels, which lie close to the energy of the top of the valence band. Therefore, it is easy to excite an electron out of the filled valence band into the new *acceptor level*. This leaves a hole in the valence band that can contribute to the extrinsic conductivity just as an electron in the conduction band does. Semiconductors that conduct through hole conduction are called *p-type semiconductors*. Semiconductors of *p* and *n* types are called *extrinsic semiconductors* as a whole, in comparison with *intrinsic semiconductors*.

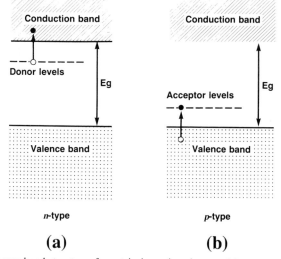

Fig. 4.7. Energy-band structures for extrinsic semiconductors: (a) *n*-type and (b) *p*-type.

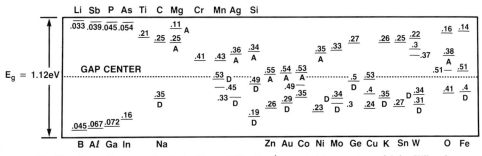

Fig. 4.8. Impurity energy levels in silicon. (After Sze.[4] Adapted by courtesy of John Wiley & Sons, Inc.)

Figure 4.8 shows the impurity energy levels for several important elements in Si.[4] The levels below the energy-gap center are measured from the top of the valence band and are acceptor levels unless indicated by D for a donor level. The levels above the energy-gap center are measured from the bottom of the conduction band and are donor levels unless indicated by A for an acceptor level. It is possible to have more than one level for a single element. In addition to chemical impurities shown in Fig. 4.8, point defects may also result in localized levels; for example, a vacancy–interstitial pair—that is, a Frenkel defect—behaves in many respects like an acceptor level.[14]

Fermi Levels in Extrinsic Semiconductors Introducing impurities into the semiconductor changes the form of the neutrality condition, which means the Fermi level must shift to preserve charge neutrality. The neutrality condition for an extrinsic semiconductor is

$$n_h + N_D^+ = n_e + N_A^- \tag{4.19}$$

where N_D^+ and N_A^- are the concentration of the ionized donors and acceptors, respectively. Dividing Eq. (4.12) by Eq. (4.13) and taking the logarithm, one obtains

$$E_F = \frac{E_C + E_V}{2} + \frac{kT}{2} \ln \frac{N_V}{N_C} + \frac{kT}{2} \ln \frac{n_e}{n_h} \tag{4.20}$$

Equations (4.18) and (4.20) lead to

$$E_F = E_{Fi} + \frac{kT}{2} \ln \frac{n_e}{n_h} \tag{4.21}$$

This equation shows that the increase in the proportion of electrons to holes, caused by introducing donors into the semiconductor, shifts the Fermi level from E_{Fi}—that is, approximately the middle of the band gap—up toward the conduction band. Figure 4.9 depicts this situation by showing (a) the

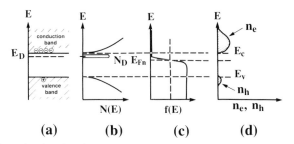

Fig. 4.9. Schematic situation in an *n*-type semiconductor: (a) simplified band diagram, (b) density of states $N(E)$, (c) Fermi–Dirac distribution function, and (d) carrier concentrations. (After Sze.[4] Adapted by courtesy of John Wiley & Sons, Inc.)

simplified band diagram, (b) the density of states $N(E)$, (c) the Fermi–Dirac distribution function $f(E)$, and (d) the carrier concentrations n_e and n_h. The reverse happens when acceptors are added. This situation is schematically shown in Fig. 4.10 as well. In these figures, E_{Fn} and E_{Fp} denote the Fermi level in the *n*-type semiconductor and that in the *p*-type semiconductor, respectively.

Concentration of Carriers   On calculating the product of $n_e n_h$ [see Eqs. (4.12) and (4.13)], the Fermi level E_F disappears:

$$n_e n_h = N_C N_V \exp[(E_V - E_C)/kT]$$
$$= N_C N_V \exp(-E_g/kT) = n_i^2 \tag{4.22}$$

Equations (4.19) and (4.22) can be combined to give the concentration of electrons n_e and holes n_p:

$$n_e = \tfrac{1}{2}\{(N_D^+ - N_A^-) + [(N_D^+ - N_A^-)^2 + 4n_i^2]^{1/2}\} \tag{4.23}$$

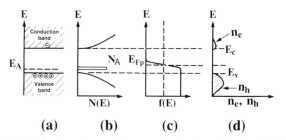

Fig. 4.10. Schematic situation in a *p*-type semiconductor: (a) simplified band diagram, (b) density of states $N(E)$, (c) Fermi–Dirac distribution function, and (d) carrier concentrations. (After Sze.[4] Adapted by courtesy of John Wiley & Sons, Inc.)

and

$$n_\mathrm{h} = \tfrac{1}{2}\{(N_\mathrm{A}^- - N_\mathrm{D}^+) + [(N_\mathrm{A}^- - N_\mathrm{D}^+)^2 + 4n_\mathrm{i}^2]^{1/2}\} \tag{4.24}$$

In an *n*-type semiconductor,

$$N_\mathrm{D}^+ - N_\mathrm{A}^- \gg n_\mathrm{i} \quad \text{and} \quad N_\mathrm{D}^+ \gg N_\mathrm{A}^-$$

and therefore

$$n_\mathrm{e} \approx N_\mathrm{D}^+ \tag{4.25}$$

and

$$n_\mathrm{h} = n_\mathrm{i}^2/n_\mathrm{e} \simeq n_\mathrm{i}^2/N_\mathrm{D}^+ \tag{4.26}$$

In a *p*-type semiconductor,

$$N_\mathrm{A}^- - N_\mathrm{D}^+ \gg n_\mathrm{i} \quad \text{and} \quad N_\mathrm{A}^- \gg N_\mathrm{D}^+$$

and therefore

$$n_\mathrm{h} \approx N_\mathrm{A}^- \tag{4.27}$$

and

$$n_\mathrm{e} = n_\mathrm{i}^2/n_\mathrm{h} \simeq n_\mathrm{i}^2/N_\mathrm{A}^- \tag{4.28}$$

For *n*-type semiconductors, the electron is called the *majority carrier* and the hole the *minority carrier*, since the electron density is the larger of the two; for *p*-type semiconductors the roles are reversed.

Generation and Recombination   The equilibrium concentration of electrons and holes in a semiconductor can be temporarily changed by several ways: by the creation of electron–hole pairs by light, by charged-particle bombardment, or by injecting carriers into the sample through a metal contact. For example, as discussed in Section 2.1.2, if light photons, whose energy $h\nu$ is greater than the bandgap, are absorbed in the crystal, the photon energy may raise an electron from the valence band to the conduction band as shown in Fig. 4.11. This process is called *generation* and produces excess carriers, which persist as long as the radiation continues. When the radiation is removed, the excess carrier concentration decays back to the equilibrium value by means of *recombination*. Such recombination may occur either directly by a band-to-band transition or by transitions through intermediate states as diagrammed in Fig. 4.11.

Deep Levels   If an impurity atom of the Group II elements, such as Zn, is added to Si, two bonds in the lattice will be missing in the vicinity of the impurity atom. This atom therefore can accept either a single electron from

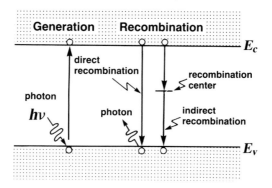

Fig. 4.11. Generation and recombination in a semiconductor.

the valence band and become singly ionized, or can accept two electrons and become doubly ionized. It may be easily understood that the energy levels that valence electrons must attain to be accepted are much higher than for a Group III acceptor such as boron. Such levels that lie in the midrange of the forbidden bandgap are called *deep levels*. Deep levels may also be created in semiconductors by lattice defects and by many transition-metal impurities such as Fe, Ni, Co, and Cu.

Thermal energy is not sufficient to excite valence electrons into the deep levels, but the deep levels can trap free electrons from the conduction band. Therefore, it is found that deep-level impurities can drastically reduce the number of free electrons for conduction. This phenomenon has been applied to some devices such as high-speed switching transistors by doping with Au, which enhances recombination and thus reduces the switching time.[4]

Resistivity   The resistivity ρ was defined as $\rho = 1/\sigma$, and thus the resistivity is derived from Eq. (4.17):

$$\rho = 1/\sigma = 1/[e(n_e\mu_e + n_h\mu_h)] \tag{4.29}$$

If $n_e \gg n_h$, as in n-type semiconductors, then

$$\rho \approx 1/en_e\mu_e \tag{4.30}$$

while if $n_h \gg n_e$, as in p-type semiconductors, then

$$\rho \approx 1/en_h\mu_h \tag{4.31}$$

Practically, the resistivity of silicon crystals doped with boron or phosphorus is calculated from the dopant concentration.[15] For boron-doped silicon,

$$\rho = 1.305 \times 10^{16}/[B]$$
$$+ 1.133 \times 10^{17}/[B][1 + (2.58 \times 10^{-19}[B])^{-0.737}] \tag{4.32}$$

where ρ is the resistivity in Ω cm and $[B]$ the concentration of boron in atoms/cm^3. For phosphorus-doped silicon,

$$\rho = (6.242 \times 10^{18}/[P]) \times 10^Z \tag{4.33}$$

where $[P]$ is the concentration of phosphorus in atoms/cm^3 and

$$Z = (A_0 + A_1 y + A_2 y^2 + A_3 y^3)/(1 + B_1 y + B_2 y^2 + B_3 y^3) \tag{4.34}$$

where

$$
\begin{aligned}
y &= (\log_{10} [P]) - 16 \\
A_0 &= -3.0769 \\
A_1 &= 2.2108 \\
A_2 &= -0.62272 \\
A_3 &= 0.057501 \\
B_1 &= -0.68157 \\
B_2 &= 0.19833 \\
B_3 &= -0.018376
\end{aligned}
\tag{4.35}
$$

Figure 4.12 shows the measured resistivity at 300 K as a function of the dopant concentration for silicon.[15,16] In the figure, the solid lines show

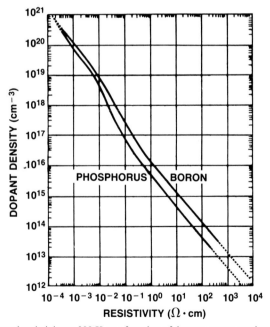

Fig. 4.12. Measured resistivity at 300 K as a function of dopant concentration for silicon. (After ASTM F723-82[15] and Irvin.[16])

conversion of the resistivity to dopant concentration for the range of actual data, while dashed lines show regions of extrapolation from the data. The conversion in silicon doped with other than boron and phosphorus has not yet been established; however, in the lightly doped range ($< 10^{17}$ atoms/cm^3), the above equations and Fig. 4.12 are expected to be reasonably accurate for other dopants. In the medium doping range between 10^{17} atoms/cm^3 and 10^{19} atoms/cm^3, the different activation energies of different dopants may cause differences in the resistivity for the same dopant concentration. In this range the differences in resistivities will be larger among acceptor dopants than among donor dopants due to the larger differences in the activation energies among the acceptor elements. In high doping ranges ($> 10^{19}$ atoms/cm^3), the formation of complexes involving dopant atoms and lattice defects may lead to a modification in the number of charged carriers. The extent of this effect will depend on the particular dopant elements and has not yet been well examined.

4.3   Electronic Device Physics

4.3.1   Surface States

As discussed in Section 3.5, all the valence electrons of the crystal surface atoms are not saturated. The unsaturated bonds of the surface atoms make the crystal surface highly reactive. Foreign atoms absorbed on the surface of a semiconductor crystal can give rise to discrete energy levels, which influence not only the surface directly but also the properties of the crystal in the region near the surface. This results at the surface in a large density of localized quantum states whose energy levels are distributed in the bandgap. Such levels are called *surface states* and may be present in densities as great as $\sim 10^{15}$/cm^2, or the order of the density of surface atoms.

Figure 4.13 illustrates the situation of negatively charged surface states on an *n*-type semiconductor. The negative charge repels free electrons from the surface, leaving positively charged ionized donors to neutralize the effect of the surface charges. The potential energy of the electron thus becomes greater in the surface region than inside the crystal. The energy-level diagram in Fig. 4.13 illustrates this behavior by bending the bands upward near the surface, and this brings about the difference in *potential energy* V_0 at surface compared with that in the bulk. Since the density of free carriers—electrons in this case—is depleted near the surface, this region is called a *depletion layer*. If the negatively charged surface states on an *n*-type semiconductor are large in number, the band bends upward to such an extent that a large number of holes are generated in valence-band states as illustrated in Fig. 4.14. These

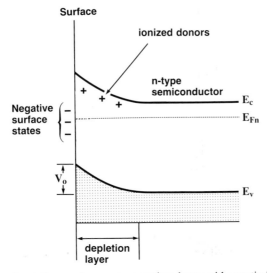

Fig. 4.13.   Energy-band diagram for an *n*-type semiconductor with negatively charged surface states.

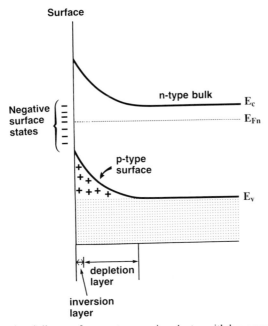

Fig. 4.14.   Energy-band diagram for an *n*-type semiconductor with large number of negatively charged surface states.

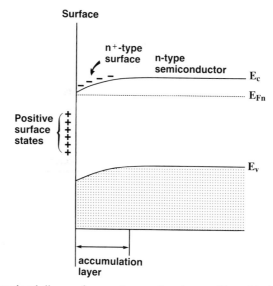

Fig. 4.15. Energy-band diagram for an *n*-type semiconductor with positively charged surface states.

holes, together with positively charged ionized donors, neutralize the negative surface charge, which results in a *p*-type region in the surface of an *n*-type semiconductor. This region is referred to as an *inversion layer*. When positively charged surface states arise on a *p*-type semiconductor, a similar situation takes place except that the bands bend downward. An *n*-type inversion layer is then produced on a *p*-type semiconductor.

When the surface states on an *n*-type semiconductor are positively charged, the charge is neutralized by the free electrons in the conduction band, and, as shown in Fig. 4.15, this produces an *accumulation layer* that may show an n^+-type surface. Negatively charged surface states on a *p*-type semiconductor generate similarly a p^+-type surface accumulation layer. The layers of depletion, inversion, and accumulation are called *space charge layers* as a whole.

4.3.2   *p-n* Junction

General Remarks   The junction between a *p*-type region and an *n*-type region in the same semiconductor crystal, a *p-n junction*, is the most important region both in modern electronic devices and in understanding the semiconductor device operation. The *p-n* junction theory is basic to the physics of semiconductor devices. It should be noted that a *p-n* junction is not the interface between two pieces of semiconductor of opposite types pressed together, but is a specific location in the semiconductor where impurity type changes from *p* to *n*, or vice versa, while the lattice structure is continuous. The structure of a *p-n* junction can be fabricated in several ways. The current important methods are diffusion, ion implantation, and epitaxy. Another method, alloying, was very important in the Ge era.

p-n Junction in Equilibrium Figure 4.16a depicts the impurity atoms together with the contributed carriers on both sides of the junction in a semiconductor crystal as it would look if the carriers did not diffuse. As discussed in the previous subsection, at any point, the neutrality condition for an extrinsic semiconductor is

$$n_h + N_D^+ = n_e + N_A^- \tag{4.19}$$

However, such a situation cannot happen realistically. As shown in Fig. 4.16b, the electrons, which are abundant in the *n*-type side, will diffuse into

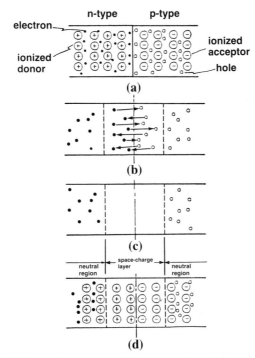

Fig. 4.16. Schematic illustrations representing a *p-n* junction: (a) impurity atoms and contributed carriers in *n*-type and *p*-type semiconductors, (b) diffusion of electrons and holes, (c) annihilation of electrons and holes by their combination, and (d) space charge layer at the *p-n* junction.

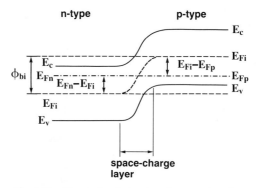

Fig. 4.17.   Energy-band structure for a *p-n* junction.

the *p*-type side, in which there are very few of them, and the holes will diffuse in the opposite direction. These electrons and holes combine with each other to lead to their annihilation (see Fig. 4.16c). That is, the flow of electrons and holes causes an immediate loss of neutrality, which is represented by Eq. (4.19); the *n*-type side is charged positively because of the ionized donors left behind, and similarly the *p*-type side is charged negatively because of the ionized acceptors, as shown in Fig. 4.16d. A potential barrier, which is called a *built-in potential* ϕ_{bi}, builds up across the resulting space charge layer and blocks any further majority carrier diffusion.

As discussed with Figs. 4.9 and 4.10, E_F is shifted toward E_C in an *n*-type semiconductor, and toward E_V in a *p*-type semiconductor. Since the Fermi level E_F is continuous throughout the crystal at equilibrium, the conduction and valence bands bend in the vicinity of the *p-n* junction as shown in Fig. 4.17.

External Potential Applied to *p-n* Junction Diode   When no external potential is applied to a *p-n* junction, the behavior of carriers and the energy-band structure are diagrammed in Figs. 4.16 and 4.17, respectively. Figures 4.18 and 4.19 represent the behavior of carriers and the energy-band structure when an external forward or an external reverse potential is applied to the junction, respectively. If a forward potential V_A is applied (Fig. 4.18), the barrier height is reduced to $\phi_{bi} - V_A$, which results in the diffusion of holes into the *n*-type region and of electrons into the *p*-type region. When a reverse potential $-V_A$ is applied (Fig. 4.19), the barrier height is increased to $\phi_{bi} + V_A$ and both the carriers are repelled away from the junction. As a result, the broadened space-charge layer is formed. This is a dielectric layer with a very low concentration of carriers.

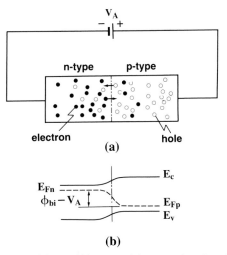

Fig. 4.18. Effect of an external forward bias potential on a *p-n* junction: (a) behavior of carriers and (b) energy-band structure.

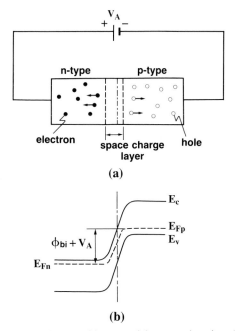

Fig. 4.19. Effect of an external reverse bias potential on a *p-n* junction: (a) behavior of carriers and (b) energy band structure.

4.4   Transistors

The term *transistor*[17] is from "transfer resistor," and transistors are unquestionably the most important semiconductor devices. The transistors discussed in this section include two types: (1) *bipolar* or *carrier injection
transistors*, and (2) *unipolar* or *field-effect transistors*. First the structure of
bipolar transistors is briefly discussed. In a bipolar transistor, the internal
currents are obtained by both majority and minority carriers, hence the name
bipolar transistors. Then field-effect transistors, which have become more
and more important in the VLSI/ULSI technology, are discussed.

4.4.1.   Bipolar Transistor

A bipolar transistor consists of a three-zone structure—*emitter* E, *collector* C,
and *base* B—with two parallel *p-n* junctions very near each other built in the
same semiconductor crystal. Two distinct types of the three-zone structure
are possible: a *p-n-p* transistor (Fig. 4.20) and an *n-p-n* transistor (Fig. 4.21).
In these figures, the basic conceptional structure (a), the cross section of a
practical silicon planar bipolar transistor (b), and the circuit symbol (c) are
shown. The operational principles of the two types are identical except for the

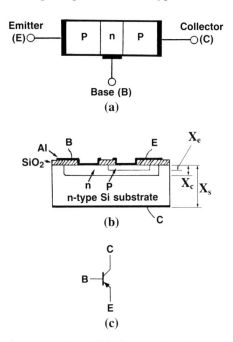

Fig. 4.20.   Structure of *p-n-p* transistor: (a) basic conceptional structure, (b) cross section of a
silicon planar transistor, and (c) circuit symbol.

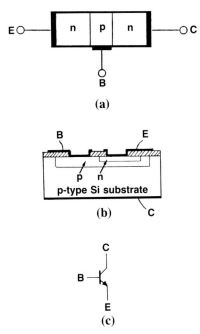

Fig. 4.21. Structure of *n-p-n* transistor: (a) basic conceptional structure, (b) cross section of a silicon planar transistor, and (c) circuit symbol.

interchange of minority and majority carrier types, and the polarity of the applied bias voltages.

The majority of discrete transistors made today are of the silicon planar type, as shown in Figs. 4.20b and 4.21b. A *p-n-p* planar silicon transistor shown in Fig. 4.20b. for example, is made by diffusing donor impurities such as phosphorus into a *p*-type Si substrate through an open window of an SiO$_2$ mask on the substrate, that is, base diffusion. Then acceptor impurities such as boron are diffused through a small window of a second mask for forming the emitter region. Note that the thicknesses of the emitter (X_e), collector (X_c), and substrate (X_s) as illustrated in Figs. 4.20b and 4.21b do not represent their realistic proportions with each other. That is, X_e and X_c are practically on the order of a few micrometers while X_s is several hundreds of micrometers or even closer to a millimeter in the case of a recent large-diameter Si substrate as described in Chapter 5. State-of-the-art bipolar ICs are commonly fabricated by planar epitaxial technology.[18] The fundamental advantage of epitaxial substrates over bulk wafers is that the structures such as high-resistivity layer/low-resistivity wafer and thin layer/wafer with opposite doping types are easily formed by epitaxial growth. Figure 4.22 shows a schematic cross section of an *n-p-n* bipolar planar transistor built on an

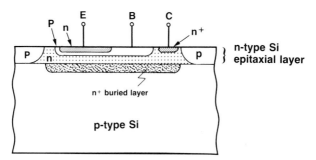

Fig. 4.22.   Cross section of an *n-p-n* bipolar planar transistor built on an epitaxial substrate.

epitaxial Si substrate. The heavily doped n^+ buried layer functions as a subcollector, which reduces the series resistance of the devices and serves for alignment in subsequent wafer processing.

4.4.2   Physics of MOS

MOS Structure   The metal oxide semiconductor field-effect transistor (MOSFET) has been realized as the most important device for VLSI/ULSI circuits such as memories and microprocessors because of its low fabrication cost, small size, and low power consumption. Figure 4.23 shows the structure of MOS or MIS (metal insulator semiconductor) and an MOS capacitor, which consists of a parallel-plate capacitor with one metallic plate, called the *gate*, and the other ohmic electrode.

C–V Characteristics   The cross sections of an MOS capacitor fabricated on *p*-type Si are schematically shown with corresponding circuit symbols in Fig. 4.24. When the gate bias voltage $V_G \ll 0$ (Fig.4.24a), the MOS capacitor has the static capacitance C_0 per unit area, which is given by

$$C_0 = K_{ox} \varepsilon_0 / t_{ox} \qquad (4.36)$$

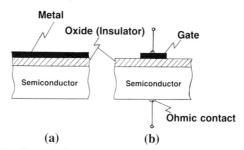

(a)   (b)

Fig. 4.23.   Cross-sectional structure: (a) MOS and (b) MOS capacitor.

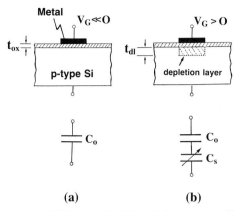

Fig. 4.24.   Capacitance in an MOS capacitor fabricated on *p*-type silicon: (a) $V_G \ll 0$, and (b) $V_G > 0$.

where K_{ox} the dielectric constant of silicon oxide and ε_0 the dielectric permittivity of vacuum, and t_{ox} the thickness of oxide layer. When $V_G > 0$, holes are repelled and a depletion layer of t_{dl} in thickness is formed under the gate, as shown in Fig. 4.24b. The thickness t_{dl} depends predominantly on V_G and the doping level. Since there are no free carriers in the depletion layer, the depletion layer can be regarded as an insulator, and then the capacitance C_s, which is nonlinearly variable with V_G, is given by

$$C_s = K_{si}\varepsilon_0/t_{dl} \tag{4.37}$$

where K_{si} is the dielectric constant of silicon. The circuit model can be assumed as two series capacitors; the total capacitance C is then given by

$$1/C = 1/C_0 + 1/C_s \tag{4.38}$$

Figure 4.25 illustrates the relationship between the MOS capacitance (C/C_0) and gate bias voltage (V_G) for a *p*-type semiconductor in an ideal

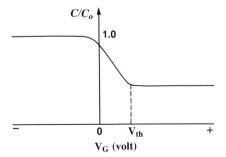

Fig. 4.25.   Ideal C–V relation in an MOS capacitor fabricated on a *p*-type semiconductor.

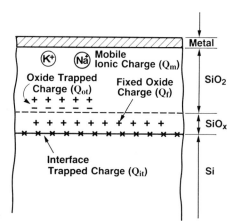

Fig. 4.26. Charges associated with thermally oxidized silicon. (After Deal.[20] ©1980 IEEE.)

situation.* Under a reverse bias condition ($V_G < 0$), C is equal to C_0 since no depletion layer exists. In the region $0 < V_G < V_{th}$ (V_{th} is referred to as the *threshold voltage*), the total capacitance C decreases with increasing t_{dl}, in turn decreasing C_s. When V_G is larger than V_{th}, C/C_0 becomes constant. However, the voltage due to the difference between the work functions of the metal and the semiconductor makes the C–V curve deviate from the ideal situation. In addition, the charges such as shown in Fig. 4.26[20] in the MOS capacitor also result in deviation of the C–V curve from the ideal situation. As shown in Fig. 4.27, when a positive charge exists at the SiO_2/Si interface, the C–V curve shifts to the left (a), while a negative charge shifts the C–V curve to the right (b). The voltage that is required on the gate of a MOS capacitor in order to achieve the ideal situation—that is, flat-band condition

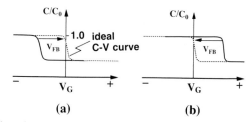

(a)   (b)

Fig. 4.27. Deviation of C–V curve from the ideal situation for a p-type semiconductor due to (a) positive charge ($V_{FB} > 0$) and (b) negative charge ($V_{FB} < 0$).

---

* Figure 4.25 shows the high-frequency C–V characteristics. At low frequencies, the capacitance goes through a minimum at V_{min} ($\sim V_{th}$) and then increases again as the inversion layer forms at the surface and C/C_0 reaches unity.[19] In the following, the high-frequency C–V plot, which is most commonly used to characterize MOS capacitors, is concerned.

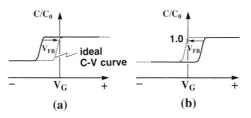

Fig. 4.28. Deviation of C-V curve from the ideal situation for an n-type semiconductor due to (a) positive charge ($V_{FB} > 0$) and (b) negative charge ($V_{FB} < 0$).

—is defined as the *flat-band voltage* and is represented with V_{FB}. For an n-type semiconductor, the C-V curve shifts similarly due to the existing states as shown in Fig. 4.28.

The flat-band voltage directly affects the threshold voltage V_{th}, which is defined as the gate bias voltage required to form an inversion layer in the surface of a semiconductor and is one of the most important factors for the operation of MOS devices. Constant V_{FB} is required to ensure high performance of MOS device operation.

4.4.3   MOS Transistor

Structure of MOSFET   Devices in which the conduction involves only one polarity are referred to as *unipolar devices*, in contrast to bipolar devices. Among the unipolar devices, the MOSFET is the most important for VLSI/ULSI circuits. Many acronyms, such as MOST (MOS transistor), IGFET (insulated-gate FET), and MISFET (metal insulator semiconductor FET), represent the same device. Figure 4.29 depicts the basic structures of MOSFETs of two different types: (a) n-channel MOSFET (NMOS) and (b) p-channel MOSFET (PMOS). Note that the vertical and horizontal scales illustrated in Fig. 4.29, as in Figs. 4.20 and 4.21, do not represent a practical relation with each other. A MOSFET consists of four essential parts: *source* (S), *gate* (G), *channel*, and *drain* D. Heavily doped polysilicon or a combination of silicide and polysilicon as well as metals such as aluminum are used as

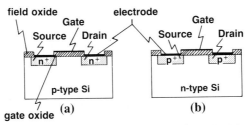

Fig. 4.29. Cross-sectional structure of enhancement-type MOSFETs: (a) n-channel MOSFET (NMOS) and (b) p-channel MOSFET (PMOS).

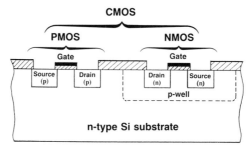

Fig. 4.30.   Cross-sectional structure of CMOS.

the gate electrode. For NMOS, the source and drain regions are fabricated by either ion implantation or diffusion of donor impurities in the surface region of the *p*-type Si substrate. For PMOS, on the other hand, they are fabricated similarly with acceptor impurities in the *n*-type Si substrate. In an IC, a MOSFET is surrounded by the *field oxide*, which is thicker than the *gate oxide*, to isolate it from adjacent devices.

Although most of today's FETs, specifically random-access memories (RAMs), are NMOS designs, complementary MOS (CMOS) RAMs are the wave of the future VLSI/ULSI devices, including memories, microprocessors, and random logic. The basic structure of CMOS fabricated on an *n*-type Si substrate is diagrammatically shown in Fig. 4.30. It has a PMOS device fabricated with source-drain diffusion into the *n*-type substrate and an NMOS device fabricated with source-drain diffusion into the *p*-well (or *p*-tub), which was formed with *p*-type impurity diffusion into the substrate. Although the CMOS fabrication process is more complicated than that for simple NMOS or PMOS processes, CMOS technology[21] permits such advantages as reduction of power consumption, much simpler circuit design resulting in a much more efficient circuit layout, and thus smaller chips. There remain, however, a few CMOS problems that must be solved. The most notorious problem is *latchup*, which becomes more difficult to manage as the circuit geometry is reduced. For solutions, it is becoming common to fabricate CMOS circuits in high-resistivity epitaxial layers on low-resistivity substrates, or to utilize a trench structure in order to effectively separate each component.

Operating Principles   Figure 4.31 diagrammatically explains the operating principle for NMOS. The n^+ source and drain regions, where the majority carrier electrons exist, are formed in the *p*-type Si substrate where the majority carrier holes dominate. When no voltage is applied to the gate (i.e., $V_G = 0$), drain current I_D does not flow even when a low voltage is applied to the drain, namely at $V_D - V_S > 0$, where V_D and V_S are the drain and source voltage, respectively, since no channel is formed between the two n^+ regions

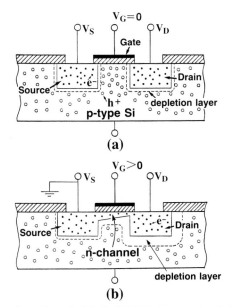

Fig. 4.31. Operating principle for NMOS: (a) $V_G = 0$ and (b) $V_G > 0$.

(see Fig. 4.31a). The only current that can flow from source to drain is the reverse leakage current. However, when a positive bias voltage V_G is applied to the gate and V_G exceeds the threshold voltage V_{th}, a surface inversion layer (i.e., *n*-channel) is formed between the source and drain, as shown in Fig. 4.31b. The two regions are then connected by a conducting *n*-channel through which a large drain current I_D can flow. The conductance of this channel can be modulated by varying the gate voltage. That is, the characteristics of the MOSFET are variable with the applied bias voltage to the gate. In the case of PMOS fabricated in *n*-type silicon, I_D is obtained similarly by applying a reverse bias voltage V_G, as shown in Fig. 4.32.

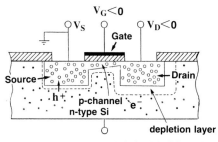

Fig. 4.32. Operating principle for PMOS at $V_G < 0$.

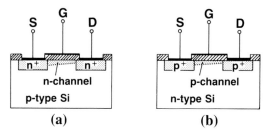

Fig. 4.33. Cross-sectional structure of depletion-type MOSFETs: (a) NMOS and (b) PMOS.

For the MOSFETs shown in Fig. 4.29, I_D does not flow as zero gate bias. These transistors are considered to be "normally off." This kind of transistor is called an *enhancement-type* device, because a gate bias voltage will enhance the conductivity of the channel. On the other hand, as shown in Fig. 4.33, MOSFETs that have a channel fabricated between the source and drain by *channel doping* are considered to be "normally on." A device that is normally on is called a *depletion-type* device, because a gate bias voltage is required to deplete the channel resulting in no I_D flow. In summary, there are basically four types of MOSFET depending on the types of channel and inversion layer: enhancement-type NMOS and PMOS, and depletion-type NMOS and PMOS. Although the operating principles of NMOS and PMOS are identical, NMOS is preferred to PMOS for devices that require high-speed switching. This is mainly because the mobility of electrons, μ_e, in an *n*-channel transistor is greater than that of holes, μ_h, in a *p*-channel transistor.

Contamination Effect on MOSFET   Although the principle of a MOSFET device had been recognized since the late 1930s, the modern practical MOSFET was made in 1960s after the bipolar transistor was invented in 1947.[4,5] This is mainly due to the unsatisfactory technology of semiconductor surface preparation and oxide growth. As discussed, the field effect is highly sensitive to surface states, and in turn to surface contamination such as positive sodium ions. The presence of positive ions in the oxide at the SiO_2–Si interface attracts electrons in the silicon. This brings about a deleterious effect on an enhancement-type NMOS and a depletion-type PMOS (see Figs. 4.29a and 4.33b). For an IC structure, a current may flow between NMOSs, which must be isolated each other, because of the unfavorable channel formed by electrons attracted to the silicon surface. That is, only an enhancement-type PMOS may be influenced little by positive ion contamination. In fact, ICs using MOSFETs were originally based on PMOS devices,[22] although NMOS is preferred to PMOS because of the higher electron mobility with respect to hole mobility as described above. The development of material and process technologies have made NMOS dominate in the IC market since the early 1970s. Again, it should be emphasized that the harmful effects of contamination on the device performance have to be minimized.

References

1. W. Schockley, "Electrons and Holes in Semiconductors." Van Nostrand-Reinhold, New York, 1950.
2. W. C. Dunlap, "An Introduction to Semiconductors." New York, 1962.
3. J. L. Moll, "Physics of Semiconductors." McGraw-Hill, New York, 1964.
4. S. M. Sze, "Physics of Semiconductor Devices," 2nd ed. Wiley, New York, 1981.
5. E. H. Nicollian and J. R. Brews, "MOS (Metal Oxide Semiconductor) Physics and Technology." Wiley, New York, 1982.
6. C. Kittel, "Elementary Solid State Physics: A Short Course." Wiley, New York, 1962.
7. J. M. Ziman, "Principles of the Theory of Solids." Cambridge Univ. Press, London and New York, 1964.
8. A. Bar-Lev, "Semiconductors and Electronic Devices," 2nd ed. Prentice-Hall, Englewood Cliffs, New Jersey, 1984.
9. Y. P. Vashni, Temperature dependence of the energy gap in semiconductors. *Physica (Amsterdam)* **34**, 149–154 (1967).
10. C. D. Thurmond, The standard thermodynamic function of the formation of electrons and holes in Ge, Si, GaAs and GaP. *J. Electrochem. Soc.* **122**, 1133–1141 (1975).
11. B. M. Welch, Advances in GaAs LSI/VLSI processing technology. *Solid State Technol.* Feb., pp. 95–101 (1980).
12. J. Bardeen and W. Schockley, Deformation potentials and mobilities in non-polar crystals. *Phys. Rev.* **80**, 72–80 (1950).
13. E. Conwell and V. F. Wesskopf, Theory of impurity scattering in semiconductors. *Phys. Rev.* **77**, 388–390 (1950).
14. L. A. Azaroff and J. J. Brophy, "Electronic Properties in Materials." McGraw-Hill, New York, 1963.
15. "Annual Book of ASTM Standards," Vol. 10.05 Electronics (II) F723–82, pp. 598–614. Am. Soc. Test. Mater. Philadelphia, Pennsylvania, 1984.
16. J. C. Irvin, Resistivity of bulk silicon and of diffused layers in silicon. *Bell Syst. Tech. J.* **41**, 387–410 (1962).
17. J. Bardeen and W. H. Brattain, The transistor, a semiconductor triode. *Phys. Rev.* **74**, 230–231 (1948).
18. H. M. Liaw, J. Rose, and P. L. Fejes, Epitaxial silicon for bipolar integrated circuits. *Solid State Technol.* May, pp. 135–143 (1984).
19. A. S. Grove, B. E. Deal, E. H. Snow, and C. T. Sah, Investigation of thermally oxidized silicon surface using metal-oxide-semiconductors. *Solid-State Electron.* 145–163 (1985).
20. B. E. Deal, Standard terminology for oxide charges associated with thermally oxidized silicon. *IEEE Trans. Electron Devices* **ED-27**, 606–608 (1980).
21. D. B. Scott, K.-L. Chen, and R. D. Davies, CMOS VLSI technology. *In* "VLSI Handbook" (N. G. Einspruch, ed.), pp. 121–150. Academic Press, New York, 1985.
22. L. C. Parrillo, VLSI process integration. *In* "VLSI Technology" (S. M. Sze, ed.). pp. 445–505. McGraw-Hill, New York, 1983.

Chapter 5

Silicon Crystal Growth and Wafer Preparation

The significant development of semiconductor materials began in December 1947 with the invention of the point-contact transistor by Bardeen and Brattain; it was followed by the invention of the junction transistor by Schockley 2 months later.[1] Very fruitful interaction between the material users and material manufacturers has continued since that time, when the necessity for "perfect and pure" crystals was recognized. The competition was often such that the crystal qualities demanded by new devices could only be met by controlling the crystal growth processes with electronic equipment built with the new devices.[2] Since dislocation-free silicon crystals already could be grown in the early 1960s using the *Dash technique*,[3] materials research and development efforts have concentrated on the purity, production yield, and problems related to device manufacturing processes.

Semiconductor devices and circuits are fabricated through many mechanical, chemical, physical, and thermal processes. A flow diagram for typical semiconductor silicon preparation processes is shown in Fig. 5.1. The preparation of silicon single-crystal substrates with mechanically and chemically polished surfaces is the first step in the long and complex device fabrication processing. Silicon is the second most abundant element on earth; more than 90% of the earth's crust is composed of silica and silicate. With a boundless supply of the raw material, the problem is to transform silicon into the usable state required by the semiconductor technology. As discussed in previous chapters, small amounts of some impurities have a strong influence on the electric characteristics of silicon. The first and absolute requirement is the extreme purity of the silicon used for electronic device fabrication. Such high-purity silicon is commonly referred to as "eleven nines" (99.999999999%). For controlling doping levels during subsequent crystal growth, the polycrystalline silicon material must have levels of impurities in

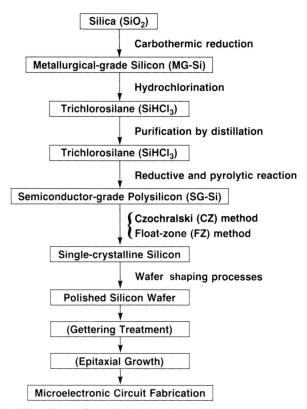

Fig. 5.1. Flow diagram for typical semiconductor silicon preparation processes.

the parts per billion atomic (ppba) range. Of primary concern is that the Group III, Group V, and metallic impurities must be less than 0.3 ppba, 1.5 ppba, and 0.1 ppba, respectively.[3] Besides the purity, the production cost and the specifications must meet the industry demands.

In this chapter, the currently used approaches to silicon materials preparation from the raw material, via single crystal ingots and polished wafers, to epitaxial silicon substrates are discussed.

5.1 Starting Materials

5.1.1 Metallurgical-Grade Silicon

The starting material for high-purity silicon single crystals is silica (SiO_2). The first step in manufacturing silicon is melting and reduction of silica. This is accomplished by mixing silica and carbon in the form of coal, coke, or

Table 5.1 Typical Impurity Concentrations found in Metallurgical-Grade Silicon[a]

Element	Concentration (ppma)
Al	1200–4000
B	37–45
P	27–30
C	NR[b]
Ca	590
Cr	50–140
Cu	24–90
Fe	1600–3000
Mg	NR
Mn	70–80
Mo	< 10
Ni	40–80
Ti	150–200
V	100–200
Zn	NR
Zr	30

[a] After McCormic.[4]
[b] NR, Not reported.

wood chips and heating the mixture to high temperatures in a submerged electrode arc furnace. Following the carbothermic reduction of silica as shown in Eq. (5.1), fused silicon is produced:

$$SiO_2 + 2C \longrightarrow Si + 2CO \tag{5.1}$$

The complex series of reactions actually occur in the furnace at temperatures ranging from 1500 to 2000°C.[3] The silicon lumps obtained in this way are called metallurgical-grade silicon (MG-Si), and its purity is about 98–99%. The remaining percent is made up of impurities such as listed in Table 5.1.[4] It is estimated that for 1 metric ton of MG-Si, 2500–2700 kg silica, 600 kg charcoal, 600–700 kg coal or coke, 300–500 kg wood chips, and 13,000–15,000 kWh of electric power are required.[5] Today, approximately 500,000 metric tons of MG-Si are produced each year in the world. Most of the products are used for aluminum alloys ($\sim 60\%$), chemical products such as silicone resin ($\sim 30\%$), and steel ($\sim 10\%$); only 1% or less of the total production of MG-Si is for manufacturing high-purity silicon for the electronics industry.[5]

5.1.2 Polycrystalline Silicon

Intermediate Chemical Compounds The next steps are to purify MG-Si to the level of semiconductor-grade silicon (SG-Si), which is used as the starting material for single-crystalline silicon. The basic concept is that powdered MG-Si is reacted with anhydrous HCl to form various chlorosilane compounds in a fluidized-bed reactor. Then the silanes are purified by distillation and chemical vapor deposition (CVD) to form SG-polysilicon.

A number of intermediate chemical compounds, such as monosilane (SiH_4), silicon tetrachloride ($SiCl_4$), trichlorosilane ($SiHCl_3$), and dichlorosilane (SiH_2Cl_2), had been considered. Among them, trichlorosilane has been dominantly used as the intermediate compound for subsequent polysilicon deposition for the following reasons: (1) it can be easily formed by the reaction of anhydrous hydrogen chloride with MG-Si at reasonably low temperatures (200–400°C); (2) it is liquid at room temperature so that purification can be accomplished using standard distillation technique; (3) it is handled easily and if dry can be stored in carbon steel tanks; (4) its liquid is easily vaporized and, when mixed with hydrogen, the gas mixtures can be transported in steel lines; (5) it can be reduced at atmospheric pressure in the presence of hydrogen; (6) its deposition can take place on heated silicon, thus eliminating contact with any foreign surfaces that may contaminate the resulting silicon; and (7) it reacts at lower temperatures (1000–1200°C) and at faster rates than does silicon tetrachloride.

Hydrochlorination of Silicon   Trichlosilane is synthesized by heating powdered MG-Si at around 300°C in a fluidized-bed reactor. That is, MG-Si is converted into $SiHCl_3$ according to the following reaction:

$$Si + 3HCl \longrightarrow SiHCl_3 + H_2 \tag{5.2}$$

The reaction is highly exothermic, and therefore heat removal is essential to assure maximum yield of trichlorosilane. In the course of converting MG-Si into $SiHCl_3$, various impurities such as Fe, Al, and B are removed by forming their halides (i.e., $FeCl_3$, $AlCl_3$, and BCl_3, respectively), besides byproducts such as $SiCl_4$ and H_2.

Distillation and Decomposition of Trichlorosilane   Distillation has been a widely used process for the purification of trichlorosilane. The trichlorosilane, which has a low boiling point (31.8°C), is fractionally distilled from the impurity halides to result in greatly increased purity with a concentration of electrically active impurities of less than 1 ppba. The high-purity trichlorosilane is then vaporized, diluted with high-purity hydrogen, and introduced into the deposition reactor. In the reactor, thin silicon rods called slim rods supported by graphite electrodes are available for the deposition of the chlorosilane to form silicon as a result of the reaction

$$SiHCl_3 + H_2 \longrightarrow Si + 3HCl \tag{5.3}$$

In addition to reaction (5.3), the following reaction occurs simultaneously in the course of polysilicon deposition and results in the formation of silicon tetrachloride, the major byproduct of the process:

$$HCl + SiHCl_3 \longrightarrow SiCl_4 + H_2 \tag{5.4}$$

The silicon tetrachloride is utilized in the production of high purity quartz, etc.

Needless to say, the purity of the silicon slim rods must be comparable to that of the deposited silicon. The slim rods are preheated to approximately 400°C at the initiation of the silicon CVD process. This preheating is required to increase the conductivity of high-purity (i.e., high-resistance) slim rods sufficiently to allow for resistive heating.[4] By deposition for 200–300 hr at around 1100°C, high-purity polysilicon rods of 150–200 mm in diameter, as shown in Fig. 5.2, are produced. For subsequent crystal growth processes, the polysilicon rods are shaped into various forms such as chunks (Fig. 5.3) for Czochralski melt growth[6] and long cylindrical rods for float-zone growth.[7] The process for hydrogen reduction of trichlorosilane on a heated silicon rod was described in the late 1950s and early 1960s in a number of process patents assigned to Siemens; therefore, this process is widely called the "Siemens method."[4]

Fig. 5.2.   As-grown polysilicon rod. (Courtesy of Monsanto Electronic Materials Company.)

Fig. 5.3.   Polysilicon chunks for Czochralski crystal growth. (Courtesy of Monsanto Electronic Materials Company.)

The major weak points of the Siemens method are its poor silicon and chlorine efficiency, relatively small batch size, and high power consumption. Poor efficiency of silicon and chlorine is associated with the large volume of silicon tetrachloride produced as the byproduct in the CVD process. Only 30% of the silicon provided to the CVD reaction is converted into high-purity polysilicon.[4] Meanwhile, the production cost of high-purity polysilicon may depend on the usefulness of the byproduct, $SiCl_4$.

Monosilane Process   A polysilicon production technology based on the production and pyrolysis of monosilane was established in the late 1960s. Monosilane potentially saves energy because it deposits as polysilicon at a lower temperature and produces higher purity polysilicon; however, it has been used little for lack of an economical way to make it and because of process problems in the deposition step.[8] Recently, with the development of economical routes to high-purity silane and the successful operation of a large-scale plant, this technology has attracted the attention of the semiconductor industry, which requires higher-purity silicon.

In the present industrial monosilane processes, magnesium and MG-Si powders are heated to 500°C in a hydrogen atmosphere to synthesize magnesium silicide (Mg_2Si), which is then made to react with ammonium chloride in liquid ammonia below $0°C$ to form monosilane according to the following reaction[9]:

$$Mg_2Si + 4NH_4Cl \xrightarrow[\text{liq. } NH_3]{0°C} SiH_4 + 2MgCl_2 + 5NH_3 \qquad (5.5)$$

High-purity polysilicon is produced by the pyrolysis of monosilane on resistively heated polysilicon filaments at $700-800°C$. In a monosilane generation process, most boron impurities are removed from silane by the chemical reaction with NH_3. The boron content of $0.01-0.02$ ppba in polysilicon has been achieved by a monosilane process. The concentration is very low compared with that in polysilicon prepared from trichlorosilane.[9] Moreover, the resulting polysilicon is less contaminated with metals through chemical transport processes because monosilane decomposition does not cause any corrosion problems.

Granular Polysilicon Deposition   Recently a significantly different process, which produces granular polysilicon, has been developed.[4,10,11] This method uses the decomposition of monosilane in a fluidized-bed deposition reactor to produce free-flowing granular polysilicon[12] and may completely replace the conventional Siemens method. Tiny silicon seed particles are fluidized in a monosilane/hydrogen mix, and polysilicon deposits to form free-flowing spherical particles that average $700 \, \mu m$ in diameter with a size distribution range from 100 to $1500 \, \mu m$, as shown in Fig. 5.4. The fluidized-bed seeds had been made by grinding SG-Si in a ball or hammer mill and leaching the product with acid, hydrogen peroxide, and water. This process was time-consuming and costly, and tended to introduce undesirable impurities through the metal grinders. In a new method, large SG-Si particles are fired at each other by a high-speed stream of gas and thereby break into particles of suitable size for a fluidized bed.[13] This process introduces no foreign materials and requires no leaching.

Because of the greater surface area, fluidized-bed reactors are much more efficient than traditional Siemens-type rod reactors. It has been reported that fluidized-bed decomposers require one-fifth to one-tenth the energy and half the capital cost of the traditional Siemens process.[11] The quality of fluidized-bed polysilicon has been proved to be equivalent to polysilicon produced by the conventional Siemens method.[14] Moreover, granular polysilicon of a free-flowing form and high bulk density enables crystal growers to obtain the most out of each production run. That is, in the Czochralski crystal growth process (see Section 5.2.1), crucibles can be quickly and easily filled to uniform loadings, which typically exceed those of randomly stacked polysilicon chunks (Fig. 5.3) produced by the Siemens method. Considering the

Fig. 5.4. Fluidized-bed granular polysilicon with an average diameter of 700 μm and a range of diameters from 100 to 1500 μm. (Courtesy of Ethyl Corporation.)

potential of moving from a batch operation to a continuous pulling, to be discussed in Section 5.4.4, the advantage of free-flowing polysilicon granules is very clear in that they provide a uniform feed into a steady-state melt.[11] Although the evaluation and test of granular polysilicon from many points of view is still under way,[12] the product seems to be a revolutionary starting material of great promise for future silicon crystal growth.

5.2   Single-Crystal Growth

5.2.1   Introduction

The growth of single crystals of silicon from high-purity polysilicon is a critical beginning step for electronic device fabrication. Although various techniques have been utilized to convert polysilicon into single crystals of silicon,[15] two techniques have dominated the production of silicon single crystals because they meet the requirements of the microelectronics device technology. One is a zone-melting method commonly called the *float-zone* (FZ) *method*,[7] and the other is a pulling method traditionally called the *Czochralski* (CZ) *method*[6] although it should be called more properly the *Teal-Little method*.[16] Figure 5.5 depicts the principles of these two crystal

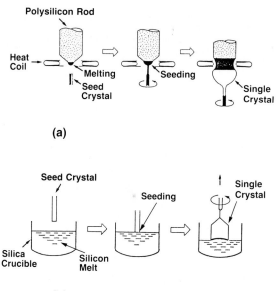

Fig. 5.5. Principles of single-crystal growth by (a) float-zone method and (b) Czochralski method.

growth methods. In the FZ method, a molten zone is passed through a polysilicon rod to convert it into a single crystal ingot; in the CZ method, a single crystal is grown by pulling from the melt contained in a quartz crucible. In both cases, the *seed crystal* plays a very important role in obtaining the single crystal with a desired crystallographic orientation.

It is estimated that about 80% of the single crystal silicon is produced by the CZ method and the rest mainly by the FZ method.[6,7] At present almost all the silicon used for microelectronic circuit fabrication is prepared by the CZ method. The silicon semiconductor industry now requires high purity and minimum defect concentration in silicon crystals in order to improve the device manufacturing yield and operation performance. These requirements are becoming increasingly stringent as the technology changes from LSI to VLSI/ULSI. Besides the quality or perfection of silicon crystals, the crystal diameter has been steadily increasing to meet the demand of device manufacturers. Since IC chips are produced through a "batch system," the diameter of silicon wafers used for device fabrication significantly affects the productivity, and in turn the production cost. Figure 5.6 shows the relation between the normalized production cost per chip with a different size and the diameter of silicon substrate.[17] The solid lines and broken lines represent the production cost of silicon substrates and that of IC device chips, respectively. Those costs

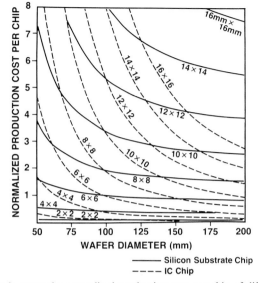

Fig. 5.6.   Relation between the normalized production cost per chip of different size and the diameter of silicon wafers. (After Takasu.[17].)

are normalized with the production cost of 6 × 6 mm chips on 100-mm-diameter silicon wafers. It is clearly recognized that silicon wafers of a larger diameter produce IC chips with less cost, and their advantage becomes more important as the IC chip size becomes larger, which is the trend of VLSI/ULSI chips. Figure 5.7 shows the number percentage trend of the diameter of silicon wafers used for electronic device fabrication.[18] It is clearly observed that the electronic device industry has favorably used larger-diameter silicon wafers in order to reduce the production cost. Recently, 200-mm-diameter and 150-mm-diameter silicon crystals have been grown by the CZ method and FZ method, respectively, and are being used experimentally for the fabrication of VLSI circuits. Technologically, it is easier and less costly to increase the crystal diameter by the CZ method than by the FZ method. Although 200-mm-diameter silicon wafers are the largest ones used for device fabrication at this moment, as far as the crystal growth technology is concerned, crystals of even larger diameter (e.g., 250 or 300 mm diameter) are available by the CZ method. The main problems that limit the practical use of the larger-diameter wafers remain in the wafer shaping technology and device processing equipment.

First the FZ method is discussed and then the CZ method. The latter will be in more detail because of its extreme importance in the VLSI/ULSI technology.

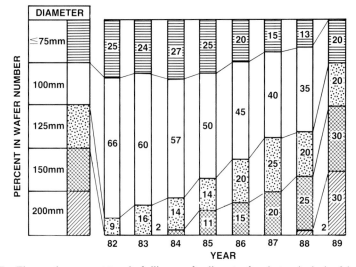

Fig. 5.7.   The number percent trend of silicon wafer diameter for electronic device fabrication. (After Kaneko.[17])

5.2.2   Float-Zone Method

General Remarks   The FZ method originated from zone melting, which was used to refine binary alloys[19] and was invented by Theuerer.[20] The inevitable reactivity of liquid silicon with crucible materials led ultimately to the development of the FZ method, which permits crystallization of silicon without contact with any crucible material, for growing crystals of requisite semiconductor purity. The first FZ silicon crystal was grown by Keck and Golay in 1953.[21] The principle of the FZ method using induction heating with radio frequency (rf) for silicon was established in late 1954.[22,23]

Operation Outline   In the FZ method, a polysilicon rod is converted into a single-crystal ingot by passing a molten zone from one end of the rod to the other. When the molten zone of silicon solidifies, polysilicon is converted into single-crystalline silicon with the help of the seed crystal.

As shown in Fig. 5.5, melting of the end of polysilicon rod to form a molten zone is first achieved by induction heating with an rf coil. The needle-eye coil has replaced the previously used conventional multiturn coil having an inner diameter larger than the feed rod. Then the tip is contacted and fused with a single-crystal seed with the desired crystal orientation. This process is called *seeding*. The seeded molten zone is passed through the polysilicon rod by simultaneously moving down the rod with the single crystal seed. As the zone travels the polysilicon rod, single-crystal silicon freezes at its end and grows as an extension of the seed crystal. After seeding, a thin neck about 2 or 3 mm in diameter and 10–20 mm long is formed. This process is called *necking*. Dislocations are usually introduced into newly grown single-crystalline silicon during the seeding operation due to thermal shock, and they are eliminated by growing a neck. This necking process, called the *Dash technique*,[2] is fundamental for growing dislocation-free crystals and is universally used for both the FZ and CZ methods. Figure 5.8 shows an X-ray topograph of the seed, neck, and conical part of a silicon single crystal grown

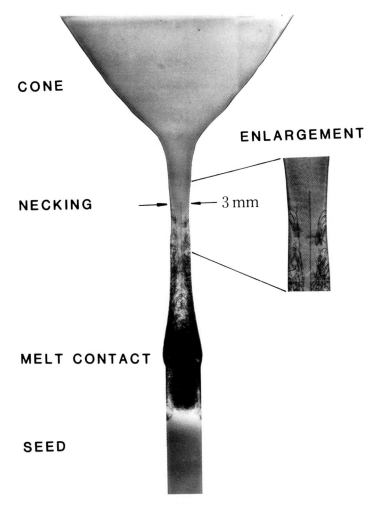

CONE

ENLARGEMENT

NECKING   →   ← 3 mm

MELT CONTACT

SEED

Fig. 5.8. X-Ray topography of seed, neck, and conical part of float-zone silicon. (After Abe.[24] Reprinted with permission of Academic Press, Inc.)

by the FZ method.[24] It is clearly observed that dislocations generated at the melt contact are perfectly eliminated through necking. After the conical part is formed, the main body with a full target diameter is grown. During the entire growth process by the FZ method, the shape of the molten zone and the ingot diameter are determined by adjusting the power to the coil and the travels rate, both of which are under computer control. The most commonly used technique to control the diameter automatically for both the FZ and CZ methods employs an infrared sensor focused on the meniscus. The shape of the meniscus on a growing crystal depends on its angle of contact at the three-phase boundary, the crystal diameter, and the magnitude of the surface tension.[25] Change in the meniscus angle, and in turn the crystal diameter, is sensed, and the information is fed back to automatically adjust the growth conditions.

In contrast with CZ crystal growth, in which the seed crystal is dipped into the silicon melt and the growing crystal is pulled upward, the thin seed crystal in the FZ method sustains the growing crystal as well as the polysilicon rod from the bottom. As a result, the rod is balanced precariously on the thin seed and neck during the entire growth process. The seed and neck can support up to a 20-kg crystal as long as the center of gravity of the growing crystal remains on the center of the growth system. If the center of gravity moves away from the center line, the seed will fracture at ingot weights above 4 kg.[7] Hence, a crystal stabilizing and supporting technique had to be invented before long and heavy FZ silicon crystals could be grown. Particularly for recent FZ crystals with a large diameter (125–150 mm), supporting the growing crystal in such a way as shown in Fig. 5.9[7] is necessary since its weight may exceed 40 kg.

Doping In order to obtain *n*- or *p*-type silicon single crystals of a required resistivity, either the polysilicon or growing crystal must be doped with the appropriate donor or acceptor impurities. Doping in the CZ method is achieved by adding the dopant to the polysilicon charged in the crucible. For FZ silicon growth, although several doping techniques have been tried,[26] the crystals are typically doped by blowing a dopant gas such as phosphine (PH_3) for *n*-type silicon or diborane (B_2H_6) for *p*-type silicon onto the molten zone. The dopant gas is usually diluted with a carrier gas, such as argon. The great advantage of this method is that the silicon producer does not need to store polysilicon sources with different resistivities.

Since the segregation coefficient (to be discussed in the next subsection) of elemental dopants for *n*-type silicon is much less than unity, FZ crystals doped by traditional methods have radial dopant gradient. Moreover, since the crystallization rate on a microscopic scale varies in the radial direction, the dopant concentrations distribute cyclically and give rise to so-called

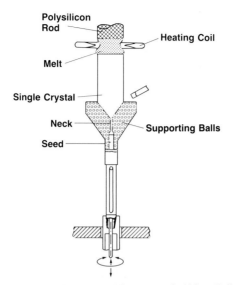

Polysilicon Rod

Heating Coil

Melt

Single Crystal

Neck

Supporting Balls

Seed

Fig. 5.9. Supporting system for float-zone silicon crystal. (After Keller and Muhlbauer.[7] Reproduced with the permission of Marcel Dekker, Inc.)

dopant striations, resulting in radial resistivity inhomogeneity. In order to obtain more homogeneously doped *n*-type silicon, neutron transmutation doping (NTD) has been applied to FZ silicon crystals.[27] This procedure involves the nuclear transmutation of silicon to phosphorus by bombarding the crystal with thermal neutrons according to the reaction

$$^{30}\text{Si}(n, \gamma) \longrightarrow {}^{31}\text{Si} \xrightarrow{2.6\,\text{hr}} {}^{31}\text{P} + \beta^- \qquad (5.6)$$

The radioactive isotope ^{31}Si is formed by ^{30}Si capturing a neutron, which then decays into the stable ^{31}P isotope (i.e., the donor atom), whose distribution is not dependent on the crystal growth parameters. Immediately after irradiation the crystals exhibit high resistivity, which is attributed to the large number of lattice defects due to radiation damage. The irradiated crystal, therefore, must be annealed in an inert ambient at temperatures around 700°C in order to annihilate the defects and to restore the resistivity to that derived from the phosphorus doping. Under the NTD scheme, crystals are grown without doping and then are irradiated in a nuclear reactor having a large ratio of thermal to fast neutrons in order to enhance neutron capture and to minimize damage to the crystal lattice. An example of the reduction of resistivity striations in NTD FZ silicon compared with standard FZ silicon is shown in Fig. 5.10.[28]

The application of NTD has been almost exclusively limited to FZ crystals because of their higher purity compared to CZ crystals. When the NTD

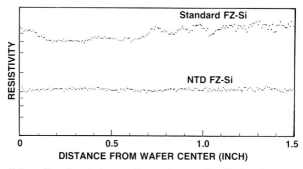

Fig. 5.10. Radial profile of typical spreading resistance distribution in standard float-zone silicon and NTD float-zone silicon. (After Stone.[28] Reproduced with the permission of North-Holland Publishing Company.)

technique was applied to CZ silicon crystals, it was found that the oxygen donor formation during the annealing process after irradiation changed the resistivity from that expected, even though phosphorus donor homogeneity was achieved.[29] The NTD has the additional shortcomings that no process is available for *p*-type dopants and that excessively long irradiation is required for low resistivities in the range of $1-10\,\Omega\,cm$.

Properties of FZ-Silicon Crystal   In FZ crystal growth, the molten silicon does not contact any substance other than the ambient gas in the growth chamber. Inherently then an FZ crystal is distinguished by its higher purity compared to a CZ crystal, which is grown from the melt involving contact with a quartz crucible. This contact gives rise to a high oxygen impurity concentration in CZ crystal. CZ silicon crystals usually contain oxygen on the order of 10^{18} atoms/cm^3, while FZ silicon contains less than 10^{16} atoms/cm^3. This higher purity allows FZ silicon to achieve high resistivities that can not be obtained in CZ silicon. The largest volume of FZ silicon consumed is in the resistivity range between 10 and $200\,\Omega\,cm$, while CZ silicon is usually prepared to resistivities of $25\,\Omega\,cm$ or less due to contamination from quartz crucibles.[30] Therefore FZ silicon is mostly used for the fabrication of semiconductor power devices that must support reverse voltages in excess of $750-1000\,V$.[31] Moreover, the combination of high-purity crystal growth and precision doping characteristics of NTD FZ-Si has been used not only for high-power devices but also for infrared detectors. By applying multiple-pass vacuum FZ processes, FZ crystals with very high resistivities from 9000 to $30,000\,\Omega\,cm$ have been grown to meet the most stringent demands for the production of infrared radiation detectors.[32]

Regarding mechanical strength, it has been recognized for many years that FZ silicon containing less oxygen impurities than CZ silicon is more vulnerable to thermal stress in the device fabrication processes.[33,34] High-temperature processing of silicon wafers during electronic device manufacturing often produces sufficient thermal stresses to generate slip dislocations and warpage.[35,36] These effects bring about yield loss due to leaky junctions, dielectric defects, and reduced lifetime, as well as reduced photolithographic yield because of the degradation of wafer flatness. Loss of geometrical planarity due to warpage can be so severe that the wafers will not be processed further.[37] Because of this, CZ silicon wafers have been used much more widely in IC device fabrication than have FZ wafers.

Deformation experiments have systematically shown that FZ silicon is more easily deformed than CZ silicon before preheating, while after preheating CZ silicon becomes more susceptible to plastic deformation.[38] A very drastic difference in the mechanical strength has been observed between FZ and CZ silicon when they contain dislocations; that is, CZ silicon is much stronger than FZ silicon against thermal stresses.[39] The difference in mechanical strength is attributed to the difference in the concentration of oxygen and associated defects.[38,40,41] This difference in mechanical stability against thermal stresses is the dominant reason why CZ silicon crystals have been exclusively used for the fabrication of ICs whose level of integration requires a large number of thermal process steps. In order to overcome these shortcomings, the growth of FZ silicon crystals with doping impurities such as oxygen[42] and nitrogen[43] has been tried. Accordingly, it was found that doping with oxygen or nitrogen in concentrations of $1-1.5 \times 10^{17}$ atoms/cm^3 or 1.5×10^{15} atoms/cm^3, respectively, into FZ silicon crystals shows a remarkable increase in mechanical strength. It has also been found that doping of a minute amount of nitrogen ($< 3 \times 10^{15}$ atoms/cm^3) into CZ silicon remarkably increases its mechanical strength.[44] Considering the oxygen concentration of $1-2 \times 10^{18}$ atoms/cm^3 present in most CZ silicon crystals, it is remarkable that nitrogen has such an extraordinarily strong effect on strengthening silicon crystal,[45] although its mechanism has not been clarified yet. Recently, FZ crystals doped with nitrogen have received increased attention for use in microelectronics circuit fabrication, where the requirement for "oxygen-free" material is stringent.[46]

5.2.3 Czochralski Method

General Remarks This method was named after J. Czochralski, who established the technique to determine the crystallization velocity of metals.[47] However, the pulling method that has been widely applied to single-crystal growth was developed by Teal and Little, who modified Czochralski's basic principle.[16] They were first successful in growing single crystals of germanium, 8 inches in length and 0.75 inches in diameter, in 1950. Subsequently they designed another apparatus for the growth of silicon at higher temperatures. The growth of single-crystal silicon of comparable size having a high degree of lattice perfection and chemical purity was first published in 1952 by Teal and Buehler.[48] Although the basic production process for single-crystal silicon used in the microelectronics industry has changed little since it was pioneered by Teal and co-workers, large-diameter silicon single crystals having a high degree of perfection that meets the VLSI/ ULSI device demands have been grown with the incorporation of the Dash technique and successive technological innovations in the apparatus.[49,50] Today, the most stringent requirement for CZ silicon crystals used for ICs is "better uniformity" of silicon characteristics such as the resistivity and oxygen concentration. Today's research and development efforts concerning silicon crystals are directed toward achieving microscopic uniformity of the crystal properties. Control of crystal uniformity, particularly at the macroscopic level, can be achieved with commercially available CZ equipment, which is equipped with computer control of the growth process. It is interesting to note, however, that most major silicon growers still add their in-house-developed "arts" to the CZ equipment, even though the development of commercial CZ equipment has been significant. Makers of crystal growth equipment have claimed "one-button" crystal growth operation (i.e., fully automated crystal growing) to improve throughput and decrease operator dependency[51,52]; however, the growth process still requires an operator's skill for some delicate steps during the growing processes.

Operation Outline   The most important three steps in CZ crystal growth are schematically shown in Fig. 5.5b. In principle, the process of CZ growth is similar to that of FZ growth: (1) melting polysilicon, (2) seeding, and (3) growing. The CZ pulling procedure, however, is more complicated than that of FZ growth and is distinguished from it by using a quartz crucible to contain molten silicon. Figures 5.11 and 5.12 show the schematic view[24] and a photograph of modern CZ crystal growth equipment, respectively. The CZ equipment shown in Fig. 5.12 allows growth of crystals up to 160 mm in diameter, 140 cm in length, and 60 kg in weight. Figure 5.13 shows some steps of actual CZ silicon crystal growth sequence:

1. Polysilicon chunks are filled in a quartz crucible and melted at temperatures higher than 1420°C (i.e., the melting point of silicon) in an inert gas or vacuum ambient.[53]

2. The melt is kept at a high temperature for a while in order to ensure complete melting and ejection of tiny bubbles, which may cause void or negative crystal defects (see Fig. 3.27), from the melt.

3. A seed crystal is dipped into the melt (seeding) until it begins to melt itself. The seed is then withdrawn from the melt so that the neck is formed (necking) by gradually reducing the diameter; this is the most

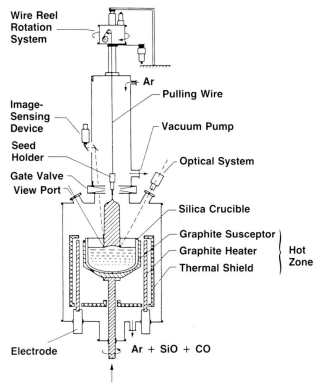

Fig. 5.11. Schematic view of typical Czochralski silicon crystal growing system. (After Abe.[24] Reproduced with the permission of Academic Press, Inc.)

delicate step and requires stable melt conditions and the operator's skill. The $\langle 111 \rangle$-oriented crystal growth requires small crystal neck diameter, but the $\langle 100 \rangle$ crystal growth requires an even thinner neck.[6] During the entire crystal growth process, inert gas (usually argon) flows downward through the pulling chamber in order to carry off reaction products such as SiO and CO. The evaporated SiO from the melt surface forms an amorphous or poorly crystallized phase by condensing on cold surfaces.[54] These deposits may cause a problem by falling back into the melt and causing the crystal to lose the single-crystal structure.

4. Increasing the crystal diameter gradually, the conical part and shoulder are grown. The diameter is increased up to the target by decreasing the pulling rate and/or the melt temperature.

5. Finally, the cylindrical part or body with a constant diameter is grown by controlling the pulling rate and the melt temperature while compensating for the melt level going down as the crystal grows. The pulling rate, in general, is reduced toward the tail end of a growing crystal,

Fig. 5.12. Appearance of modern Czochralski silicon crystal growing equipment. (Courtesy of Kokusai Electric Co., Ltd.)

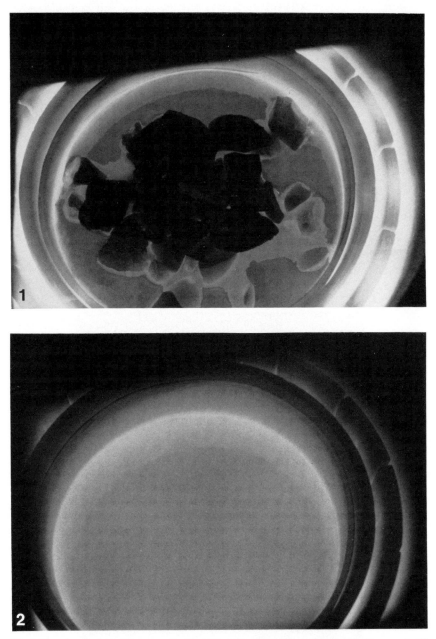

Fig. 5.13.   Czochralski silicon growth steps: (1) melting polysilicon chunks, (2) silicon melt, (3) seeding and necking, (4) pulling seed-end part, and (5) pulling body part. (Courtesy of Monsanto Electronic materials Company.) (*Figure continues.*)

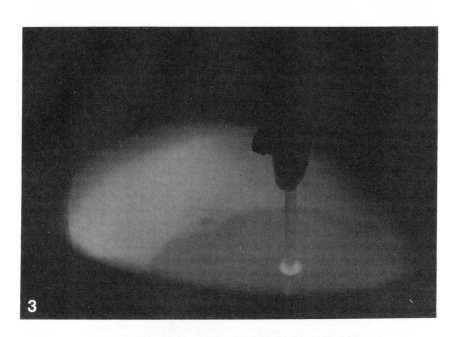

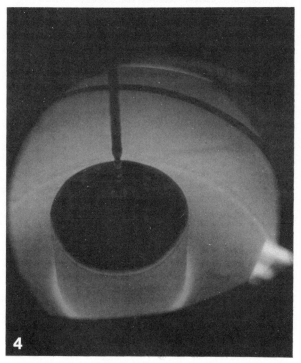

Fig. 5.13. (*Continued*)

Fig. 5.13. (*Continued*)

mainly due to the increasing heat radiation from the crucible wall as the melt level goes down and exposes more crucible wall to the growing crystal. That is, removal of heat becomes more difficult and thus more time is required to grow a certain length of the crystal.[6] Near the end of the growth process but before the crucible is completely empty of molten silicon, the crystal diameter must be reduced gradually to form an end-cone (see Fig. 1.3) in order to minimize thermal shock, which can cause slip dislocations at the tail end. When the diameter becomes small enough, the crystal can be separated from the melt without the generation of dislocations.

Figure 5.14 shows the seed-end part of an as-grown CZ silicon crystal. A neck of 2 mm in diameter is estimated to be able to hold a 150 kg suspended load. A seed-cone, which is the transition region from the seed to the cylindrical part, is usually formed rather flat for economic reasons, although crystals grown several years ago[55] were distinctly tapered. This might be desirable from the viewpoint of the crystal quality. The shoulder part and its vicinity should not be used for device fabrication because this part is considered a transition region and has considerable inhomogeneity in crystal characteristics due to the abrupt change in growth conditions.

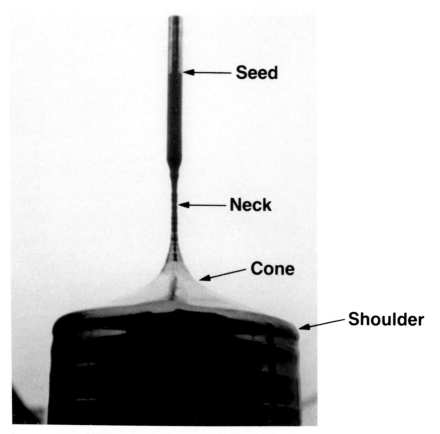

Fig. 5.14. Seed-end part of as-grown Czochralski silicon crystal. (Courtesy of Monsanto Electronic Material Company.)

Theory of Melt Growth   Crystal growth, in general, is a complex process, accompanied by a phase change. CZ crystal growth is a process in which a transformation from liquid to solid—that is, from the melt to the crystal —takes place successively. The fundamental difference between the melt and the crystal is their structural symmetry. That is, the environment of an atom in a crystal possesses crystallographic symmetry, while that in a melt does not. Therefore, the melt to crystal transformation can not proceed continuously because of the difference in symmetry.[56]

The Gibbs free energy G is defined[55] as

$$G = H - TS \qquad (5.7)$$

where H is the enthalpy, S entropy, and T temperature. At equilibrium, the Gibbs free energy of the two phases are equal:

$$G_L = G_S \tag{5.8}$$

where the subscripts L and S denote liquid and solid, respectively. Therefore,

$$\Delta G = G_L - G_S = \Delta H - T\,\Delta S = 0 \tag{5.9}$$

where

$$\Delta H = H_L - H_S \tag{5.10}$$

and

$$\Delta S = S_L - S_S \tag{5.11}$$

From Eq. (5.9),

$$\Delta S = \Delta H / T_E \tag{5.12}$$

where T_E is the equilibrium transformation temperature. The term ΔH is also known as the *heat of fusion*. Near equilibrium,

$$\Delta G = \Delta H(T_E - T)/T_E = \Delta H\,\Delta T/T_E = \Delta S\,\Delta T \tag{5.13}$$

where $\Delta T = T_E - T$ is the undercooling.

In a first-order phase transformation, which is defined by the above equations, there is a discontinuity in the internal energy, the enthalpy, and the entropy associated with the change of state. Therefore, the transformation cannot be homogeneous, and thus solidification takes place at the crystal –melt interface and will advance or recede, as a function of temperature, until it reaches its new equilibrium. Consequently, the solidification rate or crystal growth rate through the CZ process is limited by the rate of removal of the heat of fusion at the crystal–melt interface.

The CZ growth geometry and related parameters are schematically shown in Fig. 5.15.[57] The material to be crystallized is first melted by induction or resistance heating in a suitable "nonreacting" container. For a silicon melt, a quartz crucible is used, as discussed later, even though they react with each other. The melt temperature T_l is adjusted to be slightly above the melting point T_m. The CZ crystal growth involves the supply of thermal energy to the melt coupled with the removal of the latent heat of solidification from the crystal and other associated heat losses from the system. After thermal equilibrium is achieved, the seed crystal is dipped into the melt and withdrawn at a rate that gives a desired crystal diameter. As the melt solidifies and the crystal is pulled, the latent heat of fusion is transferred to the crystal. The heat is transported from the crystal–melt interface up the growing crystal.[58] Heat is lost from the crystal surface by thermal radiation and

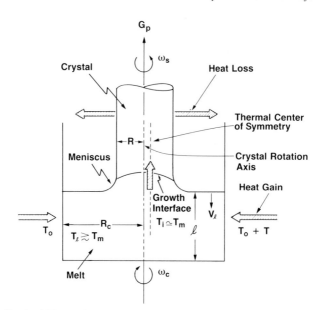

Fig. 5.15. Czochralski growth geometry and related parameters. (After Carruthers and Witt.[57] Reproduced with the permission of North-Holland Publishing Company.)

convention. Growth at a constant crystal diameter R is achieved by maintaining the solidification isotherm in a vertical position intersecting the meniscus at the point where the isotherm becomes perpendicular to the melt surface.[57] The following parameters contribute to maintaining this condition: the pulling rate G_p, the rate of melt level drop G_m, the heat fluxes gain and loss, the crystal rotation ω_s, and the crucible rotation $-\omega_c$.

Using the simplifying assumption of a flat growth interface, no radial temperature gradient, and zero melt temperature gradient, the maximum pulling rate V_{max} is given by

$$[G_p]_{max} = -\frac{K_c}{(\Delta H)\rho_c}\left(\frac{dT}{dz}\right)_c \propto \left|\left(\frac{dT}{dz}\right)_c - \left(\frac{dT}{dz}\right)_m\right| \qquad (5.14)$$

where K_c is the thermal conductivity of the crystal, ρ_c the crystal density, and $(dT/dz)_c$ and $(dT/dz)_m$ the temperature gradient in the crystal and the melt at the crystal–melt interface, respectively.[59] The negative sign in Eq. (5.14) accounts for the fact that dT/dz is a negative quantity for the usual coordinate system in which z is zero at the interface and increases positively along the crystal length. Note that the crystal growth rate under the conditions described above cannot be expected to be equal to the pulling rate. The macroscopic growth rate is always greater than the pulling rate because the melt level drops at G_m with pulling the crystal. The instantaneous microscopic growth rate differs substantially from the macroscopic growth rate due to the deviation of thermal center of symmetry from the crystal rotation axis, which results in periodic thermal fluctuations at the growth interface. Thermal asymmetry may originate from asymmetric heat gain to the crucible containing the melt (i.e., T_0 versus $T_0 + \Delta T$ as shown in Fig. 5.15), or alternatively may be associated with a lack of coincidence of the rotational and thermal axis.[57] The microscopic growth rate G_g is given by

$$G_g = G_p - \frac{2\pi \, \Delta T_i \omega}{(dT/dz)_m} (\cos 2\pi)\omega t \qquad (5.15)$$

where ΔT_i is the temperature difference experienced during a complete revolution about the axis of rotation at the growth interface, ω the crystal relative rotation ($= |\omega_s - \omega_c|$), $(dT/dz)_m$ the temperature gradient in the melt adjacent to the growth interface, and t the time.[59] Taking

$$\alpha = \frac{2\pi \, \Delta T_i \omega}{G_p (dT/dz)_m} \qquad (5.16)$$

it follows that

$$G_g = G_p[1 - \alpha(\cos 2\pi)\omega t] \qquad (5.17)$$

The minimum growth rate during each rotation is accordingly

$$[G_g]_{min} = G_p(1 - \alpha) \qquad (5.18)$$

From Eq. (5.18), it is apparent that when $\alpha > 1$, then $[G_g]_{min} < 0$; that is, local remelting takes place.[60] It has been observed that local remelting enhances the formation of point defect clusters,[61] which are due to excess interstitial silicon agglomeration.[62]

As seen in Eq. (5.14), for a one-dimensional analysis, $[G_p]_{max}$ depends only on the crystal temperature gradient at the interface; however, the temperature gradient is a very complex function of puller geometry and ambient conditions.[59] In addition to heat flow, the growth rate depends on the structure of the crystal–melt interface, that is, atomically smooth or rough, which effects the incorporation of atoms thereto. In general, the crystal growth rate is proportional to the undercooling ΔT for a rough interface[63]; for a smooth interface, where growth occurs by propagating screw dislocations, the growth rate is proportional to $(\Delta T)^2$.[64] In the case of CZ silicon crystal growth, it has been observed that the maximum practical growth rates are only 80% of the theoretical maximum.[65] These lower rates compared with the theoretical one may result from the effects of temperature fluctuations in the melt that occur near the crystal–melt interface. When a fluctuation causes a transient increase in the interface temperature, *remelt* of the grown crystal occurs and the nominal pulling rate must be reduced until solidification restarts. Subsequent decrease in temperature increases the solidification rate, leading to an increase in the crystal ingot diameter. To maintain the desired diameter, the pulling rate at that instant must be increased. Such temperature oscillations thus lead to pulling rate variations and overall lower pulling rates. A potential problem enhanced by high pulling rates in CZ silicon is that of single-crystalline structure loss due to excessive impurity concentrations at high growth rates when growing heavily doped single crystals.[66]

As the diameter of the growing crystal is increased, the maximum pulling rate decreases because the heat loss is proportional to the surface area of the crystal ingot, which increases only linearly with the diameter, but the heat gain is proportional to the volume being crystallized, which increases as the square of the ingot radius.

Growth Interface   In the CZ crystal growth process, silicon melt crystallizes successively at the crystal–melt interface, that is, the crystallization front. Therefore, heat transfer at the interface plays an important role in the growth of silicon crystal. The above condition points to the importance of maintaining thermal conditions near the interface constant during crystal growth in order to obtain uniform and high-quality material along the entire length of the crystal. The detailed modeling of the heat transfer during CZ crystal growth predicts the interrelationships among the important process variables such as diameter, pulling rate, crystal–melt interface shape, temperature gradient and its symmetry both in the crystal and the crucible, etc.[67] Among those factors, the shape of the crystal–melt interface directly influences the crystal perfections[68] and impurity distribution throughout the crystal cross section.[69,70]

Since the crystal–melt interface is generally curved in CZ-grown crystals in order to maintain thermal conditions uniform at the growth interface, a nonuniform distribution of both crystal defects and impurities is found across the transverse section of a flat wafer. It has been observed that the ingot section with a high dislocation density is associated with the highly curved portion of the interface, and this is the case regardless of whether the interface is convex or concave toward the melt.[68] The notable difference in the dislocation pattern for the concave and convex interfaces is, however, the gradual disappearance of the dislocations when growth proceeds via a concave interface. This is because of dislocations formed at the crystal–melt interface that tend to develop perpendicular to that interface. The nonuniform impurity distribution is attributed to a difference in segregation coefficient at the flat portion and at the curved portion of the growth interface, as well as to the fact that different parts of the same transverse cross section are crystallized at different times.

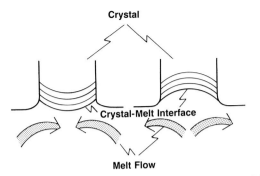

Fig. 5.16. Melt flow effect on crystal–melt growth interface shape. (After Kuroda and Kozuka.[71] Reproduced with the permission of North-Holland Publishing Company.)

Although experimental observations[6,71,72] and numerical analyses[73,74] have shown that the shape of the crystal–melt interface depends on growth conditions such as the pulling rate, rotation rate, crystal diameter, and crucible size, the melt flow in the crucible straightforwardly affects the interface shape as schematically illustrated in Fig. 5.16.[71] The melt flow pattern also dominates the incorporation of impurities such as oxygen. This impurity is dissolved in the melt as a result of the high-temperature reaction between the silicon melt and the quartz (SiO_2) crucible.

Facet Growth   It is well known that facets invariably correlated to the {111} lattice planes on the growth interface appear in crystals that have the diamond structure when they are grown from the melt in the ⟨111⟩ orientation.[75] That is, crystals pulled in the ⟨111⟩ direction have three such facets spaced symmetrically and a fourth {111} plane, which is normal to the growth direction, lying approximately parallel to the crystal–melt interface. As discussed in Section 3.3.2, the {111} planes are most closely packed in the diamond structure. Growth in a direction normal to the plane is slow, whereas growth by lateral extension of the {111} plane is favored. The preferred directions of lateral growth in the (111) plane are $[2\bar{1}\bar{1}]$, $[\bar{1}2\bar{1}]$, and $[\bar{1}\bar{1}2]$, which give rise to the three symmetrically placed habit lines commonly observed when a crystal is pulled in the [111] direction. Accordingly, the crystal–melt interface of the growing crystal along ⟨111⟩ consists of two parameters: one related to the curved solidification isotherm and the other to the facet. The growth interface of a ⟨100⟩ crystal, however, is dominated only by the isotherm, and no facet is observed.[76]

Because of this facet-related phenomenon, silicon crystal growth in a ⟨111⟩ direction is less smooth than in other directions such as ⟨100⟩. Since the facets are nearly perfect planes, their growth proceeds by a nucleation process requiring substantial undercooling.[77] Because of the normally concave solidification isotherm and the nucleation effect of the crystal edge, the growth of facet layers starts at the periphery of the crystal; once nucleated, the {111} facets grow laterally toward the center of crystal rapidly. This unsteady or abrupt facet growth causes a variation in the effective segregation coefficient, discussed in the next subsection, and results in so-called *growth striations*, which locally have different impurity/dopant concentrations. In addition, the impurities are likely to segregate in the center region of the crystal due to this lateral crystal growth mechanism.[6]

Melt Flow in CZ Crucible During CZ crystal growth, the silicon melt circulates in a very complex manner in the quartz crucible. The melt flow pattern in CZ crystal growth is a result of the combination of five basic types of melt convection, as shown in Fig. 5.17.[78] They are (a) natural thermal convection caused by the buoyant force due to a temperature gradients, (b) Marangoni convection or thermocapillary convection caused by a surface-tension gradient because of the temperature dependence of the surface tension of the melt,[79] (c) forced convection caused by crystal pulling, (d) forced convection caused by crystal rotation, and (e) forced convection by crucible rotation.

Among these five parameters, thermal convection and forced convection caused by crystal rotation play a very important role in CZ crystal growth. Thermal convection flows are always generated in fluids whenever a temperature gradient exists. The basic flow generated in a melt can be either symmetrical or asymmetrical, with the hot melt rising along the crucible wall

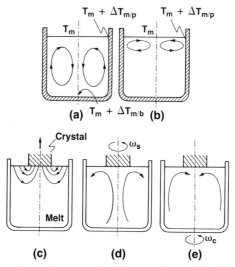

Fig. 5.17. Basic convection patterns of melt in Czochralski crucible. (After Kobayashi.[78])

and descending at the center of crucible as shown in Fig. 5.17a. The basic flow pattern is determined by the crucible geometry, aspect ratio (diameter over height of melt), and thermal boundary conditions. The driving force is given by the dimensionless *Grashof number* Gr:

$$\mathrm{Gr} = g\beta\,\Delta T_m L^3/v_K \tag{5.19}$$

where g is the acceleration due to gravity, β is the melt thermal expansion coefficient, ΔT_m the temperature difference in the melt over length L, when L is a specific length in the system (crucible diameter or melt depth), and v_K the kinematic viscosity of the melt. In the case when the melt is heated at the bottom, the *Rayleigh number* Ra is used instead of Gr:

$$\mathrm{Ra} = (\mathrm{Gr})(\mathrm{Pr}) = \frac{g\beta\,\Delta T_m L^3}{K_m v_K} \tag{5.20}$$

where Pr is the *Prandtl number* ($= v_K/K_m$) and K_m is the thermal conductivity of the melt. It follows from these equations that the thermal convection of the melt is suppressed by decreasing the value of Ra. Equation (5.20) also shows that thermal convection flows can be reduced in a low-gravity environment such as space or the *Skylab*. Since in the Czochralski process heat is supplied to the crucible laterally by the heater as shown in Fig. 5.11, the temperature difference between the periphery and the center regions of the melt, which is $\Delta T_{m/p}$ in Fig. 5.17a, becomes larger than the difference between the surface and bottom of the melt, $\Delta T_{m/b}$. Thus, in the case of crystal growth using a large-diameter crucible, a strong thermal convection occurs due to the direct contribution of the third power of L in Gr, and due to an increase in the temperature gradient in the melt. Melt turbulence due to these strong convections accelerates temperature variations in the melt, and eventually causes remelting and supercooling at the growth interface, which consequently may introduce crystal disorder and inhomogeneous impurity distribution. Thus the effect of melt flow on crystal quality becomes more significant as the diameters of crystals and crucibles become larger. In order to suppress the effect of thermal convection, a well-controlled forced convection is usually induced by crystal rotation. As illustrated in Fig. 5.17d, the crystal rotation sets up a pumping action, which draws melt up and pumps it radially outward near the surface. As a result, crucible rotation that brings about melt convection from the crucible wall toward the center, as illustrated in Fig. 5.17e, contributes both to reducing the unfavorable effect of thermal asymmetry on a growing crystal and to decreasing and stabilizing the thickness of the growth interface diffusion boundary layer (see Section 5.3.1). The degree of mixing by crystal rotation is characterized by the dimensionless *Reynolds number* Re:

$$\mathrm{Re} = \omega_s R^2/v_K \tag{5.21}$$

where R is the crystal radius. The effect of forced convection versus thermal convection has been effectively described by the ratio $Re^{2.5}/Gr$; that is, thermal convection is dominant in the melt for a low value, whereas for a high value forced convection becomes dominant.[80] The critical ratio at which the melt flow changes from thermal convection dominance to forced convection dominance is found to be on the order of 10.[80] In addition to the well-controlled forced convection that rises along the axis and flows radially outward, rotation of the crucible improves the thermal symmetry for crystal growing and homogenizes the impurity distribution.

In CZ silicon crystal growth, although some features of the surface flow can be seen, it is difficult to observe the behavior of the melt in a crucible because of the metallic character of molten silicon, that is, nontransparent with low emissivity. Therefore, in order to aid in the understanding of the melt flow pattern under realistic conditions, numerous theoretical and experimental simulations using different liquids, such as glycerine, $NaNO_3$ melt, and CaF_2 melt, which approximate those prevailing the CZ growth, have been performed.[79,81–86]

Consequently, according to the rotation conditions of crystal (ω_s) and crucible (ω_c), the basic melt flow patterns that may occur during CZ crystal growth are schematically shown in Fig. 5.18.[6] Because of the reason noted above, both the crystal and crucible are rotated, normally counter to each other, with the former at a higher rotation rate. In practical CZ crystal growth, however, a situation that makes the melt flow more complex is that the melt level gradually goes down as the crystal grows. This change in the melt depth and the length of the grown crystal significantly affects the melt and heat flow conditions, as illustrated in Fig. 5.19. For a deep melt, the forced convection caused by crystal rotation affects only the upper region of

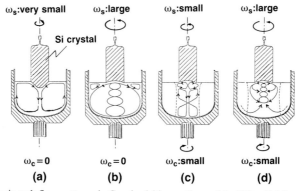

Fig. 5.18. Basic melt flow patterns in Czochralski growth crucible. (After Zulehner and Huber.[6] Reproduced by permission of Springer-Verlag.)

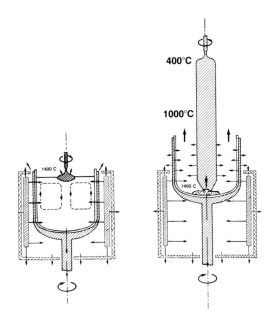

Fig. 5.19.   Thermal environment in Czochralski crystal growth at initial stage and final stage. (The arrows indicate the approximate direction of heat flow.[6])

the melt in the crucible and the rest is controlled by thermal convection. For a shallow melt, on the other hand, the forced convection affects the whole melt in the crucible, and the vertical flow extends from the crucible bottom to the melt surface.[85] Therefore, the crystal growth parameters must be changed successively as the crystal grows and the melt level goes down.

Spatial Location in Grown Crystal   As Fig. 5.19 clearly depicts, every portion of a CZ crystal is grown at a different time with different growth conditions. Thus, every portion has a different thermal history as a result of its different position along the crystal length. For example, the seed-end portion has a longer thermal history in the temperature range from the melting point of silicon to around 400°C in a puller, while the tail-end portion has a shorter history and is cooled down rather rapidly from the melting point. Eventually, every silicon wafer prepared from a different portion of a grown crystal may exhibit different physicochemical characteristics according to its location in the ingot. It has been reported that the difference in location in a grown crystal most distinctly affects the oxygen precipitation behavior,[87,88] and this will be discussed in detail in Section 7.2. This structural sensitivity is further exacerbated, as is always the case, because the crystal–melt interface is curved. Because of this, every point along a radius of a given silicon wafer has crystallized from the melt at a different time. In addition to the effect on structural sensitivity, the difference in the spatial location causes impurity distribution variation both axially and radially due to the segregation coefficient, which is usually not unity. This will be discussed in the next section.

Finally, note again that the thermal environment of the growing crystal is of immense importance to the successful production of homogeneous silicon wafers used for microelectronic device fabrication. The hot-zone characteristics of the crystal growth equipment affect the crystal–melt interface shape, concentration and distribution of impurities, and grown-in point defects, as well as strain within the crystal during growth. The diameter and length of the crystal, melt charge weight, crucible diameter, aspect ratio (the ratio of melt depth to crucible diameter), seed/crucible rotation rates, and pulling rate are several factors influencing the achievement of the desired crystal properties. Although significant research efforts are in progress to prehomogenize silicon wafers in an effort to eliminate the grown-in characteristic variations,[89] the desired goal is a high-yielding, high-performance circuit fabrication process where no adverse effects are caused by the silicon substrates.[90]

5.3   Impurities

As mentioned in Section 4.1, the importance of silicon in electronic device technology relies on semiconductor properties that are very sensitive to impurities. Because of this sensitivity, the electrical properties of silicon are precisely controlled by adding a small amount of dopant. In addition to this dopant sensitivity, contamination by impurities, particularly transition metals, harmfully affects the electrical properties of silicon and results in serious degradation of device performance. Moreover, oxygen on the order of several tens of parts per million atomic (ppma) is incorporated into Czochralski silicon crystals as a result of the reaction between the silicon melt and the quartz crucible. Regardless of how much oxygen is in the crystal, the characteristics of silicon crystals are greatly affected by the behavior of oxygen.[89] In addition, carbon is also incorporated into CZ silicon crystals either from the polysilicon raw materials or during the growth process, due to a considerable amount of graphite parts in the CZ pulling equipment. Although the concentration of carbon in commercial CZ silicon crystals is usually less than 0.1 ppma, carbon has been recognized as an impurity that greatly affects the oxygen behavior in silicon.

In this section, the incorporation and the behavior of such impurities in CZ silicon crystals is discussed.

5.3.1 Impurity Inhomogeneity

Segregation In crystallization from a melt, various impurities contained in the melt are incorporated into the growing crystal. The impurity concentration of the solid phase, in general, differs from that of the liquid phase due to a *segregation* phenomenon. The equibrium segregation behavior associated with the solidification of multiple-component systems can be determined from the corresponding phase diagram of the binary system with a *solute* (impurity) and a *solvent* (host material) as components. Most impurities in silicon lower the melting point T_m of silicon because they form *eutectic systems* with it. Figure 5.20 represents a portion of such a phase diagram near the melting point of silicon. Because of the different solubilities of impurity A in solid and liquid silicon at the same temperature T_x, the binary phase diagram consist of *solidus* and *liquidus* lines, which intersect the 100% silicon axis at T_m. Both lines can be approximated as linear for low impurity concentrations. Therefore, the ratio of the solubility of impurity A in solid silicon $[C_A]_S$ to that in liquid silicon $[C_A]_L$ remains constant over a certain concentration range. This ratio is referred to as the *equilibrium segregation* or *distribution coefficient* and is defined with k_0 as

$$k_0 = [C_A]_S/[C_A]_L \tag{5.22}$$

The segregation coefficient defined by Eq. (5.22) actually represents the ratio of the difference in thermodynamic potential due to the interaction and entropy of mixing of the impurity with the host material in the solid and liquid phase. As seen from Fig. 5.20, the impurity solubility in liquid silicon is always higher than that in solid silicon, that is, $k_0 < 1$. Figure 5.21 shows the opposite case, where impurity B raises the melting temperature and $[C_B]_S$ is always larger than $[C_B]_L$, that is, $k_0 > 1$. As listed in Table 5.2,[6,91] the only reported example for silicon in this case is oxygen, with $k_0 = 1.25 \pm 0.17$.[92]

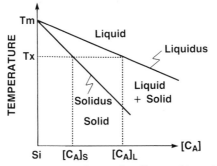

Fig. 5.20. Schematic binary phase diagram for silicon and impurity A with $k_0 < 1$.

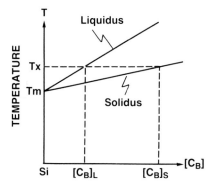

Fig. 5.21.   Schematic binary phase diagram for silicon and impurity B with $k_0 > 1$.

   Equilibrium segregation can be obtained by calculation from the binary phase diagram under the assumption that solidification proceeds very slowly, or the solidification rate can be approximated to be zero, that is, an equilibrium phase transition. The values of the equilibrium segregation coefficient for silicon have been correlated to the tetrahedral radius of the impurities, which implies that, in general, larger impurity elements have a smaller equilibrium segregation coefficient.

Normal Freezing   Solidification of a liquid from one end to the other is schematically illustrated in Fig. 5.22. Under the *normal freezing* condition, which assumes that (1) diffusion in the solid is negligible, (2) impurity concentration in the liquid is uniform, and (3) k_0 is constant, the impurity concentration in the solid $[C]_S$ as a function of the fraction solidified **g** is given by

$$[C]_S = k_0[C]_0(1 - \mathbf{g})^{k_0 - 1} \qquad (5.23)$$

where $[C]_0$ is the initial impurity concentration in the liquid.[19] The axial distributions of impurities with different k_0 calculated according to Eq. (5.20) for a Czochralski-grown crystal are plotted in Fig. 5.23.[19]

Effective Segregation Coefficient   The equilibrium segregation coefficient k_0 is applicable for solidification only at a negligibly slow growth rate. For finite or higher solidification rates, however, impurity atoms with $k_0 < 1$ are rejected by the advancing solid at a greater rate than they can diffuse into the bulk of the melt. In the CZ crystal growth process, with the start of solidification at a given seed–melt interface, segregation takes place and the rejected impurity atoms begin to accumulate in the melt layer near the

Table 5.2 Equilibrium Segregation Coefficient, Maximum Solid Solubility, and Diffusion Constant of Impurity Elements in Silicon[a]

Element	Group	Equilibrium segregation coefficient	Maximum solid solubility (atoms/cm^3)	Diffusion constant D_0 (cm^2/sec)	Diffusion constant Q (eV)
H	Ia	—	—	9.4×10^{-3}	0.48
Li	Ia	1×10^{-2}	6.5×10^{19}	2.5×10^{-3}	0.66
Cu	Ib	4×10^{-4}	1.5×10^{18}	4.7×10^{-3}	0.43
Ag	Ib	$\sim 1 \times 10^{-6}$	2.0×10^{17}	2.0×10^{-3}	1.60
Au	Ib	2.5×10^{-5}	1.2×10^{17}	2.4×10^{-4} (I)	0.39
				2.75×10^{-3} (S)	2.04
Zn	IIb	$\sim 1 \times 10^{-5}$	6×10^{16}	1×10^{-1}	1.40
B	IIIb	8×10^{-1}	1×10^{21}	9.1×10^{-2}	3.36
Al	IIIb	2×10^{-3}	5×10^{20}	1.385	3.39
Ga	IIIb	8×10^{-3}	4×10^{19}	3.74×10^{-1}	3.41
In	IIIb	4×10^{-4}	4×10^{17}	7.85×10^{-1}	3.63
Ti	IVa	2×10^{-6}	—	2.0×10^{-5}	1.50
C	IVb	7×10^{-2}	3.3×10^{17}	3.3×10^{-1}	2.92
Ge	IVb	3.3×10^{-1}	[b]	1.535×10^{3}	4.65
Sn	IVb	1.6×10^{-2}	5×10^{19}	3.2×10	4.25
N	Va	7×10^{-4}	5×10^{15}	8.7×10^{-1}	3.29
P	Va	3.5×10^{-1}	1.3×10^{21}	3.85	3.66
As	Va	3×10^{-1}	1.8×10^{21}	3.8×10^{-1}	3.58
Sb	Va	2.3×10^{-2}	7×10^{19}	2.14×10^{-1}	3.65
Bi	Va	7×10^{-4}	8×10^{17}	1.08	3.85
Cr	VIa	1.1×10^{-5}	—	1×10^{-2}	1.00
O	VIb	1.25	2.7×10^{18}	1.3×10^{-1}	2.53
S	VIb	$\sim 1 \times 10^{-5}$	3×10^{16}	9.2×10^{-1}	2.20
Mn	VIIa	4.5×10^{-5}	3×10^{16}	1.42×10^{-1}	1.30
Fe	VIII	8×10^{-6}	3×10^{16}	1.3×10^{-3}	0.68
Co	VIII	8×10^{-6}	2.3×10^{16}	9.2×10^{4}	2.80
Ni	VIII	3×10^{-5}	8×10^{17}	2×10^{-5}	0.47

[a] After Zulehner and Huber[6] and Shimura and Huff.[91]
[b] Silicon and germanium are completely miscible in both the liquid and solid phases.

growth interface and diffuse in the direction of the bulk of the melt. An impurity concentration gradient thus develops just ahead of the advancing crystal. A schematic profile in this situation is represented in Fig. 5.24.[93] The thickness of the diffusion boundary layer δ is defined as the distance from the intersection of the tangent line on the $[C]_L$ curve from $x = 0$ to the extrapolated horizontal plateau of $[C]_L$. Consequently, taking into account

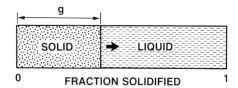

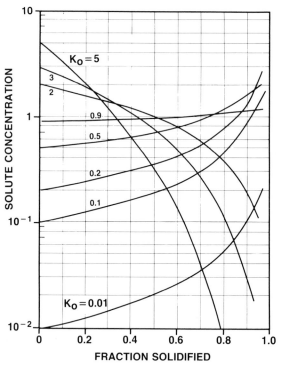

Fig. 5.22.   Solidification by normal freezing. (After Pfann.[19])

thickness δ, the solidification rate G_s, and the diffusion coefficient D of the impurity in the liquid phase, an *effective segregation coefficient* k_{eff} can be defined at any moment if stirring and convection currents in the liquid keep $[C]_L$ virtually uniform[94]:

$$k_{eff} = k_0/[k_0 + (1 - k_0)\exp(-G_s\delta/D)] \qquad (5.24)$$

The thickness δ of the diffusion boundary layer, for small values of G_s, simply depends on the relative rotation of the crystal ω and various physical properties of the liquid phase and is given by

$$\delta = 1.6D^{1/3}v_K^{1/6}\omega^{-1/2} \qquad (5.25)$$

Fig. 5.23.   Axial distribution of impurities with different k_0 for normal freezing. (After Pfann.[19])

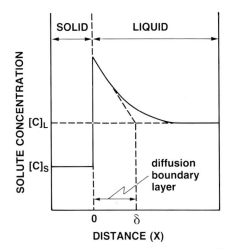

Fig. 5.24. Schematic concentration distribution of solute near the solid–liquid interface. (After Wagner.[93])

where v_K is the kinematic viscosity of the liquid. Finally, the impurity concentration $[C]_S$ in a CZ crystal can be derived by substituting k_{eff} for k_0 in Eq. (5.23):

$$[C]_S = k_{eff}[C]_0(1 - g)^{k_{eff} - 1} \qquad (5.26)$$

Consequently, it is understood that a macroscopic longitudinal impurity variation is inherent in the CZ batch growth process due to the segregation phenomenon. Moreover, the longitudinal distribution of impurities is influenced by change in the magnitude and nature of melt convection, which varies as the melt aspect ratio is decreased during crystal growth.

Striations In most crystal growth processes, there are transients in the parameters such as the instantaneous microscopic growth rate G_g and the diffusion boundary layer thickness δ, which result in variations of the effective segregation coefficient k_{eff}. The variations give rise to microscopic compositional inhomogeneities in the form of *striations* parallel to the crystal–melt interface. Striations can be delineated with several techniques, such as preferential chemical etching and X-ray topography (see Chapter 6). Figure 5.25 shows striations revealed by chemical etching for the shoulder part of a CZ silicon crystal. The gradual change in the shape of growth interfaces is clearly observed. As seen in Eqs. (5.15) and (5.24), the major causes of striations are temperature fluctuations near the crystal–melt interface induced by unstable thermal convection in the melt and crystal rotation in an

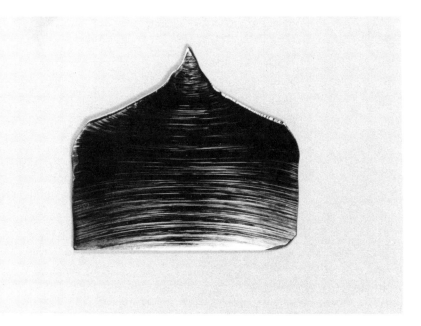

Fig. 5.25. Growth striations revealed by chemical etching for shoulder part of Czochralski silicon. (Courtesy of J. W. Moody, Monsanto Electronic Materials Company, and A. Yamaguchi, Monsanto Japan, Limited.)

asymmetric thermal environment. In addition, mechanical vibrations due to poor pulling control mechanisms in the crystal growth equipment can also cause temperature fluctuations. In CZ-grown crystals, several types of impurity striations have been observed and are customarily classified into two categories: *rotational* and *nonrotational* striations. Rotational striations are characterized by a relatively simple mechanism, while nonrotational striations are complex and of varying origins.[95]

Rotational striations show long-range periodicity, compared with nonrotational striations, and are attributed to thermal asymmetry with respect to the rotational axis. The higher thermal gradients and the more pronounced thermal convection are observed in melts with higher melting points. Therefore, the extent of thermal asymmetry is great and remelting is common during rotation and pulling of a silicon crystal because of the high melting point of silicon. With high thermal asymmetry in a crystal growth system, the microscopic growth rate [see Eq. (5.15)] will assume negative values with each rotation cycle, which causes more or less partial remelting. Consequently, rotational striations are caused by localized regions of decreased microscopic growth rate, which result in a decreased concentration for impurities

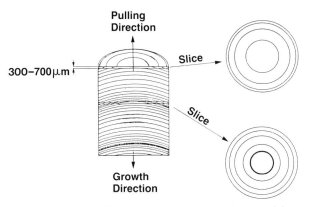

Fig. 5.26. Schematic illustration of Czochralski crystal cross section containing curved crystal–melt interfaces and planar wafers sliced at different portions.

with an equilibrium segregation coefficient $k_0 < 1$.[96] Accordingly, the periodicity d_{rs} of rotational striations is given[95] by

$$d_{rs} = G_p/\omega \tag{5.27}$$

On the other hand, nonrotational striations are generally attributed to transient growth and segregation behavior due to variations in instantaneous microscopic growth rates. As Eq. (5.15) shows, the variation is mainly caused by fluctuations either in temperature at the crystal–melt interface during growth or in pulling rate originating from poor control mechanism arrangement.[97]

Figure 5.26 schematically illustrates a CZ-grown crystal cross section containing a curved crystal–melt interface, which results in inhomogeneities on a slice surface. As each planar wafer is sliced, it contains different portions of several curved striations. Differing "phonograph rings," which are occasionally referred to as *swirl*, can then be observed across the surface of each wafer.

5.3.2 Impurity Doping

Dopants In order to obtain a desired resistivity, a certain amount of dopant is added to a silicon melt according to the resistivity–concentration calculation (see Section 4.2.3). It is common practice to add dopants in the form of highly doped silicon particles or chunks of about 0.01 Ω cm resistivity, which are called the dopant fixture, since the amount of pure dopant needed is unmanageably small, except for heavily doped silicon materials, that is, n^+ or p^+ silicon. Moreover, the physicochemical properties of pure dopant elements are often quite different from those in the silicon melt.

The criteria for selection of a dopant for a semiconductor material are that it has the following properties: (1) suitable energy level (see Fig. 4.8), (2) high solubility (see Fig. 3.31 and Table 5.2), (3) suitable or low diffusivity (see Fig. 5.27 and Table 5.2), and (4) low vapor pressure. A high diffusivity or high vapor pressure leads to undesirable diffusion or vaporization of dopants, which results in instability of device operation and in difficulty in achieving precise resistivity control. Too small a solubility limits the resistivity that can

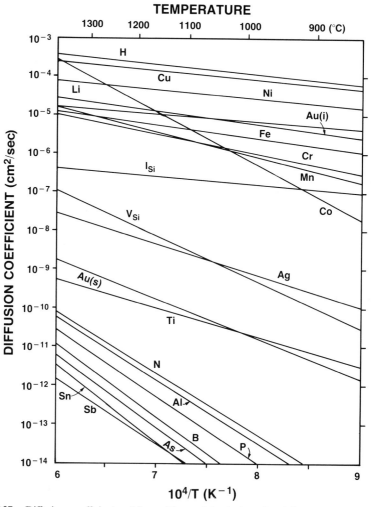

Fig. 5.27. Diffusion coefficients of impurities and intrinsic point defects in silicon. (After Shimura and Huff[91] and Tan and Gösele.[99])

be obtained. In addition to those criteria, the chemical properties (e.g., toxic or not) must be considered. A further consideration from the viewpoint of crystal growth is that the impurity has a segregation coefficient that is closer to unity in order to achieve a uniform resistivity as much as possible from the seed-end to the tail-end of the CZ crystal ingot. Consequently, phosphorus (P) and boron (B) are the most commonly used donor and acceptor dopants for silicon, respectively. For n^+ silicon, in which donor atoms are heavily doped, antimony (Sb) instead of phosphorus is usually used because of its smaller diffusivity, in spite of its small segregation coefficient and high vapor pressure, which lead to large concentration variations in both the axial and the radial directions.

Diffusion   The phenomenon of atomic migration in solids is called *solid-state diffusion* or simply *diffusion*, which includes both self-diffusion and impurity-diffusion. Control of the diffusion of dopant impurities in silicon is the basis of *p-n* junction formation and microelectronic device fabrication. Atomic diffusion in silicon can proceed by way of both direct and indirect mechanisms.

As depicted in Fig. 5.28, there are three basic mechanisms that explain atomic diffusion in crystalline materials.[98] Impurity atoms having no strong bonding interactions with silicon atoms are located exclusively at interstitial sites, and they jump directly between these sites (*interstitial mechanism*). Species such as hydrogen, helium, and many metallic impurities supposedly diffuse by this mechanism. The diffusion of substitutional impurities, such as Group III and Group V dopants, and self-interstitials require intrinsic point defects—that is, vacancies (*vacancy mechanism*) or self-interstitials (*interstitialcy mechanism*)—as diffusion vehicles.[98] It is now generally accepted that both vacancies (V_{Si}) and self-interstitials (I_{Si}) are present in silicon and that they contribute to impurity diffusion and self-diffusion.[99] In the presence of

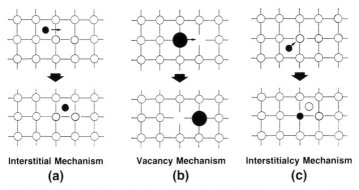

Interstitial Mechanism	Vacancy Mechanism	Interstitialcy Mechanism
(a)	**(b)**	**(c)**

Fig. 5.28.   Three basic mechanisms of atomic diffusion in crystalline material.

both vacancies and self-interstitials, the diffusivity D of a substitutional impurity is given by

$$D = D_V + D_I \tag{5.28}$$

where D_V and D_I are the diffusion components involving vacancies as a diffusion vehicle and involving self-interstitials, respectively.[100] It has been observed that oxidation enhances the diffusion of boron and phosphorus (*oxidation-enhanced diffusion*) and retards that of antimony (*oxidation-retarded diffusion*), but the diffusion of arsenic is not influenced.[101] Accordingly, based on the fact that oxidation of silicon generates excess self-interstitials at the SiO_2–Si interface[102] (see Chapter 7), it is well established that boron and phosphorus diffuse dominantly by the interstitialcy mechanism but that antimony diffuses by the vacancy mechanism, and that arsenic diffuses via both interstitials and vacancies. Referring to Table 3.9, these phenomena suggest that impurities with smaller ionic radii tend to diffuse by way of the interstitialcy mechanism, while impurities with large ionic radii tend to diffuse by the vacancy mechanism. As for an impurity such as arsenic whose radii is very close to that of silicon, diffusion proceeds via both interstitials and vacancies.[101] It follows that the diffusivity of substitutional impurities in silicon is considerably affected by the heat-treatment ambient, that is, oxidizing or nonoxidizing ambients. In addition, it is understood that the diffusivity depends also on the background impurity level in silicon, since each impurity diffuses with different point defects as already discussed.[103,104]

This effect of intrinsic point defects on impurity diffusion implies that published diffusion coefficients in a number of references will vary from each other to some extent.[91] That is, the diffusivity of impurities in a crystal depends on the quality of host crystal as well as on their intrinsic physical properties.

As just discussed, atoms move in a random motion in a crystal. Because of this randomness, no net movement of species occurs in a homogeneous crystal, in spite of many atomic jumps. In the case where a concentration gradient exists, impurity atoms migrate in a certain direction throughout the crystal, which results in a net transport of impurities. The flow of species is expressed, in the one-dimensional case, as a flux J given by

$$J = -D \frac{\partial C}{\partial x} \tag{5.29}$$

where D is the proportionality constant, which is referred to as the *diffusion coefficient*, and $\partial C / \partial x$ is the concentration gradient in the x direction. Equation (5.29) is known as *Fick's first law*, which states that the impurity flux is proportional to the negative of the concentration gradient. Thus the concentration gradient acts as a driving force that pulls impurity atoms from regions of high concentration to regions of low concentration until the concentration becomes uniform. From a physical point of view, motion of atoms in a solid usually proceeds by thermally activated jumps over potential barriers between different sites. The diffusion coefficient D can therefore be expressed in the form

$$D = D_0 \exp(-Q/kT) \tag{5.30}$$

where D_0 is the temperature-independent preexponential factor that contains the frequency for this process and the migration entropy, and Q is the activation energy of diffusion related to the barrier height. Experimentally, D_0 and Q (i.e., *diffusion constants*) are obtained from an Arrhenius plot of $(\ln D)$ versus $(1/T)$, which is linear when Eq. (5.30) is valid. The dimensions of the terms in Eq. (5.29) are given as

$$(\text{mass/length}^2 \text{ time}) = -(\text{length}^2/\text{time})(\text{mass/length}^3/\text{length}) \tag{5.31}$$

The unit of D is length^2/time, and cm^2/sec is most practically used in the field of electronics technology.

In the case of a finite volume of impurity, a case frequently encountered during a diffusion process in electronic device fabrication, the impurity concentration gradient decreases with diffusion time t. That is,

$$\frac{\partial C}{\partial t} = -\frac{\partial J}{\partial x} \tag{5.32}$$

Substituting Eq. (5.29) into Eq. (5.31), *Fick's second law* of diffusion is obtained:

$$\frac{\partial C}{\partial t} = \frac{\partial}{\partial x}\left(D \frac{\partial C}{\partial x}\right) \tag{5.33}$$

where the diffusion coefficient D is a function of concentration. For many systems of interest, D is independent of the concentration and then Eq. (5.32) reduces to

$$\frac{\partial C}{\partial t} = D \frac{\partial^2 C}{\partial x^2} \tag{5.34}$$

In one case of interest, the concentration $C(x, t)$ of an impurity is given by solving Eq. (5.34) under the following boundary conditions:

$$C(x, 0) = 0 \tag{5.35}$$

$$C(0, t) = C_0 \tag{5.36}$$

and

$$C(\infty, t) = 0 \tag{5.37}$$

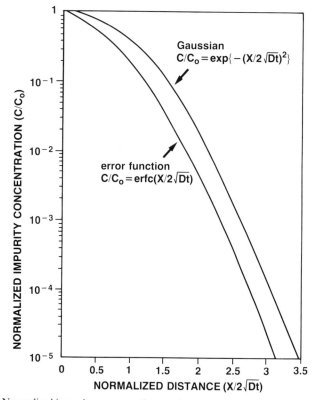

Fig. 5.29. Normalized impurity concentration as a function of normalized distance for complementary error function and for gaussian distributions.

finally

$$C(x, t) = C_0 \, \text{erfc}(x/2\sqrt{Dt}) \tag{5.38}$$

where C_0 is the surface concentration, x the distance diffused, t the time of diffusion, and erfc is the *complementary error function*. The value $\sqrt{Dt}$ is referred to as the *diffusion length*. Figure 5.29 is a plot of normalized impurity concentration (C/C_0) as a function of normalized distance ($x/2\sqrt{Dt}$) for the complementary error function (erfc) and for gaussian distributions.

Axial Distribution   As already discussed, the incorporation of impurities into CZ silicon crystals is influenced by many factors, such as the segregation coefficient, pulling rate, melt flow, and crystal–melt interface shape. Macroscopically, Fig. 5.23 shows that the axial concentration distribution of impurities is greater as the segregation coefficient deviates more from 1.

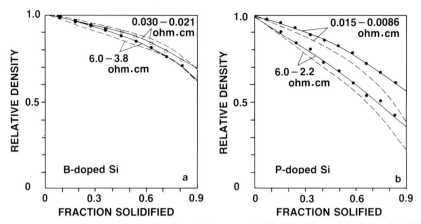

Fig. 5.30. Axial resistivity variations in (a) boron- and (b) phosphorus-doped Czochralski silicon crystals. Broken lines are calculated according to normal freezing theory and transferred into resistivity values by Irvin's curve.[105] (After Zulehner and Huber.[6] Reproduced by permission of Springer-Verlag.)

Because resistivity depends on the dopant concentration, the resistivity of a CZ silicon crystal also varies in the axial direction. Figure 5.30a and b shows the axial resistivity variations observed in boron- and phosphorus-doped CZ silicon crystals grown under the same conditions.[6] The experimentally obtained curves for boron-doping (a) fit the calculated curves well, while in the phosphorus-doping case (b), the measured resistivities are higher than the calculated ones due partially to the evaporation of phosphorus during the crystal growth process. In CZ silicon crystals doped with antimony, as shown in Fig. 5.31, it has been observed that the axial resistivity distribution varies according to pressure and gas flow conditions in the puller.[6] The variation shown in Fig. 5.31 suggests that the growth conditions are more influential than the physical segregation coefficient in controlling the resistivity distribution in antimony-doped CZ silicon crystals.

It can be understood from Figs. 5.23, 5.30, and 5.31 that the segregation phenomenon of dopants causes a low production yield of silicon crystal that meets the resistivity tolerance required by device manufacturers. For example, it has been calculated that phosphorus-doped and boron-doped silicon crystals with a resistivity $10 \pm 2 \, \Omega$ cm can be produced only by crystallization of 40% and 66% of 30 kg charged polysilicon, respectively, for 125-mm-diameter crystal growth.[24] This problem becomes even more serious for crystal growth using a dopant with a smaller segregation coefficient.

Radial Distribution   Radial impurity concentration gradients are known to affect the local electrical properties of semiconductor devices and to cause

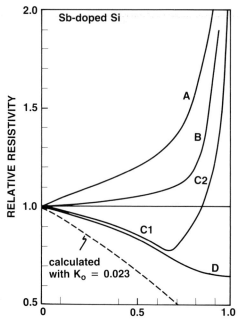

Fig. 5.31. Axial resistivity variations in antimony-doped Czochralski silicon crystals (initial resistivity is 0.018 Ω cm). Curves A and B are for same pressure of 11 mbar but different gas flow characteristics. Curves C_1 and C_2 are for different pressures of 67 mbar and 11 mbar, respectively, in the same run. Curve D is for pressure of 75 mbar. (After Zulehner and Huber.[6] Reproduced by permission of Springer-Verlag.)

differences in the rate of chemical etching, mechanochemical polishing, and so on. Therefore it is important to minimize radial fluctuations in dopant concentration for high device fabrication yields.

Radial variations in impurity concentration caused by curved growth interfaces have been schematically shown in Fig. 5.26. Microscopically, the radial impurity distribution is affected far more than the axial one by the stirring of the melt, which in turn is affected by many growth parameters as previously discussed. In the case of volatile dopants like antimony, evaporation also plays an influential role. In order to minimize the radial variations in dopant concentration, the diffusion boundary thickness δ in Eq. (5.24) must be kept uniform or more practically reduced throughout the growth interface. In general, as Eq. (5.25) shows, the values of δ can be reduced by increasing the relative crystal rotation rate ω.

The crystal orientation also strongly influences the radial impurity distribution, since it affects the shape of the growth interface significantly, as noted previously. Examples that show the radial resistivity gradient in boron- and phosphorus-doped $\langle 100 \rangle$ or $\langle 111 \rangle$ CZ silicon crystals are given in Fig. 5.32a

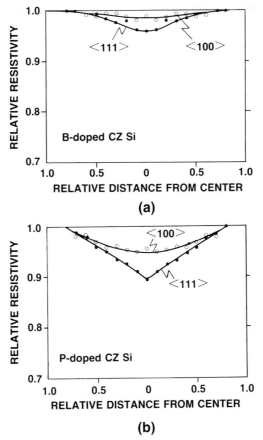

Fig. 5.32. Radial resistivity variations in ⟨100⟩ and ⟨111⟩ Czochralski silicon crystals: (a) boron-doped an (b) phosphorus-doped. (After Zulehner and Huber.[6] Reproduced by permission of Springer-Verlag.)

and b, respectively.[6] The growth conditions for all the crystals are considered to be comparable; however, larger radial variations are observed in ⟨111⟩ crystals for both dopant cases. This larger gradient is attributed to facet growth in a ⟨111⟩ crystal in which the {111} layers are built up by rapid lateral growth starting at the crystal edge and growing toward the center as previously discussed. This lateral facet growth causes segregation and enrichment of impurities with $k_0 < 1$ in the center of the crystal.

5.3.3 Oxygen and Carbon

Sources As schematically shown in Fig. 5.11, quartz or vitreous silica (SiO_2) crucibles and graphite heating elements are used in the CZ crystal growth method. The surface of the crucible that contacts the silicon melt is gradually dissolved[106] as a result of the reaction

$$SiO_2 + Si \longrightarrow 2SiO \qquad (5.39)$$

This reaction enriches the silicon melt with oxygen. Most of the oxygen atoms evaporate from the melt surface as volatile silicon monooxide (SiO), but some of them incorporate into a silicon crystal through the crystal–melt interface.[107]

Carbon, however, in CZ silicon crystals originates mainly from the polycrystalline starting material. Carbon ranging from 0.1 to 1 ppma, depending on the manufacturer, is found in the polysilicon. Sources for the carbon in polysilicon are assumed to be mainly carbon-containing impurities found in trichlorosilane used in the production of polysilicon.[108] Graphite parts in CZ pulling equipment can also contribute to carbon contamination by reacting with oxygen, which is always present in the growth ambient in a concentration of several parts per million atomic. The resulting products of CO and CO_2 are dissolved into the silicon melt and account for the carbon impurity in silicon crystals.[109] Thus, oxygen and carbon are the two major nondoping impurities that incorporate into CZ silicon crystals in the way schematically shown in Fig. 5.33. The behavior of these impurities in silicon has been the subject of intensive study since the late 1950s.[110–113]

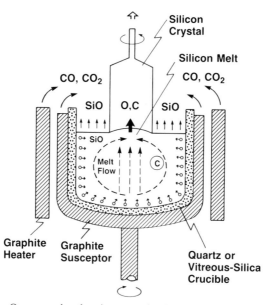

Fig. 5.33.   Oxygen and carbon incorporation into Czochralski silicon crystal.

Oxygen   Oxygen atoms incorporated into silicon occupy interstitial sites in the silicon lattice with average positions midway between two neighboring silicon atoms along the four equivalent [111], [11$\bar{1}$], [1$\bar{1}$1], and [$\bar{1}$11] bond directions.[114] Figure 5.34 shows a model for the interstitial configuration of an oxygen atom in silicon, in which the oxygen atom interrupts a normal $\langle 111 \rangle$ Si-Si valence bond.[114,115] Because of the crystal symmetry, the nonlinear Si–O–Si bridge has six equivalent positions. Transitions between those six positions occur frequently because the transition does not involve the breaking of a chemical bond: that is, it probably requires a small activation energy.[116] Oxygen atoms, although possessing strong bond interaction with silicon atoms as already noted, are believed to diffuse by direct jumping from one Si–Si line to a neighboring Si–Si line in such a way that only one Si–O bond is broken as schematically shown. That is, unlike other impurities, the oxygen diffusion process in silicon has been considered not to strongly involve vacancies or self-interstitials.[100]

Thus far, a considerable number of values of oxygen diffusion coefficients in silicon have been reported.[6,91,117] It is interesting to point out that oxygen diffusion data obtained in the period from late 1950s to late 1970s lead to a large discrepancy,[6] while those obtained after 1980 show very good agreement in spite of the fact that they were determined experimentally by means of various techniques such as secondary ion mass spectrometry (SIMS) and

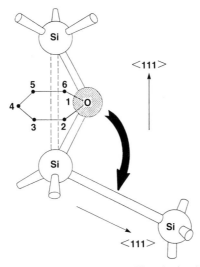

Fig. 5.34.   Interstitial configuration of oxygen atom in silicon lattice. (After Kaiser *et al.*[114] and Haas.[115])

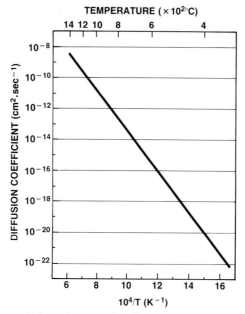

Fig. 5.35. Diffusion coefficient of oxygen in silicon as a function of temperature. (After Mikkelsen.[117])

charged particle activation analysis (CPAA).[117] Those more recent data are represented in Fig. 5.35, giving the oxygen diffusion coefficient D_{Oi} as

$$D_{Oi} \quad (cm^2 \ sec^{-1}) = 0.13 \ exp(-2.53/kT) \qquad (5.40)$$

The discrepancy in "old" oxygen diffusion data may be attributed to the fact that the diffusivity of oxygen, similar to other impurities, is to some extent affected by the quality of silicon host crystal. Moreover, enhanced oxygen diffusion has been observed in silicon crystals deliberately contaminated with metallic impurities such as copper and iron, and the results have been tentatively explained in terms of an oxygen–vacancy interaction with the vacancies being generated by site switching of the metal atoms.[118] Accordingly, a vacancy-dominant mechanism can be involved in the diffusion of interstitial oxygen.[119] In addition, anomalous oxygen diffusivity that is about four orders of magnitude higher than the diffusivity obtained from Fig. 5.35 has been observed at 450°C, which is the favorable temperature for oxygen thermal donor generation (see Section 7.2).[120] This high diffusivity has been explained in terms of a high *effective* diffusivity due to *quickly diffusing gas-like molecular oxygen* in dynamical equilibrium with interstitial oxygen.[121]

The published solubilities of oxygen in silicon are generally less consistent than the diffusivity data, although more recent data reported after 1980 are

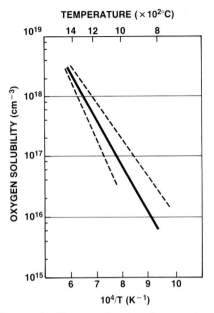

Fig. 5.36. Solubility of oxygen in silicon as a function of temperature. Broken lines represent highest and lowest variation. (After Mikkelsen.[117])

more consistent, similar to the case of diffusion data. Figure 5.36 represents a collection of oxygen solubility data obtained recently by means of infrared (IR) absorption, SIMS, and CPAA for CZ silicon and FZ silicon.[117] In Fig. 5.36, the solid line represents the generic solubility curve obtained from all the data, while the broken lines represent the highest and lowest variation in the solubility as a function of temperature. The solid line gives the solubility of oxygen $[O]_s$ in silicon as

$$[O]_s \quad (\text{atoms} \cdot \text{cm}^{-3}) = 9 \times 10^{22} \exp(-1.52/kT) \qquad (5.41)$$

Note that excess oxygen can remain in a metastable solid solution during the cooling of the as-grown silicon crystal, although the equilibrium oxygen solubility decreases rapidly with decreasing temperature, as shown in Fig. 5.36. That is, CZ silicon crystals exclusively consist of supersaturated oxygen, which is ready to precipitate if certain conditions are given, at any temperature in the range of interest in electronic device fabrication processing.

Oxygen Concentration Control   Under the assumptions that there is complete mixing in the silicon melt and that significant oxygen concentration differences exist only at the three boundaries involved, that is, (1) the crucible–melt interface, (2) the crystal–melt interface, and (3) the melt–ambient interface, the following simplified relation is established:

$$[O]_c = [O]_m - [O]_e \qquad (5.42)$$

where $[O]_c$ is the total oxygen atoms incorporated into a crystal through the crystal–melt interface, $[O]_m$ the oxygen dissolved into the melt through silica crucible dissolution, and $[O]_e$ the oxygen evaporated from the melt free surface.[122] The first term is given by

$$[O]_c = A_{s/m} k_{eff} C_m \qquad (5.43)$$

where $A_{s/m}$ is the area of crystal–melt interface, k_{eff} effective segregation coefficient of oxygen, and C_m the concentration of oxygen in the bulk melt. The second term is

$$[O]_m = A_{c/m} D(C_c - C_m)/\delta_c \qquad (5.44)$$

where $A_{c/m}$ is the area of crucible–melt interface, D the diffusion coefficient of oxygen in molten silicon, and $(C_c - C_m)$ the difference in oxygen concentration across the boundary layer of thickness δ_c at the crucible–melt interface. An increase in crucible rotation results in a decrease in δ_c; thus higher $[O]_m$ can be achieved by fast crucible rotation. The third term is

$$[O]_e = A_{m/a} D(C_s - C_m)/\delta_s \qquad (5.45)$$

where $A_{m/a}$ is the area of melt–ambient interface, that is, the area of free melt surface, and $(C_s - C_m)$ is the difference in oxygen concentration across the boundary layer of thickness δ_s.

Thus the oxygen incorporation into a growing silicon crystal involves three diffusion boundary layers and three interface areas: crucible–melt, melt–ambient, and crystal–melt. The thickness of boundary layers, as previously discussed, depends on the melt convection flows, while the interface areas are determined by the melt charge weight, size and shape of the crucible, and crystal diameter. Under these circumstances, the ratio of the crucible–melt interface area to the free melt surface area is the primary factor that determines the oxygen concentration incorporated into CZ silicon. Since the large silica crucibles used in current CZ growth system are of approximately cylindrical shape, the crucible–melt interface area constantly decreases during growth as the melt volume decreases, while the free melt surface area remains unchanged because the crystal diameter is constant over most of the growth. As a result, $[O]_m$ decreases but $[O]_e$ remains constant during growth. This phenomenon, together with the segregation coefficient of oxygen, which is larger than unity ($k_0 = 1.25$), leads to a gradual decrease in the oxygen concentration toward the tail end of a growing CZ crystal. Therefore, in order to obtain an axially uniform distribution of oxygen, the

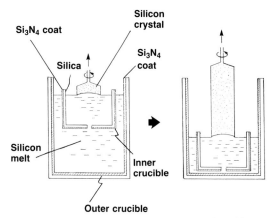

Fig. 5.37.   Concept of double-crucible Czochralski growth. (After Shimura and Kimura.[124])

ratio of $[O]_m$ to $[O]_e$ must be kept constant by controlling silicon melt flows, usually by controlling crystal/crucible rotation rates.

It follows, therefore, that a more uniform axial oxygen distribution would be obtained if the ratio of $A_{c/m}$ to $A_{m/a}$ could be maintained constant during crystal growth. This can be realized in the crystal growth using a crucible in the shape of a truncated cone.[123] However, the melt charge size accommodated in such crucibles does not meet the requirement of high production throughput. Another approach to maintain the ratio practically constant may be found in the basic concept schematically illustrated in Fig. 5.37 in which a smaller-diameter silica crucible with a silicon nitride–coated outer surface is submerged in the melt held by a larger crucible of which the inner surface is coated with silicon nitride.[124] This double-crucible CZ crystal growth arrangement may enable the ratio to remain constant and in turn may grow a silicon crystal with a uniform axial oxygen distribution at a desired concentration. However, this method has not been realized yet because of its complex operating mechanism and potential contamination with nitrogen. A more practical solution is the system in which silicon starting material is continuously fed into the crucible as the crystal is grown. In this way, a constant $A_{c/m}/A_{m/a}$ ratio can be achieved.[49] Continuous or semicontinuous CZ crystal growth techniques, which potentially have great advantages in obtaining high-quality crystals, will be discussed more in the next section.

Carbon   The properties of carbon in silicon are in considerable contrast to those of oxygen. First, carbon atoms occupy normal substitutional sites in the silicon lattice, and are electrically inactive.[125] Since the tetrahedral covalent radius of carbon is smaller than that of silicon (see Table 3.9), substitutional

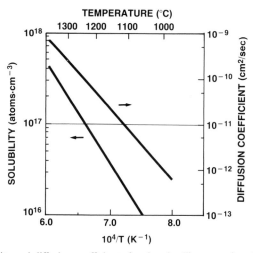

Fig. 5.38. Solubility and diffusion coefficient of carbon in silicon as a function of temperature. (After Bean and Newman[129] and Newman and Wakerfield.[130])

carbon contracts the silicon lattice, resulting in a decrease in the lattice constant,[126] while interstitial oxygen expands the silicon lattice.[127] Second, the concentration of carbon in a CZ silicon crystal increases toward the tail end because of its segregation coefficient, which is smaller than unity, that is, $k_0 = 0.07$ (see Fig. 5.23).[128]

The solubility, $[C]_s$, and diffusivity, D_C, of carbon in silicon are given by the following equations[129,130]:

$$[C]_s \quad (\text{atoms} \cdot \text{cm}^{-3}) = 4 \times 10^{24} \exp(-2.3/kT) \qquad (5.46)$$

and

$$D_C \quad (\text{cm}^{-2} \text{sec}^{-1}) = 1.9 \exp(-3.2/kT) \qquad (5.47)$$

These values as a function of temperature are shown in Fig. 5.38. As in the case of substitutionally dissolved Group III and IV dopants, which are smaller than silicon in size, the diffusion of carbon is enhanced by the presence of self-interstitials.[131]

5.3.4 Transition Metals

Sources As shown in Fig. 4.8, transition metals give rise to deep-level defects in silicon. Although the electrical activity of most of these metal-related levels disappears when they form larger complexes or precipitates, such extended defects are still detrimental for device performance (see Section 7.3). Among transition metals, $3d$ transition metals (particularly, Cr, Mn, Fe, Co, Ni, and Cu) are of great concern.[132] Table 5.3 summarizes the concentration of the

Table 5.3 Specific Transition-Metal Concentrations (ppma) for Various Parts and Materials Used in Silicon Crystal Processing[a]

Element	Quartz crucible	Graphite susceptor	Silica polishing slurry	Aluminum lapping compound	Quartz furnace tube	SiC furnace linear
Cr	$< 3.7 \times 10^{-1}$	< 1.6	4.5×10^{-1}	7.2	3.3×10^{-2}	7.51×10
Mn	$< 7.5 \times 10^{-1}$	$< 4.6 \times 10^{-1}$	3.3	—	—	1.03×10
Fe	< 5.9	$< 3.7 \times 10^{2}$	9.68×10	1.1×10^{3}	3.4	1.93×10^{4}
Co	$< 1.0 \times 10^{-2}$	$< 6.6 \times 10^{-1}$	5.7×10^{-2}	—	8.6×10^{-3}	5.11×10
Ni	$< 9.0 \times 10^{-1}$	$< 3.1 \times 10$	2.5	2.87×10	2.4×10^{-1}	1.95×10^{2}
Cu	$< 3.4 \times 10^{-1}$	—	2.6	—	2.8×10^{-1}	—

[a] After Schmidt and Pearce.[133]

above transition metals detected by neutron activation analysis (NAA) for various parts and materials commonly used in silicon processing.[133] In Table 5.4, typical levels of impurities (transition metals are marked with asterisk) are given for hydrofluoric acid (48% HF), which is one of the chemical reagents most commonly used for silicon etching and cleaning. These impurities are identified and measured as trace elements by classical chemical analysis techniques, including atomic absorption, gas chromatography, and infrared and ultraviolet spectroscopy.[134] It is obvious that transition metals may contaminate silicon at any process from polysilicon starting material preparation to final cleaning procedure.

Diffusivity   Transition metals are typically fast diffusers in silicon, as shown in Fig. 5.27. This fast diffusivity is primarily because they exist dominantly in interstitial sites in the silicon lattice, and diffuse via an interstitial mechanism.[132] On the other hand, substitutional dopant elements forming covalent bonds with silicon atoms diffuse via interstitialcy or vacancy mechanisms, as discussed in Section 5.3.2. For these processes, the activation energy Q consists of the formation energy of an intrinsic point defect and the energy of migration of the diffusant as well. In contrast to those substitutional atoms, since interstitial transition atoms have no strong bond interaction with silicon and the interstitial diffusion is independent of the presence of intrinsic point defects, the activation energies Q for interstitial diffusion, except for Co, are found to be considerably lower than those of substitutional dopants (see Table 5.2). Why Co has a large activation energy remains unexplained. Moreover, interstitial diffusion in the diamond lattice is much easier than that in a more closely packed metal lattice where geometric barriers to diffusion exist and considerable dilation of the lattice is needed in order for an impurity to move from one interstitial site to another. The

Table 5.4 Concentration of Dissolved Trace Impurities in Semiconductor-Grade Hydrofluoric acid[a]

Element[b]	Concentration (ppma)
Aluminum (Al)	< 0.05
Antimony (Sb)	< 0.02
Arsenic (As)	< 0.05
Barium (Ba)	< 0.1
Beryllium (Be)	< 0.02
Bismuth (Bi)	< 0.1
Boron (B)	< 0.05
Cadmium (Cd)	< 0.02 (typically < 0.01)
Calcium (Ca)	< 0.5 (typically 0.20)
*Chromium (Cr)	< 0.02 (typically < 0.01)
*Cobalt (Co)	< 0.02 (typically < 0.01)
*Copper (Cu)	< 0.02 (typically < 0.01)
Gallium (Ga)	< 0.02
*Gold (Au)	< 0.1 (typically 0.05)
Indium (In)	< 0.02
*Iron (Fe)	< 0.5 (typically 0.23)
Lead (Pb)	< 0.05 (typically 0.03)
Lithium (Li)	< 0.02 (typically < 0.01)
Magnesium (Mg)	< 0.2 (typically 0.06)
*Manganese (Mn)	< 0.05 (typically < 0.01)
*Molybdenum (Mo)	< 0.05
*Nickel (Ni)	< 0.02 (typically < 0.01)
*Platinum (Pt)	< 0.2
Potassium (K)	< 0.1 (typically 0.04)
Silver (Ag)	< 0.02 (typically < 0.01)
Sodium (Na)	< 0.2 (typically 0.08)
Strontium (Sr)	< 0.02 (typically < 0.01)
Thallium (Tl)	< 0.06
*Titanium (Ti)	< 0.1
Tin (Sn)	< 0.1
Vanadium (V)	< 0.05
Zinc (Zn)	< 0.1 (typically 0.02)
Zirconium (Zr)	< 0.1

[a] After Juleff *et al.*[134]
[b] Transition metals indicated with asterisk.

activation energy for interstitial diffusion in silicon is, therefore, considered to be determined mainly by Coulombic interaction between the impurity and host atoms due to the imperfect screening of nuclear charges by electrons.[135] Evidence to support this is the experimental observation that shows the diffusivity of 3*d* transition elements tends to increase as the atomic number increases; that is, the larger atomic radius of the transition metal, the smaller the Coulombic effect of the nucleus.

Solubility The solubility curves of some transition metals were shown in Fig. 3.31. It has been observed that the behaviors of transition metals during the cooling process are not identical. That is, Cr, Mn, and Fe atoms can be quenched in interstitial sites of tetrahedral symmetry in the silicon lattice, while Co, Ni, and Cu vanish out of the interstitial sites during quenching and form preferably silicide precipitates.[132] The variance in behavior results in different transition metal species generating different types of surface micro-defects as discussed in Section 7.2.3.

5.4 New Crystal Growth Methods

Silicon crystals used for microelectronic circuit fabrication must meet many requirements of device manufacture. In addition to requirements for silicon *wafers*, whose criteria for VLSI/ULSI electronics will be discussed in Chapter 8, the following crystallographic demands have become obvious from the view of high-yield and high-performance microelectronic device manufacturing:

1. large diameter,
2. low defect density,
3. uniform and low radial resistivity gradient, and
4. optimum initial oxygen concentration and its precipitation.

It is clear that silicon crystal manufacturers must not only meet the above requirements but also produce those crystals economically and with high manufacturing yield. The main concerns of silicon crystal growers are crystallographic perfection and axial distribution of dopants in CZ silicon. In order to overcome some problems that the conventional CZ crystal growth method inevitably has, several new crystal growth methods have been developed. In this section, those methods that have been successfully operated or are expected to be feasible in the near future are discussed.

5.4.1 Magnetic-Field-Applied Czochralski Method

Background As discussed in Section 5.2.3, melt convection flow in a crucible strongly affects the crystal quality of CZ silicon. In particular, unfavorable growth striations are induced by unsteady melt convections resulting in temperature fluctuations at the growth interface. The ability of a magnetic field to inhibit thermal convection in electrically conducting fluids[136] was first applied in 1966 to crystal growth of indium antimonide by using the horizontal boat technique[137] and horizontal zone-melting technique[138]. Through these investigations, it was confirmed that a magnetic field of sufficient strength can suppress the temperature fluctuations accompanying melt convection, and can dramatically reduce growth striations. In 1970, a transverse magnetic field of 4000 G was successfully applied to the Czochralski growth of indium antimonide crystals.[139]

The effect of the magnetic field on reducing growth striations is explained by its ability to decrease turbulent thermal convection of a melt and in turn decrease temperature fluctuations at the crystal–melt interface. The fluid flow damping caused by the magnetic field is due to the induced magnetomotive force when the flow is orthogonal to the magnetic flux lines, which results in an increase in the effective kinematic viscosity of the conducting melt.[136] This increase is proportional to the *Hartman number* Ha,

$$Ha = BL(\sigma/\rho_m \nu_K)^{1/2} \tag{5.48}$$

where B is the magnetic flux density, σ the electrical conductivity, and ρ_m the melt density. An increase in the melt viscosity brings about suppression of the thermal convection [see Eq. (5.20)] and results in a decrease in temperature fluctuations at the growth interface. Numerical simulations of melt flow in a CZ crucible when a magnetic field is applied have been performed by several investigators.[140–142]

Silicon crystal growth by the magnetic-field-applied CZ (MCZ) method was reported for the first time in 1980.[143] Originally MCZ was aimed at growing CZ silicon crystals that contain a low oxygen concentration and then have a high resistivity with a low radial variation. That is, MCZ silicon was expected to replace FZ silicon almost exclusively used for power-device fabrication. Since then, various types of magnetic field configurations, in terms of the magnetic field direction (horizontal or vertical) and the type of magnets used (normal conductive or superconductive), have been developed.[144] Recently, MCZ silicon, which contains a wide range oxygen concentration at a desired level from low to high, has been of great interest for different device application. In addition, both horizontal[145] and vertical[146] magnetic fields have also been applied to FZ silicon crystal growth, and a dramatic decrease in growth striations has been achieved in both cases.

Magnetic Field Direction   The primary concern in applying a magnetic field to CZ silicon growth is the direction of the magnetic field: that is, perpendicular (horizontal or transverse) or parallel (vertical or axial) to the growth axis. Hereafter the CZ growth with a horizontal and a vertical magnetic field are referred to as HMCZ and VMCZ, respectively. Figure 5.39 shows the magnetic flux lines calculated for superconductive magnets without interruption due to the furnace for both VMCZ and HMCZ cases.[144] A vertical magnetic field can be applied to the melt, by surrounding the growth chamber with a solenoid coil, and is much easier and less costly than a horizontal magnetic field, which requires heavy electromagnets. Moreover, a

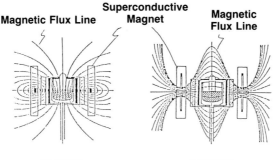

Fig. 5.39. Magnet flux lines caculated for VMCZ and HMCZ cases with superconductive magnets. (After Ohwa *et al.*[144])

magnetic field parallel to the crystal growth axis offers the advantage that the axial symmetry of the configuration is not disturbed, thus minimizing azimuthal variations in the crystal.[147] However, it has been found that the variation of radial dopant distribution becomes more pronounced in VMCZ growth.[148] In addition, it has been observed that the oxygen concentration in VMCZ silicon crystals tends to be higher than in CZ silicon grown without a magnetic field,[149] and occasionally it exceeds the solubility limit obtained from Fig. 5.36.[144]

The primary interest in MCZ silicon growth is its expected advantage in microscopic control of impurities, particularly oxygen. In Fig. 5.40, the effect of a magnetic field on melt convection, which strongly affects oxygen incorporation, is schematically illustrated for both VMCZ and HMCZ cases.[145] As a result of the magnetic field damping, forced convection caused by crystal rotation becomes dominant in the VMCZ case, while in the

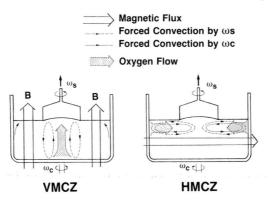

Fig. 5.40. Effect of vertical and horizontal magnetic field on damping melt flow in Czochralski crucible. (After Hoshi *et al.*[150])

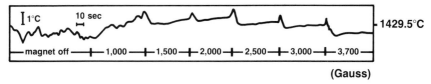

(Gauss)

Fig. 5.41. Effect of horizontal magnetic field on suppressing temperature fluctuations in silicon melt. (After Suzuki et al.[157] Reproduced by permission of The Electrochemical Society, Inc.)

HMCZ case forced convection caused by both crystal and crucible rotation contributes to the melt flow. In order to achieve an axially uniform oxygen distribution in a crystal, the crystal rotation must be changed, usually to slower rotation, as the melt level goes down in the VMCZ growth, while the oxygen concentration in HMCZ can be controlled by changing crucible rotation. It is well known that changing crystal rotation during CZ crystal growth usually causes detrimental variation in the radial gradients of dopants and oxygen [see Eq. (5.24)].

There are several publications that report that homogeneous or high-quality silicon crystals have been grown by the VMCZ method.[148,151-154] However, it should be noted that those VMCZ silicon crystals reported are limited in size to smaller than 80 mm in diameter, and were grown from a small melt charge less than 4 kg. For large VMCZ silicon crystal growth (i.e., 100-125 mm in diameter and 22-30 kg melt charge size), axial oxygen concentration control has been found to be very difficult.[144] In contrast to the VMCZ growth, the HMCZ method has achieved the growth of large-diameter (100-125 mm) high-quality silicon crystals grown from large melt

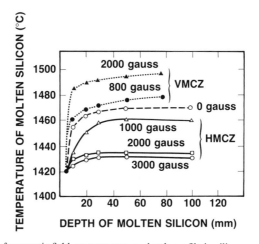

Fig. 5.42. Effect of magnetic field on temperature depth profile in silicon melt. (After Ohwa et al.[144] Reproduced with the permission of The Electrochemical Society, Inc.)

charges (20–30 kg),[144,155] and no problem is anticipated for even larger crystals. From a practical point of view, the HMCZ method seems to be more successful than the VMCZ in growing high-quality silicon crystals with both an axial and radial uniformity of oxygen at a desired level.

MCZ Silicon Melt It is usually observed that the surface of silicon melt in a CZ growth chamber vibrates due to thermal convection. By applying a horizontal magnetic field larger than a certain critical value, surface vibration is eliminated.[156] Figure 5.41 shows the effect of a horizontal magnetic field on suppressing temperature fluctuations in a silicon melt close to the surface.[155] Temperature fluctuations of about 1.5°C appear in the melt at zero magnetic field, and as the field increases, the fluctuations gradually decrease to less than 0.1°C at 1500 G. This is attributed to reduced vibration of the melt surface due to the suppression of thermal convection. The spikes observed in Fig. 5.41 may be caused by the induction current generated when the magnetic field strength was changed stepwise. Since a magnetic field suppresses thermal convection, the temperature depth profile in the melt is significantly affected by applying a magnetic field. Figure 5.42 shows the effect of a magnetic field on the vertical temperature profile measured at the center of a silicon melt in a 400-mm-diameter quartz crucible.[144] It has been obvious that the vertical and horizontal magnetic fields influence temperature profiles differently. That is, the radial temperature gradient and the average melt temperature increase with application of a vertical field, while they decrease with a horizontal magnetic field. This difference in effect may be explained qualitatively with Fig. 5.40 in which a magnetic field dampens the melt flow perpendicular to the magnetic flux.

Growth Rate of HMCZ Silicon It has been observed that HMCZ silicon can be grown at a rate about twice the usual rate at zero magnetic field.[155] As shown in Eq. (5.14), larger temperature gradients in a growing crystal $[(dT/dz)_c]$ and smaller gradients in the melt $[(dT/dz)_m]$ result in a higher growth rate. The small $(dT/dz)_m$ achieved in an HMCZ melt shown in Fig. 5.42 contributes to a higher growth rate. In order to make the temperature gradient in a growing crystal $(dT/dz)_c$ larger, it is necessary to modify the hot zone in a puller by, for example, constructing a heat-shield assembly that screens the growing crystal from heat radiation.[155]

Properties of HMCZ Silicon The value of HMCZ silicon lies in its high quality and controlled oxygen concentration over a wide range. Figure 5.43 shows growth striations in HMCZ silicon grown without any rotation and the axial distribution profile of phosphorus dopant concentrations.[156] No growth striation is found in the region grown with a magnetic field; however,

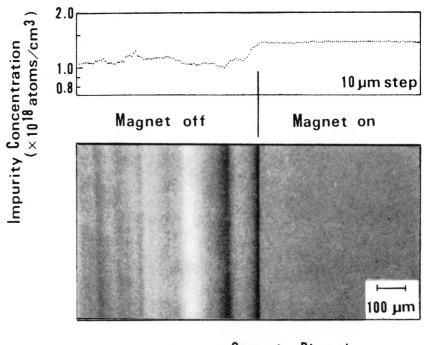

Fig. 5.43.   Growth striations in boron-doped HMCZ silicon crystal grown without rotation. Magnetic field applied is 3700 G. (After Hoshi *et al.*[156] Reprinted with the permission of The Electrochemical Society, Inc. Courtesy of K. Hoshi and N. Isawa, Sony Corporation.)

nonrotational striations with irregular intervals are clearly observed in the region grown without a magnetic field. Phosphorus concentrations vary in correspondence with the growth striations in the area, while the axial distribution of phosphorus is very uniform in the region grown with the magnetic field. Growth striations and axial distribution of antimony dopant concentrations for HMCZ silicon grown with crystal and crucible rotations are shown in Fig. 5.44.[155] Both rotational and nonrotational striations are observed in the regions grown without a magnetic field. The center region grown with a magnetic field shows no striation; however, rotational striations due to thermal asymmetry as discussed in Section 5.3.1 are observed at the periphery of the crystal. These experimental observations suggest that a horizontal magnetic field is effective in reducing growth striations, particularly nonrotational striations, but that thermal asymmetry must be removed in order to eliminate rotational striations completely. In addition to the uniform distribution of dopants in HMCZ silicon, it should be noted that the impurity concentrations in the region grown with a magnetic field tend to be higher

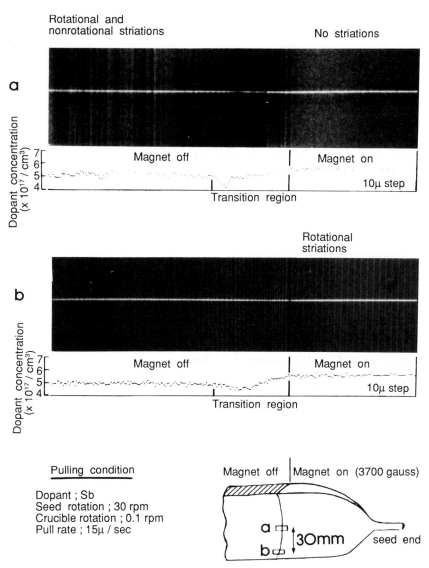

Fig. 5.44. Growth striations and axial distribution of antimony dopant concentration in HMCZ silicon crystal growth with rotation; $\omega = 30$ rpm and $\omega_c = 0.1$ rpm. (After Suzuki *et al.*[155] Reproduced with the permission of The Electrochemical Society, Inc. Courtesy of K. Hoshi and N. Isawa, Sony Corporation.)

than those in the region without a magnetic field, as shown in Figs. 5.43 and 5.44. This increase in impurity concentrations is attributed to the increase in the effective segregation coefficient due to the effect of the magnetic field on increasing kinematic viscosity of molten silicon [see Eqs. (5.24) and (5.25)].

HMCZ silicon, which contains a wide range of oxygen concentration at a desired level from low to high, has been of great interest for different device application, since the obtainable oxygen level is limited to the range of 10–18 ppma* for conventional CZ silicon. It has been recently reported that a wide range of oxygen concentration in silicon, 5–25 ppma, can be achieved by the HMCZ method.[156,157]

In addition to the above-mentioned advantages, defects and oxygen behave more predictably in HMCZ silicon than in conventional CZ silicon. The predictable behavior is valuable in terms of their application to microelectronic device fabrication. The behavior of oxygen and defects will be discussed further in Chapter 7.

Future of HMCZ Silicon   As far as crystal quality is concerned, there is no doubt that the HMCZ method provides the silicon crystals most favorable to the semiconductor device industry. At the present time, the production cost of HMCZ silcon is higher than that of conventional silicon crystals because the MCZ method consumes more electrical power and requires additional equipment and operating space for the electromagnets. However, taking into account the higher growth rate of HMCZ, and adopting superconductive magnets that need smaller space and consume less electrical power compared with conductive magnets, the production cost of HMCZ silicon crystals may become comparable to that of conventional CZ silicon crystals. In addition, the improved crystal quality of HMCZ silicon may increase production yields and lower the barrier concerning its production cost.

5.4.2 Multiple Czochralski Growth Method

Background   Crystal production costs depend to a large extent on costs for materials, in particular for quartz crucibles.[49,158] In the conventional CZ process, called a *batch process*, a crystal is pulled from a single crucible

---

* As discussed in Section 6.2.2, the interstitial oxygen concentration $[O_i]$ is most commonly obtained by the following equation:

$$[O_i] = f_c \alpha_{03} \quad (\times 10^{17} \text{ atoms/cm}^3)$$
$$= 2f_c \alpha_{03} \quad (\text{ppma})$$

where f_c is the conversion factor and α_{03} is the infrared absorption coefficient at 1106 cm^{-1} due to interstitial oxygen (in cm^{-1}). Most confusingly, three different conversion factors (f_c = 4.81, 3.03, and 2.45) have been used as one likes in the silicon industry.[91] Although the confusion may be resolved by using a unified value of f_c = 3.03 in the near future, f_c = 2.45 is used in this book.

charge, and the quartz crucible is only used once and then is discarded. This is because the crucible with a small amount of remaining silicon cracks as it cools from a high temperature at each growth run. The multiple CZ growth method,[158] which spreads the crucible cost over several crystal pulls, was originally studied in order to reduce the production cost of solar-grade silicon crystals, which can be of lower quality than those used for microelectronic circuit fabrication. In addition to cost reduction performed by multiple use of quartz crucibles, this method is also attractive for growing high-quality CZ crystals with a tight resistivity range. That is, the pulling of a crystal is stopped before the lower resistivity specification is exceeded, and then another crystal can be grown after the crucible is replenished with silicon, which allows the growth of a second crystal at the target resistivity.

Operation Outline The operational procedure for multiple CZ growth is shown schematically in Fig. 5. 45. A silicon crystal is pulled according to the usual CZ growth procedure until the resistivity reaches ρ_2 ($< \rho_1$), which is the lower limit of the target value. Upon removal of the crystal, a new polysilicon rod is recharged and the reseeding is performed for the second growth. When the crucible is replenished, the resistivity requirement is achieved with the new polysilicon rod and by adding the appropriate amount of dopant. By repeating this entire process, multiple CZ silicon ingots that hit the target resistivity can be grown.

Advantages and Disadvantages As previously noted, the production yield of silicon crystals that meet the resistivity range is low due to the phenomenon of dopant segregation. This low yield problem is most serious for *n*-type crystals doped with phosphorus or antimony, which has a small segregation coefficient. By the multiple CZ growth method, efficient production of *n*-type silicon crystals can be achieved. One of the disadvantages of this method is that the growth chamber must be extensively modified, which results in very

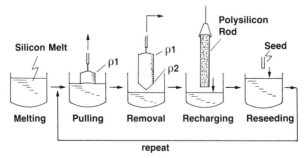

Fig. 5.45. Operation outline of multiple Czochralski growth method.

complicated growth equipment. The overall equipment design requires the following criteria to meet a multiple crystal growth run[158]:

1. The crucible must always be kept hot to prevent breakage.
2. The hot zone must always be under vacuum or low argon pressure to reduce the buildup of silicon monoxide deposits.
3. The ingots must be removed from the furnace without admitting any air.
4. Recharge polysilicon must be added to the hot crucible without contaminating the silicon melt or the hot-zone environment.

Another disadvantage can be impurity buildup in the residual melt due to segregation.[159] From Eq. (5.23), the concentration of impurity in the melt $[C]_L$ is given as

$$[C]_L = [C]_0 (1 - g)^{k_0 - 1} \tag{5.49}$$

Repeated application of Eq. (5.49) coupled with feedstock addition of impurity concentration $[C]_0$ gives the impurity concentration in the melt at the beginning of the nth pull, $[C]_L^n$, as

$$[C]_L^n = [C]_0 P^{n-1} + [C]_0 g(P^{n-2} + P^{n-3} + \cdots + 1)$$
$$= [C]_0 \{ P^{n-1} + g[(P^{n-1} - 1)/(P - 1)] \} \tag{5.50}$$

where $P = (1 - g)^{k_0}$. At the end of the nth growth run, the impurity concentration in the melt $[C]_L^{n+1}$ increases according to the normal freeze law:

$$[C]_L^{n+1} = (P/1 - g)[C]_L^n \tag{5.51}.$$

If $k_0 \ll 1$, then $P \approx 1$ and Eqs. (5.50) and (5.51) can be approximated by

$$[C]_L^n \approx [C]_0 [1 + g(n - 1)] \tag{5.52}$$

and

$$[C]_L^{n+1} \approx [C]_0 [1 + ng(1 - g)] \tag{5.53}$$

respectively. Thus, this impurity buildup in the residual melt may limit multiple growth of high-quality CZ crystals.

5.4.3 Continuous Czochralski Growth Method

Advantages and Disadvantages Another strategy to replenish a quartz crucible with melt economically would be to continuously add feed as the crystal is grown and thereby maintain the melt at a constant volume. In addition to saving crucible costs, the continuous-charging Czochralski (CCZ) method provides an ideal situation for silicon crystal growth. As already discussed, many of the inhomogeneities in crystals grown by the conventional CZ batch process are a direct result of the nonsteady kinetics due primarily to the melt volume change during crystal growth. The CCZ method aims not only to reduce the production cost of crystals but also to grow crystals under steady conditions. By maintaining the melt volume constant, steady thermal and melt flow conditions can be achieved. See Fig. 5.19, which shows the change in thermal environments during conventional CZ growth.

Second, concerning the control of oxygen incorporation into a silicon crystal, the ratio of $A_{c/m}$ (the area of crucible–melt interface) to $A_{m/a}$ (the area of free-melt surface) can be kept constant during the crystal growth. That is, the total oxygen atoms incorporated into a crystal through the crystal–melt interface can be maintained constant [see Eq. (5.42)]. Moreover, by continuous charging of polysilicon and dopant, the resistivity of the growing crystal is tightly controlled. The resistivity ρ of crystal grown by the CCZ method is given by

$$\rho = \rho_0 \frac{k_0[C]_0}{[C]_r - ([C]_r - k_0[C]_0) \exp(-k_0 V_s/V_0)} \tag{5.54}$$

where ρ_0 is the initial crystal resistivity at $V_s = 0$, $[C]_0$ the initial melt dopant concentration, $[C]_r$ the incoming melt replenishment concentration, V_s the crystal volume grown, and V_0 the primary crucible melt volume.[49] Equation (5.54) shows that the crystal resistivity will continuously decrease as more crystals are grown, due to the exponential term, without any special precaution. However, if $[C]_r = k_0[C]_0$, the exponential term is eliminated and the resistivity will remain constant and independent of crystal volume grown. Thus, the axial crystal resistivity can be maintained constant by controlling the resistivity of incoming polysilicon continuously charged as $k_0[C]_0$.

The disadvantages of the CCZ method would be, as in the case of the multiple CZ method, the impurity buildup in the residual melt, and the complexity of growth equipment. As far as the impurity is concerned, however, it has been shown that the CCZ procedure leads to a much slower buildup of impurities than does sequential replenishment.[160] The impurity concentration equation for the CCZ process is given by

$$\frac{[C]_L}{[C]_0} = \frac{1}{k_0} \left[1 - (1 - k_0) \exp\left(-\frac{k_0 V_s}{V_0} \right) \right] \tag{5.55}$$

If $k_0 V_s/V_0 \ll 1$, Eq. (5.55) can be approximated by

$$[C]_L = [C]_0 [1 + (1 - k_0)V_s/V_0] \tag{5.56}$$

Thus, a lower impurity buildup and greater axial uniformity can be achieved by the CCZ method when compared with both the sequential and conventional CZ methods.

Structure of Continuous-Charging Crystal Puller   Continuous charging can be performed by either liquid or solid feeding. Figure 5.46 shows a continuous liquid feed system where the charge is fed continuously from the meltdown chamber through the liquid transfer tube into the growth chamber.[159] The liquid transport is controlled by hydrostatic pressure. A continuous solid feed system, such as shown in Fig. 5.47,[161] has also been developed. This system consists basically of a hopper for storing the polysilicon raw material and a vibratory feeder that transfers polysilicon chunks to the crucible. In the crucible that contains the silicon melt, a quartz baffle may be required in order to prevent melt turbulance, caused by feeding solid chunks, around the growth interface. Free-flowing polysilicon granules such as shown in Fig. 5.4 are obviously advantageous for the solid-feed CCZ method.

Future of Continuous Czochralski Growth   As just discussed, the CCZ method will certainly solve most of the problems related to inhomogeneities in crystals grown by the conventional CZ method. In spite of the great attraction of this method, there has been no publication yet that reports the commercial use of this method. This situation may suggest that further development is required to establish the CCZ method, or that publication has been reserved simply because of the developer's policy. In any case, there is no doubt that the appearance of free-flowing granular polysilicon with high purity may accelerate the development of the CCZ method and

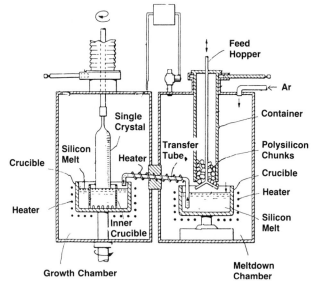

Fig. 5.46. Continuous liquid-feed Czochralski growth furnace. (After Lorenzini *et al.*[159])

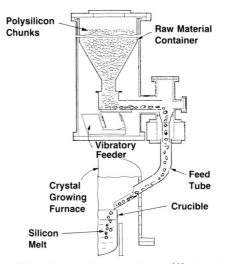

Fig. 5.47. Continuous solid-feed Czochralski growth furnace.[161] (Reproduced with the permission of *Electronics*).

that the CCZ method will become the leading technology of silicon crystal growth in the near future. Moreover, it is anticipated that the combination of MCZ and CCZ—namely, the magnetic-field-applied continuous CZ (MCCZ) method —will be the ultimate crystal growth method and may provide ideal silicon crystals for the wide application of microelectronic technology.

Finally, it should be emphasized that the difference in thermal history of the crystal parts from the seed to tail ends of a grown crystal, as shown in Fig. 5.19, must be considered even when the crystal was grown by the ideal growth method. In order to homogenize the grown crystal or to obtain "real" axial uniformity, some posttreatment would be necessary for the crystal.[89]

5.5 Wafer Shaping Process and Wafer Properties

Semiconductor electronic devices are exclusively fabricated on polished wafers. Thus, the first step in device fabrication is the preparation of smooth, clean, and damage-free surfaces. The requirements for geometrical tolerance of the polished wafer as well as their bulk crystal properties have become more stringent as the complexity of device design has increased. As the semiconductor device industry approaches the ULSI era, microelectronic circuits and their fabrication processes are proving more sensitive to the characteristics of starting material, that is, polished wafers. In this section, the wafer shaping processes and some important properties of silicon wafers are discussed.

5.5.1 Wafer Shaping Process

General Remarks Polished wafers are prepared through a complex sequence of shaping, polishing, and cleaning steps after a single crystal ingot has been grown. Most of these processes, referred to as *wafer shaping*, are somewhat similar among silicon manufacturers; however, the details strongly depend on the know-how developed by each manufacturer. Therefore, the details of wafer shaping processes have not been publicized since the competition among silicon suppliers is as extreme as that among device manufactures. Thus, the wafer shaping procedure presented below, according to Fig. 5.48, shall be generic. The cleaning process undertaken after each step will be discussed separately in the next subsection.

Trimming The single-crystal ingot first passes through routine evaluation of properties such as resistivity and crystal perfection. The seed and tail ends are cut off, and then the ingot is ground to a cylindrical shape of precise

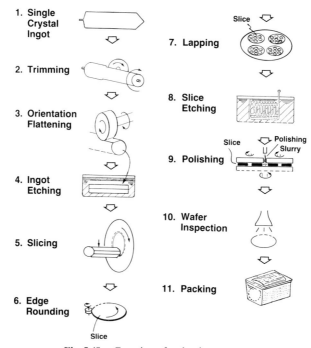

Fig. 5.48. Generic wafer shaping processes.

diameter, since the ingot does not grow perfectly round nor with sufficiently uniform diameter. At this point the diameter is slighty larger than the final one, since the diameter is decreased during subsequent steps such as chemical etching.

Orientation Flattening One or more flats that show the crystallographic orientation are ground along the length of the cylindrical ingot. The largest flat, called the *primary flat*, is usually positioned perpendicular to the ⟨110⟩ orientation. The primary flat is used to correctly align the wafers during the device fabrication processes with automated wafer handling equipment. Since devices fabricated on the wafer must be oriented to specific crystallographic orientations, the primary flat is also used as the reference. The smaller flat, or *secondary flat*, is utilized to identify the wafer plane orientation and the conductive type as shown in Fig. 5.49.

Ingot Etching The ground crystal ingot is dipped into chemical etchant in order to remove mechanical damage induced by grinding and flattening, since the surface of the ingot will eventually become the edge of a polished wafer. Although the primary purpose of this etching process is to remove the mechanical damage, this operation also has a direct impact on the diameter tolerance. The composition of the etchant varies, but most etchants are based on the well-known HNO_3–HF system. Etchant modifiers such as acetic acid

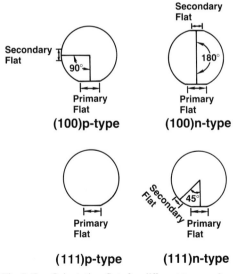

Fig. 5.49. Orientation flats for different type wafers.

are also used.[162] Chemical etching with this etchant system proceeds with the following two-step reaction:

$$Si + 4HNO_3 \longrightarrow SiO_2 + 4NO_2 + 2H_2O \tag{5.57}$$

and

$$SiO_2 + 6HF \longrightarrow H_2SiF_6 + 2H_2O \tag{5.58}$$

Since the reactions are exothermic, temperature control is critical for uniform etching. The reaction product NO_x is toxic, and safety controls must be used.

Slicing   The slicing step produces silicon slices from the shaped ingot and critically defines the important factors of a wafer: the surface orientation, thickness, taper, bow, and so on. The ingot must be rigidly mounted to maintain exact crystallographic orientation, obtained with X-ray diffraction, during the slicing process. The most common way to slice large-diameter crystals is *inner diameter* (*ID*) *slicing*, which uses stainless steel blades with diamond particles bonded on the inner edges.[163] Continuous monitoring with a blade deflection sensor is critically important to assure that slices are sawn with a minimum of bow, taper, and flucture.

Edge Rounding   This step can come either before or after lapping. The edges of silicon slices are rounded by an edge grinder. Figure 5.50 illustrates typical shapes of rounded edges. Edge rounding substantially reduces mechanical defects, such as edge chips and cracks, due to handling with tweezers or loading into furnace boats. The higher mechanical strength of the wafer edge ensures a lower incidence of process-induced plastic deformation (i.e., slip dislocations, which are preferentially nucleated at edge chips and cracks), and increases the wafer yield due to lower frequency of wafer breakage during subsequent processes. Moreover, edge rounding minimizes occurrence of

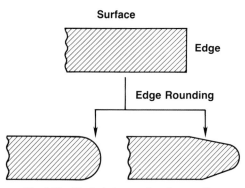

Fig. 5.50.   Typical shapes of wafer rounding.

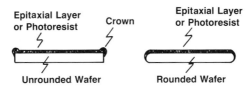

Fig. 5.51. Epitaxial crown or photoresist pile-up phenomenon.

epitaxial crown in the epitaxial deposition process and a pile-up of photoresist at the periphery of the slice. The epitaxial crown or photoresist pile-up is schematically illustrated in Fig. 5.51. It is easily understood that this pile-up phenomenon causes serious problems in device patterning, especially when contact masks are used.

Lapping Lapping is performed on both sides of slices, primarily to remove the nonuniform damage left by slicing, and to attain a high degree of parallelism and flatness of the slice.[164] The lapping abrasive slurry is typically a mixture of alumina or silicon carbide and glycerine.

Slice Etching The mechanical damage induced during the previous shaping steps is removed entirely by chemical etching. The etching process is mostly accomplished with the chemical systems noted above [see Eqs. (5.57) and (5.58)]. Recently, a caustic etching system has been attractive as an alternative to the acid-etching system.[165] The system employs one of the alkaline hydroxides, such as potassium hydroxide, with certain stabilizers. The general reaction is

$$Si + 4OH^- \longrightarrow SiO_4^{-4} + 2H_2 \tag{5.59}$$

One of the advantages of this system is that the reaction system avoids the production of toxic NO_x. However, the etched surface of silicon tends to be somewhat rougher than that produced by acid etching, since the caustic etching is basically crystal orientation-dependent.[166]

Polishing The polishing process is to produce a highly reflective and damage-free surface on one side, and sometimes both sides, of the silicon slice. Polishing of silicon is accomplished by a mechanochemical process in which a polishing pad and a polishing slurry of sodium hydroxide and fine silica particles are involved. Removal rate and resulting flatness of the wafer depend on various operating factors such as temperature, pressure, pad material, rotation rate, and slurry composition. A slower polishing rate, in general, results in a smoother surface.

A recent decrease in the device geometry requires polished silicon wafers whose surface deviates only a few micrometers from the highest point to the lowest point when the wafer is held on a flat vacuum chuck. The polishing process, along with crystal growing, has always been one of the most proprietary areas for the silicon wafer producers.

5.5.2   Cleaning

General Remarks   As repeatedly mentioned, surface contamination affects the electronic device performance and yields more severely as device geometry shrinks. Thus, a primary concern in microelectronic circuit fabrication is the level of contamination on the wafers and the removal of such contamination before further processing. Cleaning of wafers is absolutely required at many steps in device fabrication processes as well as in wafer shaping processes. The types of contaminants range from organic compounds to metallic impurities that are encountered by handling or processing. These contaminants attach to the wafer surface either chemically or physically. A variety of wafer cleaning techniques have been used to remove various different types of contamination. In this subsection, the cleaning procedure after polishing, that is, the final stage before shipping to silicon users, is discussed.

Types of Contaminants   Contaminants on silicon surfaces can be classified broadly as molecular, ionic, and atomic.[167] Typical molecular contaminants are waxes, resins, and oils, which are commonly used in the processes after sawing. They may also include organic compounds from human skin[168] and plastic containers used for interim storage of slices. Such molecular contaminants are usually held to the silicon surface by weak electrostatic forces. Since water-insoluble organic compounds tend to prevent the effective removal of adsorbed contaminants, the molecular contaminants must be removed at the first stage in cleaning processes.

Ionic contaminants, such as Na^+, Cl^-, F^-, and I^-, are present after etching of slices in HF-containing or caustic etchants.[169] They may deposit on the silicon surface by physical adsorption or by chemical absorption (chemisorption). The removal of chemisorbed ions is much more difficult than that of physically attached ions, and a chemical reaction must be used to remove these contaminants.

The atomic contaminants present on silicon surfaces that are of the most serious concern are transition metals such as gold, iron, copper, and nickel. They originate mainly from acid etchants (see Table 5.4). These transition metals, as well as ionic impurities, seriously degrade device performance. The removal of these transition metals requires reactive agents, which dissolve them and form metal complexes to prevent redeposition from the solution.

Chemical Cleaning   Many chemical cleaning processes that aim to eliminate contaminants have been developed. The procedure most widely used in the semiconductor industry is the so-called RCA method (see Appendix XII),[167] which is a sequence of baths including H_2O-H_2O_2-HN_4OH solution (solution 1) and H_2O-H_2O_2-HCl (solution 2). Solution 1, typically 5-1-1 to 7-2-1 parts by volume of H_2O-H_2O_2-NH_4OH, removes organic contaminants by both the solvating action of ammonium hydroxide (NH_4OH) and the powerful oxidizing action of hydrogen peroxide (H_2O_2). The ammonium hydroxide also serves to complex some of the Group I and Group II metals such as copper, gold, nickel, cobalt, and cadmium. Solution 2, typically 6-1-1 to 8-2-1 parts by volume of H_2O-H_2O_2-HCl, removes alkali and transition metals, and prevents redeposition from the solution by forming soluble metal complexes. The chemical cleaning procedure based on the RCA method but slightly modified may be used by every silicon supplier and device manufacturer in the semiconductor industry.

Although various modifications in the RCA method may have been made, very little has been published. The following is an example of the modification recently performed by the originator of the RCA method.[170] Since the presence of a natural thin oxide layer on the silicon surface is suspected to hinder surface cleaning, the elimination of the oxide layer prior to further cleaning may further increase the purification efficiency.[171] Brief etching in dilute hydrofluoric acid (HF) solution after solution-1 cleaning was added to the original RCA procedure. The etching of silicon oxide proceeds according the reaction shown by Eq. (5.58). It has been suggested that this brief etching should be performed with a very dilute high-purity HF solution and for a very short period of time to avoid reinducing the contaminants from the HF solution. Experiments shows that a 10-sec immersion in 1:50 HF-H_2O is sufficient to remove the oxide layer. Subsequent rinsing with deionized (DI) water should also be kept very brief, (e.g., 30 sec), to only remove HF solution from the wafer assembly and to minimize regrowth of a new hydrous oxide layer. It has been observed that the elimination of DI water rinse between HF and subsequent cleaning solution immersion gives high cleaning efficiencies and will not cause any silicon surface attack as long as only small amounts of HF remain on the silicon surface.[171]

Two very important comments on chemical cleaning of silicon wafers have been made by the developer of the RCA method.[170] First, the wafers must never be dried during the processes, because dried residues are difficult to redissolve and may mask the surface during subsequent treatments. Second, Pyrex glassware should not be used with solutions 1 and 2 because substantial amounts of sodium, potassium, boron, and other impurities are leached out of the glass by the hot solutions. Fused silica beakers should be used instead; high-quality opaque fused silica is much less costly than clear fused quartz and is acceptable for wafer-cleaning vessels. Rinse tanks and vessels for HF solution should be constructed of high-grade polypropylene plastic.

High-Purity Water   It is unquestionable that the high-purity water is the most important material to assure purity for any cleaning processes. The quality of water used significantly impacts the purity of wafers and the yield in manufacturing of complex microelectronic devices. As feature dimensions have shrunk to a micrometer or less, these devices become susceptible to particles one-tenth of a micrometer in size.[172] Accordingly, the demands for the water used in the semiconductor industry have become more stringent with advancing technology as shown in Table 5.5.[173] Significant attention is also being focused on the yield loss problem due to microbes in DI water. Water produced by a sequence of deionization, filtration, and ultraviolet irradiation has not been able to avoid the contamination due microbes. A recently developed technique that purifies DI water by oxidation using ozone may provide ultra-high-purity DI water to meet the stringent requirements.[173]

Hydrophilic or Hydrophobic Surface   Silicon dioxide (SiO_2) is *hydrophilic* and easily wet by water, while a surface of silicon is *hydrophobic*. A hydrophobic surface on a silicon wafer can be achieved by dipping into a hydrofluoric acid (HF) solution. A hydrophobic surface that emerges from an HF etch is very reactive and picks up contaminants and particulates, while a hydrophilic surface reduces this pick-up phenomenon.[91] Therefore, a hydrophilic surface is preferred. It has been observed that HF etching, which makes

Table 5.5   Trend of Pure Water Specifications for Microelectronic Device Fabrication[a]

Specification	1980	1983	1984	1985
DRAM bits	16K	64K	256K	1M
Resistivity (MΩ cm at 25°C)	15	17–18	18	18
Particles (μm)	0.2	0.2–0.1	0.1	0.1–0.05
Particles/cm^3	200–300	50–150	20–50	–
Total organic carbon (mg/l)	1	0.5–1	0.05–0.2	0.05
Bacteria/cm^3	1	0.5–1	0.02	0.01
SiO_2 (ppb)	–	20–30	10	10
Dissolved oxygen (mg/l)	8	0.1–0.5	0.1	0.1

[a] After C. Nebel and Nezgod.[173]

the surface hydrophobic, significantly enhances both the rate of impurity, particularly carbon, adsorption on a silicon surface,[174,175] and enhances the oxide growth both at room temperature[175] and high temperatures.[176] Surface roughening in an HF-etched surface that occurs with 1100°C–vacuum heating has also been reported.[174] From these observations, it is clear that etching with HF as a final step should be avoided.

According to the results of a recent investigation on the surface chemistry,[177] a hydrophilic surface is caused by chemical bonding with singular and associated OH groups on the surface, such as units of the type Si–[O$_3$H] with an Si–H bond. The OH groups stabilize the oxide during heating. As a result, the oxide on hydrophilic wafers grows mainly while the wafer is in the cleaning agent and consists of hydrated SiO$_2$ through all the stages of the growth. On the other hand, the hydrophobic state is mainly characterized by Si–H and SiCH$_x$, and to a lesser extent Si–F, groups on the surface. On a hydrophobic surface, the oxidation begins with the formation of a lower oxide state, which turns into SiO$_2$ on storage in air.

Wafer Scrubbing and Sonic Cleaning   Wafer cleaning with scrubbing either by brush or high-pressure fluid jet can enhance yields, particularly when it is applied during device fabrication processing, by removing dirt and residual impurities left on the wafer by processing and during storage.[178] Since brush bristles are usually more than several micrometers in diameter, they are not able to penetrate into dense device topography that has micrometer-order dimensions. High-pressure jet scrubbing, however, can be used to remove contaminants lodged into dense topography. Ultrasonic cleaning, where wafers are immersed in a solution and are agitated by sonic energy, also has advantages similar to those of high-pressure jet scrubbing.

However, a serious disadvantage of these techniques, particularly of brush scrubbing, is that they may induce some mechanical damage that causes stacking faults during subsequent thermal cycles on the wafer surface. For this reason, the application of these cleaning techniques may be limited to cleaning processes for device chips and packages.

Finally, it should be emphasized that avoiding contamination rather than removing contaminants should be achieved, since the former may be less costly and less difficult than the latter.

5.5.3   Mechanical Properties of the Wafer

General Remarks   Mechanical and dimensional properties of wafers directly affect the performance of lithography. For example, wafer flatness in the focal plane of the electron beam delineator or optical printer is crucial for uniform imaging in the electron beam-lithographic or photolithographic processes. In this section, the dimensional specifications and the wafer deformation, which is related to the specification in general, are discussed. The specific criteria of silicon wafers for VLSI/ULSI electronics will be discussed in the final chapter of this book.

Dimensional Specification   Silicon wafers must be dimensionally specified with various parameters for proper use for device fabrication. The primary parameters are the diameter and the thickness. As Figs 5.6 and 5.7 show, 150-mm-diameter wafers have begun to impact the silicon marketplace in order to attain more economical production of microelectronic device chips. The thickness of a silicon wafer becomes greater with the diameter (e.g., 525 μm for a 100-mm-diameter wafer and 625 μm for a 150-mm-diameter wafer). In addition to these primary parameters, silicon polished wafers are characterized with other dimensional parameters.

Flatness is one of the most critical wafer parameters required today. It directly impacts device linewidth capability, process latitude, and throughput. The process and equipment used to fabricate devices define the flatness required. Processes utilizing 1 × 1 imaging systems are sensitive to *global flatness*, whereas stepper imaging systems are sensitive to *local site flatness*. The most common parameter analyzed is *total indicating reading* TIR for a vacuum-chucked wafer. As Fig. 5.52 depicts, TIR concerns a frontside-referenced measurement. An imaginary reference plane is superimposed on the surface, approximating a least-squares fit of that surface. The TIR is defined as the sum of the maximum positive and negative deviation from this reference plane. If the reference is chosen to coincide with the focal plane of mask aligner, the *focal-plane deviation* FPD is defined as the largest deviation, positive or negative, from this plane. In order to determine a specified local site flatness, it is necessary to simulate wafer stepper equipment. Figure 5.53 shows a wafer divided into an array that represents exposure areas typically

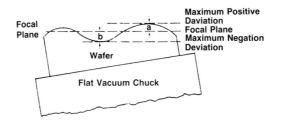

$$\text{TIR} \;=\; a + b$$
$$\text{FPD} \;=\; a \;(\text{if } a > b)$$
$$\phantom{\text{FPD} \;} =\; b \;(\text{if } b > a)$$

Fig. 5.52.   Global flatness with characteristic TIR and FPD.

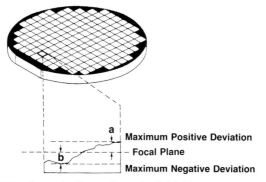

Fig. 5.53.   Local-site flatness for wafer divided into exposure arrays.

15×15 mm or 20×20 mm. Viewing the site cross section of a square chip, TIR and FPD can be defined for each exposure site.

Bow relates to the concave or convex deformation of a wafer and is independent of thickness variation because it is applied to both polished and unpolished silicon slices in a free unclamped state. Reading the two values **a** and **b** shown in Fig. 5.54, the wafer bow is defined as half of the measured value, that is, $(\mathbf{a} - \mathbf{b})/2$.[179] As shown in Fig. 5.55, bow may not affect the photolithographic processes if the wafer surface is sufficiently flat. In extreme cases, however, the clamping action by vacuum chucking cannot remove all the bow. The presence of excessive thickness variation and warpage of a silicon wafer will affect the ability to perform high-quality line imaging. A wafer warped by thermal processing may cause an additional problem of difficulty in vacuum chuck mounting.

Warp is defined as the difference between the maximum and minimum distances of the median surface of the wafer from a reference plane encountered during scanning pattern, while *total thickness variation* TTV is defined as the difference between the maximum and minimum values of thickness

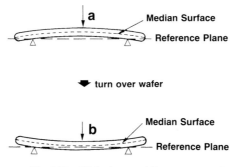

Fig. 5.54.   Wafer bow and its measurement.

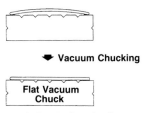

Fig. 5.55.   Removal of wafer bow by flat vacuum chucking.

encountered in the wafer.[180] Figure 5.56 depicts some deformed wafer shapes with related warp and TTV.

The definition of bow and warp is not necessarily unified in the silicon industry, particularly between the United States and Japan. For example, bow and warp or warpage are occasionally defined in Japan as follows[181]: bow = initial wafer deformation caused by mechanical wafer shaping process, and warpage = wafer deformation caused by heat treatment.

This discrepancy between the two countries may originate from the difference in the language; that is, there is only one word, *sori*, in Japanese for

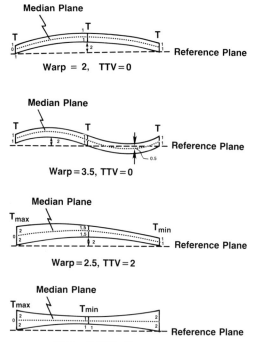

Fig. 5.56.   Deformed wafer shapes with related warp and TTV values. (After ASTM F657.[180] © ASTM. Reproduced with permission.)

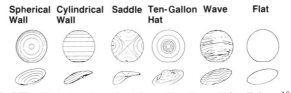

Spherical   Cylindrical   Saddle   Ten-Gallon   Wave   Flat
Wall   Wall   Hat

Fig. 5.57.   Variations in wafer deformation shape. (After Takasu.[181])

English words "bow" and "warp." The confusion occurs when *sori* is translated into English and vice versa.

Wafer Deformation   Silicon polished wafers may vary in their microscopic shape, which deviates from the ideal shape of a flat disk. Commonly observed deformed wafer shapes are schematically illustrated in Fig. 5.57.[181] Note that the illustrations shown in Fig. 5.57 very much exaggerate the real situation. The deviation from the ideal shape (i.e., a flat wafer) is on the order of several tens of micrometers at the worst case. This corresponds to several tens of millimeters of deviation from the flat plane of 125 mm in diameter in the case of a 125-mm-diameter polished wafer.

The previous discussion considers only the appearance of wafers; however, in order to investigate the origin of the deformation and to find the way of correction, it is important to separate the elastic and plastic deformation that occurs in an internally and/or externally deformed wafer. Table 5.6 models the appearance and the crystal lattice of deformed wafers and their characteristics.[181] The ideal wafer (a) has neither elastic nor plastic deformation. The appearance of wafer (d) looks the same as wafer (a); however, it has elastic deformation due to the curved lattice. On the other hand, the appearance of wafer (b) is deformed but is perfect as far as the crystal lattice is

Table 5.6   Types of Bow and Warp and Their Characteristics[a]

Type of bow and warp	Surface appearance	Lattice curvature	Elastic deformation	Plastic deformation	Comments
a	Flat	Flat	No	No	Ideal
b	Curved	Flat	No	No	
c	Curved	Curved	Yes	No	
d	Flat	Curved	Yes	No	
e	Curved	Flat	(Yes)	Yes	Slips
f	Curved	Curved	Yes	(Yes)	Backside damage
g	Curved	Curved	Yes	(Yes)	Backside film

[a] After Takasu.[181]

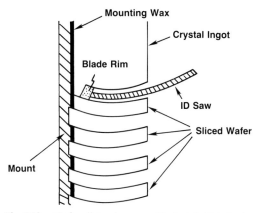

Fig. 5.58.   Wafer slicing by saw with deviated blade rim.

concerned; therefore the wafer has neither elastic nor plastic deformation as in the case of wafer (a). Simple elastic deformation is observed in wafer (c). Case (e) may take place by slip motion toward the wafer surface on {111} planes, and elastic strain, which is much smaller than for other cases, may be induced by slips as well as plastic deformation. Backside damage on wafer (f) and a backside film such as polysilicon and/or silicon nitride for wafer (g), which are commonly utilized for *extrinsic gettering sinks* (see Section 7.4), may not induce plastic deformation at the initial stage but will generate lattice defects locally close to the back surface during subsequent thermal cycles. These lattice defects result in plastic deformation.

Origin of Wafer Deformation   Wafer deformation is induced most dominantly during the slicing process when the saw-blade rim deviates from the

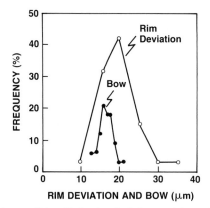

Fig. 5.59.   Relation between slicing blade rim deviation and bow of as-sliced wafer. (After Takasu et al.[182])

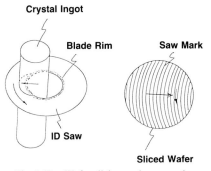

Fig. 5.60.   Wafer slicing and saw mark.

ideal flat position. Deformation is not related to the crystal quality such as growth striations.[181] The blade rim deviation causes curved slicing, as schematically illustrated in Fig. 5.58, which results in wafer bow. The relation between the blade rim deviation and bow in an as-sliced wafer is shown in Fig. 5.59.[182] Saw marks caused by blade rim bending such as illustrated in Fig. 5.60 usually result in material damage as deep as 10–40 μm. These saw marks can also cause bow during thermal cycles if the damage is not removed by subsequent lapping and etching operations.

Deformation found in wafers (b) and (e) in Table 5.6 can be removed by subsequent lapping, but it is difficult to correct or remove the deformation for most cases by lapping. Therefore careful operation of the slicing process, to minimize deformation, should be carried out. Operational experience has shown that the key to a good slicing operation is skilled handling and attaching a thin-blade saw to the slicer.

Thermal cycles induce further wafer deformation due to defect generation such as oxygen precipitation. Thus, the crystal quality or growth striations may strongly affect wafer bow. The deformation induced during thermal cycles will be discussed in Section 7.2.

5.6   Silicon Epitaxy

The term *epitaxy* is applied to the process of growing a thin crystalline layer on a crystalline substrate. When the thin layer is grown on a substrate of the same material, such as silicon on silicon or gallium arsenide on gallium arsenide, the process is termed *homoepitaxy*. If the thin layer and the substrate are of different materials, such as gallium arsenide on silicon or silicon carbide on silicon, the process is termed *heteroepitaxy*. The word "epitaxy" is composed of two parts: *epi* (Greek for "on" or "upon") and *taxis* (Greek for "arrangement" or "ordered"). Thus, epitaxy refers to the orienting of crystallizing atoms of one species by a single-crystalline substrate of another species, so that the two atomic planes in contact have planar unit cells that match each other in shape and size.[183] Although the adjective form of epitaxy should be *epitaxic* or *epitaxical*, the form *epitaxial* is used in this book following the custom in the semiconductor industry.

Silicon epitaxy was originally developed to attain high performance of bipolar transitors and later bipolar integrated circuits. For example, by growing a lightly doped epitaxial layer over a heavily doped substrate, the bipolar device can be optimized for high breakdown voltage of a collector-substrate junction while maintaining a low collector resistance, which provides high device operating speeds at moderate currents. Recently, the equivalent structure but with a thinner epitaxial layer ($< 5 \mu$m), has proved to minimize latch-up problems in CMOS ICs. This problem becomes more serious with shrinking device features.[184] The structure of a high-resistivity layer on a low-resistivity substrate can be achieved by epitaxy, but not by conventional diffusion. This is the most significant advantage of epitaxy. Other advantages of fabricating electronic devices in an epitaxial layer are that the doping concentration profile can be sharply controlled and that the layer can be free from impurities such as oxygen. Moreover, the technology of silicon epitaxy on insulator (SOI) is anticipated to become more and more important for even more sophisticated electronic devices.[185] However, the epitaxial process has some disadvantages. The increased wafer cost is the most serious one. It is not scientific, but the price of the substrate is an important factor for large-scale fabrication of electronic devices, particularly memory devices, which must be produced with a low cost. At this point, the price of silicon epitaxial substrate is about three times that of a substrate without epitaxy. This higher price is mainly due to the low throughput of epitaxial growth equipment. It is anticipated that epitaxial silicon substrates will be used for most device fabrication if the price drops to less than one-and-a-half times that of a nonepitaxial wafer. The key to more consumption of epitaxial substrates is certainly an increase in production throughput.

Epitaxial growth can be achieved either from the vapor phase (VPE), liquid phase (LPE), or solid phase (SPE). For silicon epitaxy, the chemical vapor deposition (CVD) technique has been widely accepted because of its excellent purity control and crystalline perfection achieved. A different approach recenty developed is molecular-beam epitaxy (MBE), which allows epitaxial growth in the temperature range of 600–900°C. These temperatures are sufficiently low to eliminate *autodoping* of impurities, a phenomenon that is a shortcoming of CVD. A discussion of the fundamentals of silicon epitaxial growth by CVD and MBE is given next. For more information on this subject, the reader is recommended to study Refs. 186–190.

5.6.1 Chemical Vapor Deposition

Source and Chemistry Epitaxial silicon is grown by the introduction of volatile reactants together with a carrier gas, most commonly hydrogen, into a suitable reactor. The system should be clean, with oxygen content below 1 ppma. Table 5.7 lists the candidate reactants: silicon tetrachloride ($SiCl_4$), trichlorosilane ($SiHCl_3$), dichlorosilane (SiH_2Cl_2), and monosilane (SiH_4).[191] Silicon tetrachloride requires high deposition temperatures, which result in significant autodoping and outdiffusion of dopants from the substrate. The high-temperature deposition of $SiCl_4$, however, results in a major advantage in that very little deposition occurs on the reactor walls, thereby reducing the frequency of reactor cleaning. This source is mostly used for thick epitaxial layers ($> 10 \ \mu$m) that can tolerate the disadvantage of high-temperature deposition. Trichlorosilane has similar characteristics to those of $SiCl_4$ but is used less because of the greater difficulty in handling, in spite of its slightly lower deposition temperature. Dichlorosilane is widely used for thin-layer deposition at lower temperatures, which reduce outdiffusion and autodoping. It has been reported that silicon epitaxial films deposited from SiH_2Cl_2 have lower defect densities[192] and thereby increase device performance[193] compared to other sources. Monosilane offers the lowest deposition temperature, and accordingly is used for very-thin-layer deposition ($< 2 \ \mu$m). The low-temperature deposition substantially decreases autodoping and outdiffusion; however, it develops the most significant disadvantage because of its low decomposition temperatures, causing heavy deposits on reactor walls. Thus, the use of SiH_4 requires frequent cleaning of the reactor, reducing its throughput. As a consequence, silicon tetrachloride ($SiCl_4$) has been most widely used for silicon epitaxy in spite of its higher deposition temperature. However, recent demands for thinner epitaxial layers and lower-temperature deposition have lead to increased use of SiH_2Cl_2 and SiH_4.

Table 5.7 Source for Silicon Epitaxial Growth by CVD[a]

Source	Normal growth rate (μm/min)	Temperature range (°C)	Allowed oxidizer level (ppm)	Phase at room temperature
$SiCl_4$	0.4–1.5	1150–1250	5–10	Liquid
$SiHCl_3$	0.4–2.0	1100–1200	5–10	Liquid
SiH_2Cl_2	0.4–3.0	1050–1150	< 5	Gaseous
SiH_4	0.2–0.3	950–1050	< 2	Gaseous

[a] After Hammond.[191]

The overall reactions for those epitaxial silicon reactants are[194]

$$SiCl_4 + 2H_2 \longrightarrow Si + 4HCl \tag{5.60}$$

$$SiHCl_3 + H_2 \longrightarrow Si + 3HCl \tag{5.61}$$

$$SiH_2Cl_2 \longrightarrow Si + 2HCl \tag{5.62}$$

$$SiH_4 \longrightarrow Si + 2H_2 \tag{5.63}$$

However, a number of intermediate and competing reactions are considered to exist for the resulting overall reactions. For example, epitaxial deposition of silicon from the hydrogen reduction of $SiCl_4$ has been estimated to proceed through the following competing reactions resulting in intermediate products[195]:

$$SiCl_4 + H_2 \longleftrightarrow SiHCl_3 + HCl \tag{5.64}$$

$$SiHCl_3 + H_2 \longleftrightarrow SiH_2Cl_2 + HCl \tag{5.65}$$

$$SiH_2Cl_2 \longleftrightarrow SiCl_2 + H_2 \tag{5.66}$$

$$SiHCl_3 \longleftrightarrow SiCl_2 + HCl \tag{5.67}$$

$$SiCl_2 + H_2 \longleftrightarrow Si + 2HCl \tag{5.68}$$

The sequence of $SiCl_4$ reaction shows the presence of $SiHCl_3$ and SiH_2Cl_2 as intermediate species. For the silicon deposition from SiH_4, on the other hand, the reaction proceeds with only two steps:

$$SiH_4 \longrightarrow SiH_2 + H_2 \tag{5.69}$$

$$SiH_2 \longrightarrow Si + H_2 \tag{5.70}$$

Growth Rate   The growth rate of epitaxial silicon depends primarily on parameters such as source gas, deposition temperature, pressure, and concentration. Figure 5.61 shows the temperature dependence of the silicon epitaxial growth rate for various silicon sources.[194] Observe that the growth rate strongly depends on the temperature in a low temperature range, region I. At higher temperatures, region II, the temperature dependence becomes small. It follows that the growth process in region I can be characterized as reaction rate- or kinetic-limited, whereas the reaction is transport-limited in region II.

It has been observed that the dependence of the growth rate on the input concentration of reactant is different for monosilane and the others that produce HCl during silicon deposition [see Eqs. (5.60)–(5.63)]. That is, a linear relationship between the growth rate and input concentration is observed until the onset of gas-phase decomposition for monosilane, whereas a certain concentration gives the maximum growth rate for chlorosilanes. This phenomenon is most pronounced in the case of $SiCl_4$, as shown in Fig. 5.62.[196] It is seen that the deposition rate initially increases with increasing

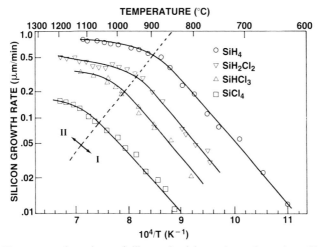

Fig. 5.61. Temperature dependence of silicon epitaxial growth rate for various silicon sources. (After Eversteyn.[194])

$SiCl_4$ concentration, peaks at a mole fraction of 0.10, and then decreases linearly with concentration. This reduction in growth rate is attributed to the reaction product HCl, which etches silicon by reaction (5.68) when the amount of the product exceeds a critical value. The vapor-phase etching of the substrate with anhydrous HCl can be regarded as *in situ cleaning* in addition to the chemical cleaning of the substrate prior to the epitaxial process.

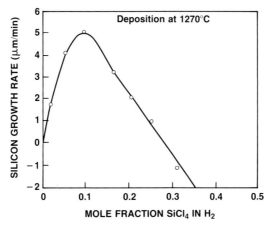

Fig. 5.62. Effect of $SiCl_4$ concentration on epitaxial silicon growth rate. (After Theuerer.[196] Reproduced with the permission of The Electrochemical Society, Inc.)

Table 5.8   Commonly Used Doping Gases for Silicon Epitaxy

Gas	Dopant	Properties
Diborane (B_2H_6)	B	Very toxic, flammable
Phosphine (PH_3)	P	Very toxic, flammable
Arsine (AsH_3)	As	Very toxic, flammable

The diffusion rate of reactants is inversely proportional to the total pressure of the reacting system. Therefore, in the transport-limit growth range, the total pressure of the reacting system has great influence on the growth rate.

Doping   In order to control the conductivity type and electrical resistivity of epitaxial layers, a dopant is added to the CVD system during the entire growth process. Typically, hydrides of the dopants are used as the source. The hydride gases commonly used for doping in epitaxial silicon are listed with their properties in Table 5.8. The hydrides are usually diluted in hydrogen to 10–1000 ppm. As shown in Table 5.8, the doping gases are very toxic and flammable, and are relatively unstable above room temperature, resulting in reaction with air to form solid products.

The incorporation of dopant atoms from the hydride gas phase into epitaxial silicon depends on many parameters, including the dopant gas concentration, temperature, deposition rate, substrate orientation, and substrate doping level. This complex situation has resulted in no simple rule to relate the incorporation of dopant atoms into the epitaxial silicon layer. Nevertheless, the doping concentrations in the range of 10^{14}–10^{20} atoms/cm^3 have been routinely achieved with well-controlled doping operation.

Autodoping and Outdiffusion   Most microelectronic circuit fabrications that use epitaxial wafers require a lightly doped epitaxial layer (10^{14}–10^{17} atoms/cm^3) on a heavily doped substrate (10^{19}–10^{21} atoms/cm^3). The substrate may also have heavily doped isolated areas or a buried layer in the surface region. During the epitaxial deposition, dopants are transported from the substrate to the epitaxial layer in two distinct ways: through *vapor-phase autodoping* and *solid-state outdiffusion*.[197,198] Figure 5.63 depicts these phenomena. Dopant is released from the substrate through solid-state outdiffusion and evaporation. These dopant atoms are reincorporated into the growing epitaxial layer by diffusion through the interface and vapor-phase deposition. The impurities released from the susceptor and the reactor

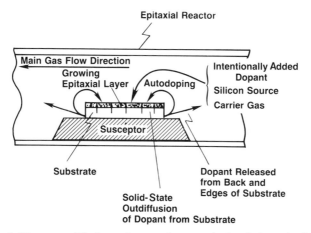

Fig. 5.63.   Solid-state outdiffusion and vapor-phase autodoping during epitaxial growth.

may also cause autodoping. These phenomena are enhanced by a high-temperature epitaxial deposition on a heavily doped substrate. This unintentional doping softens the ideally abrupt junction between the buried layer or substrate and the epitaxial layer, therefore limiting the vertical shrinking of device geometry. Vapor-phase autodoping and solid-state outdiffusion cannot be prevented during the epitaxial growth but can be minimized by choosing low growth temperatures and high growth rates for short times. In addition, it has been reported that epitaxial growth under low-pressure conditions (< 50 torr), low-pressure CVD (LPCVD), dramatically reduced autodoping, and results in the abrupt dopant profile at junctions[199] This reduction in autodoping is attributed to decrease in deposition temperature and in dopant partial pressure at the surface by decreasing the total pressure.

Pattern Shift and Pattern Distortion   When bipolar devices are fabricated, an epitaxial layer is deposited over buried layers in which one or more diffusions are applied to the substrate to create the necessary isolation, collector, emitter, or base functions (see Fig. 4.27). The presence of buried layers complicates subsequent epitaxial processes. It is sometimes found that the device pattern in the epitaxial layer is misplaced and misshaped relative to its original configuration in the substrate. These phenomena are termed *pattern distortion* and *pattern shift*. Pattern distortion is defined as the change in lateral dimension of a given feature after epitaxy, relative to the dimension of the photomask used for patterning, while pattern shift concerns the displacement of a surface feature relative to the underlying buried layer.[200] The crystal orientation has a significant effect on pattern shift, as briefly discussed in Section 3.4.[200,201] For the case of (100) wafers the pattern shift is at a minimum when they are oriented on the exact $\langle 100 \rangle$ axis. On the other hand, for (111) wafers pattern shift is minimized when the surface is misoriented by 2–5° from the exact $\langle 111 \rangle$ axis toward the nearest $\langle 110 \rangle$. Current practice is to misorient (111) wafers normally by 4° off-orientation for silicon epitaxy. In general, pattern shift increases with growth rate and reduced deposition temperature. It has been reported that the pattern shift is substantially reduced as the reaction pressure is reduced.[202] Pattern distortion exhibits an opposite relationship to the parameters previously mentioned. An explanation of pattern shift and pattern distortion is not yet complete.

Epitaxial Reactor Geometry   Consideration of the principles of CVD and the demands for high-yield production of silicon epitaxial substrates leads to a series of guidelines for optimum reactor design for a specific purpose. Three principal geometries have been widely recognized for silicon epitaxy: (1) horizontal reactor with induction heating, (2) vertical pancake reactor with induction heating, and (3) cylinder or barrel reactor with induction or radiant heating. Figures 5.64a, b, and c depict these respectively. Each design has its relative advantages and disadvantages.

The horizontal reactor offers high throughput and is suitable for the production of many different device types. However, controlling the deposition process over its entire susceptor length is not easy. The vertical pancake reactor achieves excellent uniformity, a lower defect density, and reduced autodoping because the gas flow is perpendicular to the substrate and the susceptor rotates. The major disadvantage is its very low throughput. The cylinder reactor, where wafers are placed in niches along the slightly sloping vertical cylindrical wall, also offers high throughput. This reactor design is essentially an expanded adaptation of the horizontal reactor in which the gas flows approximately parallel to the substrate surfaces. The uniformity is improved with a shorter susceptor length and rotating cylinder.

Reactor heating methods, including radient heating, rf induction heating, and hot-wall furnace, strongly affect the silicon epitaxy process. The hot-wall furnace is usable only for deposition in the medium temperature range (600–900°C); therefore it is suitable for polysilicon deposition but not for silicon epitaxy, which requires higher deposition temperatures. Radio-frequency induction is a very convenient way to heat the SiC-coated substrate susceptor without heating the quartz reactor vessel. This method, therefore, has been most widely accepted for heating epitaxial reactors. With increasing substrate diameters, it becomes more difficult to prevent slip dislocations during epitaxial growth. These dislocations originate from the steep vertical and lateral temperature gradients common in rf induction heating. Radiant heating significantly prevents the onset of slip dislocations, and thus has largely supplanted rf induction heating.

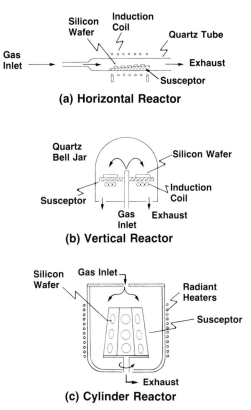

Fig. 5.64. Three principal geometries of silicon epitaxial reactors: (a) horizontal reactor with induction heating, (b) vertical pancake reactor with induction heating, and (c) cylinder or barrel reactor with induction or radiant heating.

5.6.2 Molecular-Beam Epitaxy

Principle Silicon molecular-beam epitaxy (MBE) differs from CVD in involving the evaporation of silicon and the desired dopants under the ultra-high vacuum conditions, for example, 10^{-8}–10^{-11} torr. The evaporated species are transported at relatively high velocity in vacuum to the substrate, where they deposit on the substrate since they are in a supersaturated state and so readily nucleate and grow. This process requires neither diffusion nor chemical reactions, and is typically carried out in a temperature range between 600 and 900°C.[203] These temperatures are sufficiently low to eliminate solid-state dopant outdiffusion from the substrate.

MBE Silicon Silicon epitaxial layer grown by the MBE method promises several remarkable improvements for future device fabrication.[203] Combining the low growth temperature as noted above, direct transport of the material on the growing surface, and immediate control by *in situ* measurement instruments, this enables the precise control of the thickness and doping profile on a submicrometer scale. Moreover, a wide variety of materials, including metals, insulators, and other semiconductors, can be evaporated under MBE conditions. Because of this high flexibility of material combinations and layer numbers, the design of novel devices, such as complex heteroepitaxial structures, should be stimulated. Thus, silicon MBE may influence the frabrication of microelectronic circuits in an evolutionary or revolutionary way.[203]

References

1. G. K. Teal, Single crystals of germanium and silicon—Basic to the transistor and integrated circuit. *IEEE Trans. Electron Devices* **ED-23**, 621–639 (1976).
2. W. C. Dash, Silicon crystals free of dislocations. *J. Appl. Phys.* **29**, 736–737 (1958).
3. L. D. Crossman and J. A. Baker, Polysilicon technology. *In* "Semiconductor Silicon 1977" (H. R. Huff and E. Sirtl, eds.), pp. 18–31. Electrochem. Soc., Princeton, New Jersey, 1977.
4. J. R. McCormic, Polycrystalline silicon–1986, *In* "Semiconductor Silicon 1981" (H. R. Huff, R. J. Kriegler, and Y. Takeishi, eds.), pp. 43–60. Electrochem. Princeton, New Jersey, 1981.
5. "Research Report of High-Purity Silicon Issue Study Group." Shin-Kinzoku Kyokai, 1985 (in Japanese).
6. W. Zulehner and D. Huber, Czochralski-grown silicon. *In* "Crystals 8: Silicon, Chemical Etching" (J. Grabmaier, ed.), pp. 1–143. Springer-Verlag, Berlin and New York, 1982.
7. W. Keller and A. Muhlbauer, "Floating-Zone Silicon." Dekker, New York, 1981.
8. P. A. Taylor, Silane: Manufacture and applications. *Solid State Technol.* July, pp. 53–59 (1987).
9. A. Yusa, Y. Yatsurugi, and T. Takaishi, Ultrahigh purification of silane for semiconductor silicon. *J. Electrochem. Soc.* **122**, 1700–1705 (1975).
10. R. Lutwack and A. Morrison, eds., "Silicon Material Preparation and Economical Wafering Methods." Noyes Publication, Park Ridge, New Jersey, 1984.
11. G. Parkinson, New ways to make crystals for semiconductor uses. *Chem. Eng. (N.Y.)* May 25, pp. 14–17 (1987).
12. S. K. Iya, R. N. Flagella, and F. S. DiPaolo, Heterogeneous decomposition of silane in a fixed bed reactor *J. Electrochem. Soc.* **129**, 1531–1535 (1982).
13. G. Hsu, N. Rohatgi, and J. Houseman, Silicon particle growth in a fluidized-bed reactor. *AIChE J.* **33**, 784–791 (1987).
14. Monsanto Electronic Material Company and Ethyl Corporation, unpublished.
15. J. Grabmaier, ed., "Crystals 5: Silicon." Springer-Verlag, Berlin and New York, 1981.
16. G. K. Teal and J. B. Little, Growth of germanium single crystals. *Phys. Rev.* **78**, 647 (1950).
17. S. Takasu, Wafer. *In* "Microdevices," pp. 88–93. Nikkei-McGraw-Hill, Tokyo, 1983 (in Japanese).
18. Y. Kaneko, ed., "Report on Large-Diameter Wafer Market." Fuji Economy, Tokyo, 1985 (in Japanese).
19. W. G. Pfann, Principles of zone-melting. *Trans. Am. Inst. Min. Metall. Eng.* **194**, 747–753 (1952).

20. H. C. Theuerer, Method of processing semiconductive materials. U.S. Patent 3,060,123 (1962).
21. P. H. Keck and M. J. E. Golay, Crystallization of silicon from a floating liquid zone. *Phys. Rev.* **89**, 1297 (1953).
22. P. H. Keck, W. Van Horn, J. Soled, and A. MacDonald, Floating zone recrystallization of silicon. *Rev. Sci. Instrum.* **25**, 331–334 (1954).
23. S. Muller, Siliciumreinigung durch tiegelfreies Zonenschmelzen. *Z. Naturforsch., B: Anorg. Chem., Org. Chem., Biochem., Biophys., Biol.* 9B, 504–505 (1954) (in German).
24. T. Abe, Crystal fabrication. *In* "VLSI Electronics Microstructure Science" (N. G. Einspruch and H. R. Huff, eds.), Vol. 12, pp. 3–61. Academic Press, New York, 1985.
25. D. T. J. Hurle, Analytical representation of the shape of the meniscus in Czochralski growth. *J. Cryst. Growth* **63**, 13–17 (1983).
26. F. F. Wang, ed., "Impurity Doping Processes in Silicon." North-Holland Publ., Amsterdam, 1981.
27. J. M. Meese, ed., "Neutron Transmutation Doping in Semiconductors." Plenum, New York, 1979.
28. B. D. Stone, Neutron transmutation doping of silicon. *In* "Impurity Doping Processes in Silicon" (F. F. Wang, ed.), pp. 217–257. North-Holland Publ., Amsterdam, 1981.
29. H. M. Liaw and C. J. Varker, Phosphorus donor homogenization of Czochralski crystals by thermal neutron irradiation over doping. *In* "Semiconductor Silicon 1977" (H. R. Huff and E. Sirtl, eds.), pp. 116–125. Ellectrochem. Soc., Princeton, New Jersey, 1977.
30. J. H. Matlock, Advances in single crystal growth of silicon. *In* "Semiconductor Silicon 1977" (H. R. Huff and E. Sirtl, eds.), pp. 32–51. Electrochem. Soc., Princeton, New Jersey, 1977.
31. S. K. Ghandhi, "Semiconductor Power Devices." Wiley, New York, 1977.
32. E. L. Kern, L. S. Yaggy, and J. A. Baker, Process improvements in detector grade silicon. *In* "Semiconductor Silicon 1977" (H. R. Huff and E. Sirtl, eds), pp. 52–60. Electrochem. Soc., Princeton, New Jersey, 1977.
33. S. M. Hu and W. J. Patrick, Effect of oxygen on dislocation movement in silicon *J. Appl. Phys.* **46**, 1869–1874 (1975).
34. S. M. Hu, Dislocation pinning effect of oxygen atoms in silicon. *Appl. Phys. Lett.* **31**, 53–56 (1977).
35. B. Leroy and C. Plougonven, Warpage of silicon wafers. *J. Electrochem. Soc.* **127**, 961–970 (1980).
36. S. Takasu, H. Otsuka, N. Yoshihiro, and T. Oku, Wafer bow and warpage. *Jpn. J. Appl. Phys., Suppl.* **20** (1), 25–30 (1981).
37. C.-O. Lee and P. J. Tobin, The effect of CMOS processing on oxygen precipitation, wafer warpage, and flatness. *J. Electrochem. Soc.* **133**, 2147–2152 (1986).
38. Y. Kondo, Plastic deformation and preheat treatment effect in CZ and FZ silicon crystals. *In* "Semiconductor Silicon 1981" (H. R. Huff, R. J. Kriegler, and Y. Takeishi, eds.), pp. 220–231. Electrochem. Soc., Princeton, New Jersey, 1981.
39. K. Sumino, H. Harada, and I. Yonenaga, The origin of the difference in the mechanical strengths of Czochralski-grown and float-zone-grown silicon. *Jpn. J. Appl. Phys.* **19**, L49–L52 (1980).
40. J. R. Patel and A. R. Chaudhuri, Oxygen precipitation effects on the deformation of dislocation-free silicon. *J. Appl. Phys.* **33**, 2223–2224 (1962).
41. I. Yonenaga, K. Sumino, and K. Hoshi, Mechanical strength of silicon crystals as a function of the oxygen concentration. *J. Appl. Phys.* **56**, 2346–2350 (1984).
42. K. Sumino, I. Yonenaga, and A. Yusa, Mechanical strength of oxygen-doped float-zone silicon crystals. *Jpn. J. Appl. Phys.* **19**, L763–L766 (1980).

43. T. Abe, K. Kikuchi, S. Shirai, and S. Muraoka, Impurities in silicon single crystals—A current view. *In* "Semiconductor Silicon 1981" (H. R. Huff, R. J. Kriegler, and Y. Takeishi, eds.), pp. 54–71. Electrochem. Soc., Princeton, New Jersey, 1981.

44. H.-D. Chiou, J. Moody, R. Sandfort, and F. Shimura, Effects of oxygen and nitrogen on slip in CZ silicon wafers. *In* "VLSI Science and Technology/1984" (K. E. Bean and G. A. Rozgonyi, eds.), pp. 59–65. Electrochem. Soc., Princeton, New Jersey, 1984.

45. K. Sumino, I. Yonenaga, M. Imai, and T. Abe, Effect of nitrogen on dislocation behavior and mechanical strength in silicon crystals. *J. Appl. Phys.* **54**, 5016–5020 (1983).

46. L. Jastrzebski, G. W. Cullen, R. Soydan, G. Harbeke, J. Lagowski, S. Vecrumba, and W. N. Henry, The effect of nitrogen on the mechanical properties of float zone silicon and on CCD device performance. *J. Electrochem. Soc.* **134**, 466–470 (1987).

47. J. Czochralski, Ein neues Verfahren zur Messung der Kristallisationsgeschwindigkeit der Metalle. *Z. Phys. Chem.* **92**, 219–221 (1918) (in German).

48. G. K. Teal and E. Buehler, Growth of silicon single crystals and of single crystal silicon p-n junctions. *Phys. Rev.* **87**, 190 (1952).

49. G. Fiegl, Recent advances and future directions in CZ-silicon crystal growth technology. *Solid State Technol.* Aug., pp. 121–131 (1983).

50. J. W. Moody and R. A. Frederick, Development in Czochralski silicon crystal growth. *Solid State Technol.* Aug., pp. 221–225 (1983).

51. G. Fiegl, Bulk properties improvement of CZ silicon from large melts by thermal convective flow control. *In* "Defects in Silicon" (W. M. Bullis and L. C. Kimerling, eds.), pp. 527–538. Electrochem. Soc., Princeton, New Jersey, 1983.

52. R. Iscoff, Crystal growing and fabrication equipment. *Semicond. Int.* Nov., pp. 68–69 (1983).

53. C. P. Chartier and C. B. Sibeley, Czochralski silicon crystal growth at reduced pressures. *Solid State Technol.* Feb., pp. 31–33 (1975).

54. L. Brewer and R. K. Edwards, The stability of SiO solid and gas. *J. Chem. Phys.* **58**, 351–358 (1954).

55. G. K. Teal, Single crystals of germanium and silicon—Basic to the transistor and integrated circuit. *IEEE Trans. Electron Devices* **ED-23**, 621–639 (1976).

56. K. A. Jackson, Theory of melt growth. *In* "Crystal Growth and Characterization" (R. Ueda and J. B. Mullin, eds.), pp. 21–32. North-Holland Publ., Amsterdam, 1975.

57. J. R. Carruthers and A. F. Witt, Transient segregation effects in Czochralski growth. *In* "Crystal Growth and Characterization" (R. Ueda and J. B. Mullin, eds.), pp. 107–154. North-Holland Publ., Amsterdam, 1975.

58. J. C. Brice, Analysis of the temperature distribution in pulled crystals. *J. Cryst. Growth* **2**, 395–401 (1968).

59. S. N. Rea, "Czochralski silicon pull rate limits. *J. Cryst. Growth* **54**, 264–274 (1981).

60. K. Morizane, A. Witt, and H. C. Gatos, Impurity distribution in single crystals. III. Impurity heterogeneities in single crystals rotated during pulling from the melt. *J. Electrochem. Soc.* **114**, 738–742 (1967).

61. A. J. R. de Kock, P. J. Rocksnoer, and P. G. T. Boonen, Microdefects in swirl-free silicon crystals. *In* "Semiconductor Silicon 1973" (H. R. Huff and R. R. Burgess, eds.), pp. 83–94. Electrochem. Soc., Princeton, New Jersey, 1973.

62. H. Föll and B. O. Kolbesen, Formation and nature of swirl defects in silicon. *Appl. Phys.* **8**, 319–331 (1975).

63. J. W. Cahn, Theory of crystal growth and interface motion in crystalline materials. *Acta Metall.* **8**, 554–562 (1960).

64. F. C. Frank, The influence of dislocations on crystal growth. *Discuss. Faraday Soc.* **5**, 48–54 (1949).

65. T. G. Digges and R. Shima, The effect of growth rate, diameter and impurity concentration on structure in Czochralski silicon crystal growth. *J. Cryst. Growth* **50**, 865–869 (1980).

66. R. H. Hopkins, R. G. Seidensticker, J. R. Davis, P. Rai-Choudhury, and P. D. Blais, Crystal growth considerations in the use of 'solar grade' silicon. *J. Cryst. Growth* **42**, 493–498 (1977).

67. R. K. Srivastava, P. A. Ramachandran, and M. P. Duduković, Czochralski growth of crystals. *J. Electrochem. Soc.* **133**, 1009–1015 (1986).

68. F. D. Rosi, Effect of crystal growth variables on electrical and structural properties of germanium. *RCA Rev.* **19**, 349–387 (1958).

69. J. A. M. Dikhoff, Cross-sectional resistivity variations in germanium single crystals. *Solid-State Electron.* **1**, 202–210 (1960).

70. M. G. Mil'vidskii, The crystallization front and the distribution of impurities across the transverse section of monocrystals grown from a melt. *Sov. Phys.—Crystallogr. (Engl. Transl.)* **6**, 647–648 (1962).

71. E. Kuroda and H. Kozuka, Influence of temperature oscillations in silicon Czochralski growth. *J. Cryst. Growth* **63**, 276–284 (1983).

72. M. G. Mil'vidskii and B. I. Golovin, The form of the crystallization front in semiconducting single crystals grown from the melt by the Czochralski method. *Sov. Phys.—Solid State (Engl. Transl.)* **3**, 737–739 (1961).

73. N. Kobayashi and T. Arizumi, The numerical analysis of the solid–liquid interface shapes during the crystal growth by the Czochralski method. *Jpn. J. Appl. Phys.* **9**, 361–367 (1970).

74. R. K. Srivastava, P. A. Ramachandran, and M. P. Duduković, Interface shape in Czochralski grown crystals: Effect of conduction and radiation. *J. Cryst. Growth* **73**, 487–504 (1985).

75. E. Billig, Growth of monocrystals of germanium from an undercooled melt. *Proc. R. Soc. London* **229**, 346–363 (1955).

76. T. Abe, The growth of Si single crystals from the melt and impurity incorporation mechanisms. *J. Cryst. Growth* **24/25**, 463–467 (1974).

77. J. C. Brice, Facet formation during crystal growth. *J. Cryst. Growth* **6**, 205–206 (1970).

78. N. Kobayashi, Melt flow in a Czochralski melt. *Proc. 84th Meet. Cryst. Eng. (Jpn. Soc. Allied Phys.* pp. 1–8 (1984) (in Japanese).

79. D. Schwabe and A. Scharmann, Marangoni convection in open boat and crucible. *J. Cryst. Growth* **52**, 435–449 (1981).

80. N. Kobayashi, Effect of fluid flow on the formation of gas bubbles in oxide crystals grown by the Czochralski method. *J. Cryst. Growth* **54**, 414–416 (1981).

81. D. S. Robertson, A study of the flow patterns in liquids using a model Czochralski crystal growing system. *Br. J. Appl. Phys.* **17**, 1047–1050 (1966).

82. J. R. Carruthers and K. Nassau, Nonmixing cells due to crucible rotation during Czochralski crystal growth. *J. Appl. Phys.* **39**, 5205–5214 (1968).

83. K. Shiroki, Simulations of Czochralski growth on crystal rotation rate influence in fixed crucibles. *J. Cryst. Growth* **40**, 129–138 (1977).

84. N. Kobayashi, Computational simulation of the melt flow during Czochralski growth. *J. Cryst. Growth* **43**, 357–363 (1978).

85. N. Kobayashi and T. Arizumi, Computational studies on the convection caused by crystal rotation in a crucible. *J. Cryst. Growth* **49**, 419–425 (1980).

86. A. D. W. Jones, An experimental model of the flow in Czochralski growth. *J. Cryst. Growth* **61**, 235–244 (1983).

87. H. Tsuya, F. Shimura, K. Ogawa, and T. Kawamura, A study on intrinsic gettering in CZ silicon crystals: Evaluation, thermal history dependence and enhancement. *J. Electrochem. Soc.* **129**, 374–379 (1982).

88. G. Fraundorf, P. Fraundorf, R. A. Craven, R. A. Frederick, J. W. Moody, and R. W. Shaw, The effect of thermal history during growth on O precipitation in Czochralski silicon. *J. Electrochem. Soc.* **132**, 1701–1704 (1985).

89. F. Shimura, Behavior and role of oxygen in silicon wafers for VLSI. *In* "VLSI Science and Technology/1982" (C. J. Dell'Oca and W. M. Bullis, eds.), pp. 17–32. Electrochem. Soc., Princeton, New Jersey, 1982.

90. H. R. Huff and F. Shimura, Silicon material for VLSI electronics. *Solid State Technol.* Mar., pp. 103–118 (1985).

91. F. Shimura and H. R. Huff, VLSI silicon material criteria. *In* "VLSI Handbook" (N. G. Einspruch, ed.), pp. 191–269. Academic Press, New York, 1985.

92. Y. Yatsurugi, N. Akiyama, Y. Endo, and T. Nozaki, Concentration, solubility, and equilibrium distribution coefficient of nitrogen and oxygen in semiconductor silicon. *J. Electrochem. Soc.* **120**, 975–979 (1973).

93. C. Wagner, Theoretical analysis of diffusion of solutes during the solidification of alloys. *Trans. Am. Inst. Min. Metall. Eng.* **196**, 154–160 (1954).

94. J. A. Burton, R. C. Prim, and W. P. Slichter, The distribution of solute in crystals grown from the melt. Part I. Theoretical. *J. Chem. Phys.* **21**, 1987–1996 (1953).

95. A. F. Witt and H. C. Gatos, "Impurity distribution in single crystals. II. Impurity striations in InSb as revealed by interference contrast microscopy. *J. Electrochem. Soc.* **113**, 808–813 (1966).

96. A. F. Witt and H. C. Gatos, Microscopic rate of growth in single crystals pulled from the melt: Indium antimonide. *J. Electrochem. Soc.* **115**, 70–75 (1968).

97. K. Morizane, A. F. Witt, and H. C. Gatos, Impurity distribution in single crystals. I. Impurity striations in nonrotated InSb crystals. *J. Electrochem. Soc.* **113**, 51–54 (1966).

98. L. A. Girifalco, "Atomic Migration in Crystals." Ginn (Blaisdell), Boston, Massachusetts, 1964.

99. T. Y. Tan and U. Gösele, Point defects, diffusion processes, and swirl defect formation in silicon. *Appl. Phys. [Part] A* **A37**, 1–17 (1985).

100. U. Gösele and T. Y. Tan, The influence of point defects on diffusion and gettering in silicon. *In* "Impurity Diffusion and Gettering in Silicon" (R. B. Fair, C. W. Pearce, and J. Washburn, eds.), pp. 105–116. Mater. Res. Soc., 1985.

101. S. Mizuo and H. Higuchi, Retardation of Sb diffusion in Si during thermal oxidation. *Jpn. J. Appl. Phys.* **20**, 739–744 (1981).

102. S. M. Hu, Formation of stacking faults and enhanced diffusion in the oxidation of silicon. *J. Appl. Phys.* **45**, 1567–1573 (1974).

103. M. Miyake, Oxidation-enhanced diffusion of ion-implanted boron in heavily phosphorus doped silicon. *J. Appl. Phys.* **58**, 711–715 (1985).

104. R. B. Fair, M. L. Manda, and J. J. Wortman, The diffusion of antimony in heavily doped and n- and p-type silicon. *J. Mater. Res.* **1**, 705–711 (1986).

105. J. C. Irvin, Resistivity of bulk silicon and of diffused layers in silicon. *Bell Syst. Tech. J.* **41**, 387–410 (1962).

106. R. E. Chaney and C. J. Varker, The dissolution of fused silica in molten silicon. *J. Cryst. Growth* **33**, 188–190 (1976).

107. W. K. Kaiser and P. H. Kech, Oxygen content of silicon single crystals. *J. Appl. Phys.* **28**, 882–887 (1957).

108. B. O. Kolbesen and A. Muhlauer, Carbon in silicon: Properties and impact on devices. *Solid-State Electron.* **25**, 759–775 (1982).

109. H. M. Liaw, Oxygen and carbon in silicon crystals. *Semicond. Int.* Oct., pp. 71–82 (1979).

110. H. R. Huff and E. Sirtl, eds., "Semiconductor Silicon 1977." Electrochem. Soc., Princeton, New Jersey, 1977.

111. H. R. Huff, R. J. Kriegler, and Y. Takeishi, eds., "Semiconductor Silicon 1981." Electrochem. Soc., Princeton, New Jersey, 1981.

112. H. R. Huff, T. Abe, and B. Kolbesen, eds., "Semiconductor Silicon 1986." Electrochem. Soc., Princeton, New Jersey, 1986.

113. J. C. Mikkelsen, Jr., S. J. Pearton, J. W. Corbett, and S. J. Pennycook, eds., "Oxygen, Carbon, Hydrogen and Nitrogen in Crystalline Silicon." Mater. Res. Soc., Pittsburgh, 1986.

114. W. Kaiser, P. H. Keck, and C. F. Lange, Infrared absorption and oxygen content in silicon and germanium. *Phys. Rev.* **101**, 1264-1268 (1956).

115. C. Haas, The diffusion of oxygen in silicon and germanium. *J. Phys. Chem. Solids* **15**, 108-111 (1960).

116. J. W. Corbett and G. D. Watkins, Stress-induced alignment of anisotropic defects in crystals. *J. Phys. Chem. Solids* **20**, 319-320 (1961).

117. J. C. Mikkelsen, Jr., The diffusivity and solubility of oxygen in silicon. *In* "Oxygen, Carbon, Hydrogen and Nitrogen in Crystalline Silicon" (J. C. Mikkelsen, Jr., S. J. Pearton, J. W. Corbett, and S. J. Pennycook, eds.), pp. 19-30. Mater. Res. Soc., Pittsburgh, 1986.

118. R. C. Newman, A. K. Tipping, and J. H. Tucker, "The effect of metallic contamination on enhanced oxygen diffusion in silicon at low temperatures. *J. Phys. C* **18**, L861-L866 (1985).

119. D. Heck, R. E. Tressler, and J. Monkowski, The effects of processing conditions on the out-diffusion of oxygen from Czochralski silicon. *J. Appl. Phys.* **54**, 5739-5743 (1983).

120. P. Gaworzewski and G. Ritter, On the out-diffusion of oxygen from silicon. *Phys. Status Solidi A* **67**, 511-516 (1981).

121. U. Gösele and T. Y. Tan, Oxygen diffusion and thermal donor formation in silicon. *Appl. Phys. [Part] A* **A28**, 79-92 (1982).

122. K. Hoshikawa, H. Hirata, H. Nakanishi, and K. Ikuta, Control of oxygen in CZ silicon growth. *In* "Oxygen, Carbon, Hydrogen and Nitrogen in Crystalline Silicon" (J. C. Mikkelsen, Jr., S. J. Pearton, J. W. Corbett, and S. J. Pennycook, eds.), pp. 101-112. Mater. Res. Soc., Pittsburgh, 1986.

123. F. Shimura, Growth method of semiconductor single crystals. Japanese Patent 54-146665 (1979) (in Japanese).

124. F. Shimura and M. Kimura, Growth method and equipment of semiconductor single crystals. Japanese Patent 56-190318 (1981) (in Japanese).

125. R. C. Newman and J. B. Willis, Vibrational absorption of carbon in silicon. *J. Phys. Chem. Solids* **26**, 373-379 (1965).

126. J. A. Baker, T. N. Tucker, N. E. Mover, and R. C. Buschert, Effect of carbon on the lattice parameter of silicon. *J. Appl. Phys.* **39**, 4365-4368 (1968).

127. Y. Takano and M. Maki, Diffusion of oxygen in silicon. *In* "Semiconductor Silicon 1973" (H. R. Huff and R. R. Burgess, eds.), pp. 469-481. Electrochem. Soc., Princeton, New Jersey, 1973.

128. T. Nozaki, Y. Yatsurugi, and N. Akiyama, Concentration and behavior of carbon in semiconductor silicon. *J. Electrochem. Soc.* **117**, 1566-1568 (1970).

129. A. R. Bean and R. C. Newman, The solubility of carbon in pulled silicon crystals. *J. Phys. Chem. Solids* **32**, 1211-1219 (1971).

130. R. C. Newman and J. Wakerfield, The diffusivity of carbon in silicon. *J. Phys. Chem. Solids* **19**, 230-234 (1961).

131. L. A. Ladd, J. P. Kalejs, and U. Gösele, Enhanced carbon diffusion in silicon during 900°C annealing. *In* "Impurity Diffusion and Gettering in Silicon" (R. B. Fair, C. W. Pearce, and J. Washburn, eds.), pp. 89-94. Mater. Res. Soc., Pittsburgh, 1985.

132. E. R. Weber, Transition metals in silicon. *Appl. Phys. [Part] A* **A30**, 1-22 (1983).

133. P. F. Schmidt and C. W. Pearce, A neutron activation analysis study of transition group metal contamination in the silicon device manufacturing process. *J. Electrochem. Soc.* **128**, 630-637 (1981).

134. E. M. Juleff, W. J. McLeod, E. A. Hulse, and S. Fawcett, Advances in contamination control of processing chemicals in VLSI. *Solid State Technol.* Sept., pp. 82-86 (1982).

135. R. A. Swalin, Diffusion of interstitial impurities in germanium and silicon. *J. Phys. Chem. Solids* **23**, 154-155 (1962).

136. S. Chandrasekhar, On the inhibition of convection by a magnetic field. *Philos. Mag.* [7] **43**, 501–532 (1952).

137. H. P. Utech and M. C. Flemings, Elimination of solute banding in indium antimonide crystal by growth in a magnetic field. *J. Appl. Phys.* **37**, 2021–2024 (1966).

138. H. A. Chedzey and D. T. Hurle, Avoidance of growth-striae in semiconductor and metal crystals grown by a zone-melting techniques. *Nature (London)* **210**, 933–934 (1966).

139. A. F. Witt, C. J. Herman, and H. C. Gatos, Czochralski-type crystal growth in transverse magnetic fields. *J. Mater. Sci.* **5**, 822–824 (1970).

140. W. E. Langlois and K.-J. Lee, Digital simulation of magnetic Czochralski flow under various laboratory conditions for silicon growth. *IBM J. Res. Dev.* **27**, 281–284 (1983).

141. G. M. Oreper and J. Szekely, The effect of an externally imposed magnetic field on buoyancy driven flow in a rectangular cavity. *J. Cryst. Growth* **64**, 505–515 (1983).

142. M. Mihelcic and K. Wingrath, Numerical simulations of the Czochralski bulk flow in an axial magnetic field: Effects on the flow and temperature oscillations in the melt. *J. Cryst. Growth* **71**, 163–168 (1985).

143. K. Hoshi, T. Suzuki, Y. Okubo, and N. Isawa, *Ext. Abstr., Electrochem. Soc. Meet., 157th* p. 811 (1980).

144. M. Ohwa, T. Higuchi, E. Toji, M. Watanabe, K. Homma, and S. Takasu, Growth of large diameter silicon single crystal under horizontal or vertical magnetic field. *In* "Semiconductor Silicon 1986" (H. R. Huff, T. Abe, and B. Kolbesen, eds.), pp. 117–128. Electrochem, Soc., Princeton, New Jersey, 1986.

145. N. de Leon, J. Guldberg, and J. Salling, Growth of homogeneous high resistivity FZ silicon crystals under magnetic field bias. *J. Cryst. Growth* **55**, 406–408 (1981).

146. H. Kimura, M. F. Harvey, D. J. O'Conner, G. D. Robertson, and G. C. Valley, Magnetic field effects on float-zone Si crystal growth. *J. Cryst. Growth* **62**, 523–531 (1983).

147. K. M. Kim and W. E. Langlois, Computer simulation of boron transport in magnetic Czochralski growth of silicon. *J. Electrochem. Soc.* **133**, 2586–2590 (1986).

148. K. Hoshikawa, H. Kohda, and H. Hirate, Homogeneous dopant distribution of silicon crystal grown by vertical magnetic field-applied Czochralski method. *Jpn. J. Appl. Phys.* **23**, L37–L39 (1984).

149. K. G. Barraclough, R. W. Series, G. J. Rae, and D. S. Kemp, Axial magnetic Czochralski silicon growth. *In* "Semiconductor Silicon 1986" (H. R. Huff, T. Abe, and B. Kolbesen, eds.), pp. 129–141. Electrochem. Soc., Princeton, New jersey, 1986.

150. K. Hoshi, N. Isawa, and T. Suzuki, Up-to-date trend of MCZ-Si. *ULSI* Aug., pp. 51–56 (1985), (in Japanese).

151. K. Hoshikawa, Czochralski silicon crystal growth in the vertical magnetic field. *Jpn. J. Appl. Phys.* **21**, L545–L547 (1982).

152. H. Hirata, K. Hoshikawa, and N. Inoue, Improvement of thermal symmetry in CZ silicon melts by the application of a vertical magnetic field. *J. Cryst. Growth* **70**, 330–334 (1984).

153. K. M. Kim and P. Smetana, Striations in CZ silicon crystals grown under various axial magnetic field strengths. *J. Appl. Phys.* **58**, 2731–2735 (1985).

154. H. Hirata and N. Inoue, Macroscopic axial dopant distribution in Czochralski silicon crystals grown in a vertical magnetic field. *Jpn. J. Appl. Phys.* **24**, 1399–1403 (1985).

155. T. Suzuki, N. Isawa, K. Hoshi, Y. Kato, and Y. Okubo, MCZ silicon crystals grown at high pulling rates. *In* "Semiconductor Silicon 1986" (H. R. Huff, T. Abe, and B. Kolbesen, eds.), pp. 142–152. Electrochem. Soc., Princeton, New Jersey, 1986.

156. K. Hoshi, N. Isawa, T. Suzuki, and Y. Ohkubo, *J. Electrochem. Soc.* **132**, 693–700 (1985).

157. M. Futagami, K. Hoshi, N. Isawa, T. Suzuki, Y. Okubo, Y. Kato, and Y. Okamoto, CMOS static RAM devices fabricated on a high oxygen-content MCZ wafer. *In* "Semiconductor Silicon 1986" (H. R. Huff, T. Abe, and B. Kolbesen, eds.), pp. 939–948. Electrochem. Soc., Princeton, New Jersey, 1986.

158. R. L. Lane and A. H. Kachare, Multiple Czochralski growth of silicon crystals from a single crucible. *J. Cryst. Growth* **50**, 437–444 (1980).

159. R. E. Lorenzini, A. Iwata, and K. Lorenz, Continuous crystal growing furnace. U.S. Patent 4,036,595 (1977).

160. R. H. Hopkins, R. G. Seidensticker, J. R. Davis, P. Rai-Choudhury, P. D. Blais, and J. R. McCormick, Crystal growth considerations in the use of 'solar grade' silicon. *J. Cryst. Growth* **42**, 493–498 (1977).

161. *Electronics* June 22, pp. 44–46 (1978).

162. H. Robbins and B. Schwartz, Chemical etching of silicon. II. The system of HF, HNO_3, H_2O, and $HC_2H_3O_2$. *J. Electrochem. Soc.* **107**, 108–111 (1960).

163. R. L. Lane, ID slicing teechnology for large diameters. *Solid State Technol.* July, pp. 119–123 (1985).

164. C. Murray, Techniques of wafer lapping and polishing. *Semicond. Int.* July, pp. 94–103 (1985).

165. J. A. Moreland, The technology of crystal and slice shaping. *In* "VLSI Electronic Microstructure Science" (N. G. Einspruch and H. R. Huff, eds.), Vol. 12, pp. 63–87. Academic Press, New York, 1985.

166. E. D. Palic, V. M. Dermudez, and O. J. Glembocki, Ellipsometric study of orientation-dependent etching of silicon in aqueous KOH. *J. Electrochem. Soc.* **132**, 871–884 (1985).

167. W. Kern and D. A. Puotinen, Cleaning solutions based on hydrogen peroxide for use in silicon semiconductor technology. *RCA Rev.* **31**, 187–206 (1970).

168. J. A. Lange, Sources of semiconductor wafer contamination. *Semicond. Int.* Apr., pp. 124–128 (1983).

169. W. Kern, Radiochemical study of semiconductor surface contamination. I. Adsorption of reagent compounds. *RCA Rev.* **31**, 207–233 (1970).

170. W. Kern, Purifying Si and SiO_2 surfaces with hydrogen peroxide. *Semicond. Int.* Apr., pp. 94–99 (1984).

171. K. D. Beyer and R. H. Kastl, Impact of deionized water rinses on silicon surface cleaning. *J. Electrochem. Soc.* **129**, 1027–1029 (1982).

172. D. L. Tolliver, Contamination control: New dimensions in VLSI manufacturing. *Solid State Technol.* Mar., pp. 129–137 (1984).

173. C. Nebel and W. W. Nezgod, Purification of deionized water by oxidation with ozone. *Solid State Technol.* Oct., pp. 185–193 (1984).

174. R. C. Henderson, Silicon cleaning with hydrogen peroxide solutions: A high energy electron diffraction and Auger electron spectroscopy study. *J. Electrochem. Soc.* **119**, 772–775 (1972).

175. S. I. Raider, R. Flitsch, and M. J. Palmer, Oxide growth on etched silicon in air at room temperature. *J. Electrochem. Soc.* **122**, 413–418 (1975).

176. F. J. Grunthaner and J. Maserjian, Experimental observations of the chemistry of the SiO_2/Si interface. *IEEE Trans. Nucl. Sci.* **NS-24**, 2108–2112 (1977).

177. M. Grunder and H. Jacob, Investigations on hydrophilic and hydrophobic silicon (100) wafer surfaces by X-ray photoelectron and high-resolution electron energy loss-spectroscopy. *Appl. Phys.* [*Part*] *A* **A39**, 73–82 (1986).

178. A. D. Weiss, Wafer cleaning update. *Semicond. Int.* Apr., pp. 82–85 (1984).

179. ASTM F534, Standard test method for bow of silicon slices. *Annu. Book ASTM Stand.* pp. 496–499 (1985).

180. ASTM F657, Standard method for measuring warp and total thickness valiation on silicon slices and wafers by a noncontact scanning method. *Annu. Book ASTM Stand.* pp. 555–563 (1985).

181. S. Takasu, Bow and warpage of silicon wafers. *Oyo Butsuri* **49**, 83–89 (1980) (in Japanese).

182. S. Takasu, H. Otsuka, N. Yoshihiro, and T. Oku, Wafer bow and warpage. *Jpn. J. Appl. Phys., Suppl.* **20** (1), 25–30 (1981).

183. C. B. Duke, Epitaxical or epitaxial? *J. Mater. Res.* **1**, vii-viii (1986).
184. J. G. Posa, Is epitaxy right for CMOS? *Electronics* Feb. 10, pp. 93–94 (1981).
185. H. W. Law, P. F. Pinizzotto, and A. F. Tasch, Jr., Silicon-on-insulator for VLSI applications. *In* "VLSI Handbook" (N. G. Einspruch, ed.), pp. 503–514. Academic Press, New York, 1985.
186. S. B. Kulkarni, Epitaxial silicon: Material characterization. *In* "VLSI Handbook" (N. G. Einspruch, ed.), pp. 305–325. Academic Press, New York, 1985.
187. T. L. Chu, Growth of epitaxial films for VLSI applications. *In* "VLSI Handbook" (N. G. Einspruch, ed.), pp. 285–304. Academic Press, New York, 1985.
188. C. W. Pearce, Epitaxy. *In* "VLSI Technology" (S. M. Sze, ed.), pp. 51–92. McGraw-Hill, New York, 1983.
189. S. Wolf and R. N. Tauber, "Silicon Processing for the VLSI Era," Vol. 1. Lattice Press, Sunset Beach, California, 1986.
190. J. Bloem and L. J. Giling, Epitaxial growth of silicon by chemical vapor deposition. *In* "VLSI Electronic Microstructure Science" (N. G. Einspruch and H. R. Huff, eds.), Vol. 12, pp. 89–139. Academic Press, New York, 1985.
191. M. L. Hammond, Silicon epitaxy. *Solid State Technol.* Nov., pp. 68–75 (1978).
192. H. R. Chang, Defect control for silicon epitaxial process using silane, dichlorosilane, and silicon tetrachloride. *In* "Defects in Silicon" (W. M. Bullis and L. C. Kimerling, eds.), pp. 549–557. Electrochem. Soc., Princeton, New Jersey, 1983.
193. S. B. Kulkarni, Defect reduction by dichlorosilane epitaxial growth. *In* "Defects in Silicon" (W. M. Bullis and L. C. Kimerling, eds.), pp. 558–567. Electrochem. Soc., Princeton, New Jersey, 1983.
194. F. C. Eversteyn, Chemical reaction engineering in the semiconductor industry. *Philips Res. Rep.* **19**, 45–66 (1974).
195. J. Nishizawa and M. Sato, Growth mechanism of chemical vapor deposition in silicon. *Proc. Int. Conf. Chem. Vap. Deposition, 8th, 1981* pp. 317–328 (1981).
196. H. C. Theuerer, Epitaxial silicon films by the hydrogen reduction of $SiCl_4$. *J. Electrochem. Soc.* **108**, 649–653 (1961).
197. P. H. Lange and J. I. Goldstein, Boron autodoping during silane epitaxy. *J. Electrochem. Soc.* **124**, 591–598 (1977).
198. G. R. Srinivasan, Autodoping effects in silicon epitaxy. *J. Electrochem. Soc.* **127**, 1334–1342 (1980).
199. E. Krullmann and W. L. Engl, Low-pressure silicon epitaxy. *IEEE Trans. Electron Devices* **ED-29**, 491–497 (1982).
200. S. P. Weeks, Pattern shift and pattern distortion during CVD epitaxy on (111) and (100) silicon. *Solid State Technol.* Nov., pp. 111–117 (1981).
201. C. M. Drum and C. A. Clark, Anisotropy of macrostep motion and pattern edge-displacement during growth of epitaxial silicon on silicon wafer {100}. *J. Electrochem. Soc.* **117**, 1401–1405 (1970).
202. R. B. Herring, Advances in reduced pressure silicon epitaxy. *Solid State Technol.* Nov., pp. 75–80 (1979).
203. E. Kasper and K. Worner, Application of Si-MBE for integrated circuits. *In* "VLSI Science and Technology/1985" (K. E. Bean and G. A. Rozgonyi, eds.), pp. 429–447. Electrochem. Soc., Princeton, New Jersey, 1984.

Chapter 6

Crystal Characterization

Semiconductor crystals have been characterized by numerous kinds of electrical and physicochemical techniques.[1-4] The goal of silicon material characterization is the improvement of device performance and yield during microelectronic circuit fabrication processing. The goal can only be achieved by symbiotic use of several existing and emerging techniques. From the viewpoint of electronic device fabrication, characterization of semiconductor silicon may be focused ultimately on electrical properties, particularly of silicon wafer surfaces, since they directly reflect the performance of device operation and manufacturing yield. However, it is true that the electrical properties of semiconductor silicon are complex manifestations of the chemical and physical properties of the silicon material. Therefore, when the desired electrical properties are not observed, the more fundamental chemical and physical properties of the silicon material must be characterized. Moreover, physicochemical characterization of silicon surely helps to control the electrical properties of the material. Meanwhile, specification of silicon wafers is becoming much more sophisticated and stringent as the relationship between microelectronic circuit device yield and specific silicon substrate parameters is being conclusively demonstrated.

The various properties of concern for characterizing silicon wafers are summarized in Table 6.1. In this chapter, the characterization of silicon

Table 6.1 Properties for Characterizing Silicon Crystals

Electrical	Chemical	Physical/structural
Conductivity type Resistivity Lifetime	Impurity content	Lattice defects

wafers from the electrical, chemical, and physical/structural points of view is discussed with respect to the diagnostic techniques for characterizing those properties listed in Table 6.1.

6.1   Electrical Characterization

6.1.1   Conductivity Type and Resistivity

Conductivity Type Determination   Conductivity type is routinely determined with quite simple equipment. Among several basic methods, the hot-probe thermal electromotive force (EMF) method is most widely used for silicon in various forms such as sawn, ground, lapped, etched, and polished wafers. Figure 6.1 schematically illustrates the experimental arrangement for the determination of conductivity type by hot-probe thermal EMF. Placing both the hot and cold probes on a silicon material, one finds a *p*-type silicon will deflect the needle toward the positive (+) side of the null indicator, and an *n*-type toward the negative (−) side. This method is applicable to type determination on silicon samples with room-temperature resistivities between 0.002 and 1400 Ω cm.[4]

Resistivity Measurement   As discussed in Chapter 4, the resistivity and the net carrier concentration are interrelated according to the well-known *Irvin's curve* (Fig. 4.12). A wide variety of resistivity measurement techniques has been established. The most widely used technique for the measurement of resistivity in the semiconductor industry is the *four-point probe method*. The usual geometry is to place four probes in a line on the sample. Current is passed through two probes and the potential developed across the other two

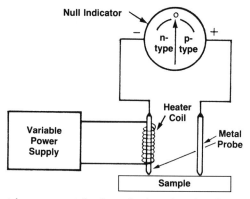

Fig. 6.1.   Experimental arrangement for determination of semiconductor conductive type by hot-probe thermal EMF.

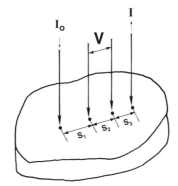

Fig. 6.2.   Principle of four-point probe resistivity measurement.

probes is measured. Any of six combinations of current and voltage probes can in principle be used with different correction factors.[5] Figure 6.2 shows the case in which current is passed through the outer two probes and the potential developed across the inner two probes is measured. The resistivity ρ is given by

$$\rho = 2\pi(V/I)/[1/S_1 + 1/S_3 - 1/(S_1 + S_2) - 1/(S_2 + S_3)] \qquad (6.1)$$

When the probes are spaced with an equal distance S ($= S_1 = S_2 = S_3$), Eq. (6.1) reduces to

$$\rho = 2\pi S(V/I) \qquad (6.2)$$

Usually the probe spacing S is preset as 0.159 cm so that $2\pi S$ equals 1. The four-point probe method is applicable to wide resistivity ranges of 0.0008–2000 Ω cm for p-type silicon and 0.0008–6000 Ω cm for n-type silicon.[6]

Contactless Resistivity Measurement   The four-point probe method has been satisfactory for determining the resistivity of semiconductors. In order to minimize electrical contact problems, however, it is customary to apply relatively high pressures on these contacts, and this in turn produces pitting and structural damage on the sample surface. The existence of this problem has led to the development of various contactless resistivity measurement methods that enable the resistivity measurement without inducing probe damage.[4] In addition, contact-resistance effects can be removed, although variations in sample size can change the coupling between sample and measuring circuit. Among the contactless resistivity measurement methods, the method widely used for silicon is related to *eddy-current* conductivity.[7] This technique usually employs a bridge system, which is imbalanced by the presence of currents in the sample induced by a magnetic field.[8] Coupling a

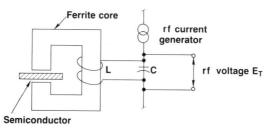

Fig. 6.3. Experimental arrangement of noncontact resistivity measurement.

silicon wafer to an rf tank circuit via a high-permeability ferrite core of circular cross section as indicated in Fig. 6.3, one finds the oscillating magnetic field causes eddy currents in the silicon wafer. In order to implement the regime of the form of Fig. 6.3, the total parallel resistance R_P, which consists of the circuit resistance R_c and sample resistance R_s, is considered:

$$1/R_P = 1/R_c + 1/R_s \qquad (6.3)$$

Holding the rf level constant,

$$iR_P = \text{constant} \qquad (6.4)$$

where i is the magnitude of the rf drive current. Therefore,

$$i \propto 1/R_P \qquad (6.5)$$

With no sample loading, the value of i has its minimum value i_0, which corresponds to $R_P = R_c$. Consequently, Eqs. (6.3) and (6.5) yield

$$(i - i_0) \propto 1/R_s \qquad (6.6)$$

In order to determine the resistivity of the sample, the measuring apparatus is first calibrated using standard specimens of known resistivities.

Spreading Resistance   The spatial resolution of standard potential probe methods is measured in millimeters. For instance, the most commonly used small four-point probe spacing[9] is about 0.5 mm and involves a minimum volume of about 5×10^{-8} cm^3.[10] The spreading resistance method allows microresistivity variation measurements across a semiconductor specimen. The resolving power of the spreading resistance method in depth turns out to be about 0.5 μm; and horizontally it amounts to 25 μm,[11] allowing local resistivity measurements on silicon with a sampling volume of about 10^{-10} cm^3.[9] It has been reported that resistivity profiles of about 0.5 μm resolution length can be achieved by the measurement of angle-lapped n/n^+ epitaxial layers.[9]

For the case of a perfectly conducting hemisphere embedded in a homogeneous semiinfinite solid of resistivity ρ, the spreading resistance R_{sp} in the semiinfinite solid is given by

$$R_{sp} = \rho/2\pi r \qquad (6.7)$$

where r is the radius of the embedded hemisphere. This relationship has been widely used in making rough estimates of the spreading resistance. For brittle semiconductor materials, a more realistic geometry can be arranged. The contact interface between a hemispherically surfaced probe and a flat solid is assumed to be a circular disk of radius r_d. The radius is given by

$$r_d = 1.1\{(Fr/2)[(1/E_1) + (1/E_2)]\}^{1/3} \qquad (6.8)$$

where F is the force applied to the contact, and E_1 and E_2 are the Young's moduli of the materials involved in the contact.[9] Then the spreading resistance of a semiinfinite solid of ρ is given by[12]

$$R_{sp} = \rho/4r_d \qquad (6.9)$$

Figure 6.4 shows a schematic diagram for a spreading resistance measurement for an angle-lapped sample.[13] Because the spreading resistance probe senses the resistivity in a microscopic sampling volume immediately under the probe tip, a profile of resistivity versus depth can be obtained easily. However, it should be noted that the spreading resistance measurement is sensitive to surface preparation.

6.1.2 Lifetime

General Remarks As was briefly discussed in Section 4.2.3, the recombination lifetime τ_r applies when excess carriers introduced by light or a forward-biased *p-n* junction, for instance, decay exponentially toward their equilibrium values as a result of recombination with charges of the opposite

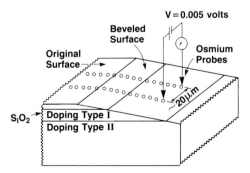

Fig. 6.4. Spreading resistance measurement for angle-lapped semiconductor sample. (After Brennan and Dickey.[13] Reproduced with the permission of *Solid State Technology*.)

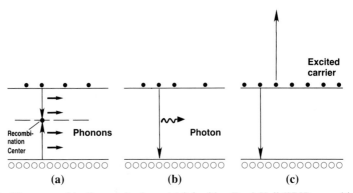

Fig. 6.5. Three recombination mechanisms: (a) Schockley–Read–Hall (SRH) or multiphonon recombination, (b) radiative recombination, and (c) Auger recombination. (After Schroder.[14] ©1982 IEEE.)

sign [see Eqs. (4.30) and (4.31)]. The recombination rate R_r for minority carriers depends nonlinearly on the excess carrier density Δ_n and can be written as

$$R_r = A \Delta n + B(\Delta n)^2 + C(\Delta n)^3 \tag{6.10}$$

The usual definition of $\tau_r = n/R_r$ gives

$$1/\tau_r = A + B \Delta n + C(\Delta n)^2$$

$$= 1/\tau_{\text{SRH}} + 1/\tau_{\text{rad}} + 1/\tau_{\text{Auger}} \tag{6.11}$$

where τ_{SRH} is the Schockley–Read–Hall (SRH) lifetime, τ_{rad} the radiative lifetime, and τ_{Auger} lifetime.[14] The physical mechanisms giving rise to these three lifetimes are schematically shown in Fig. 6.5. As was shown in Fig. 4.11, SRH recombination is dominant in indirect-bandgap semiconductors such as Si, while radiative recombination dominates direct-bandgap ones such as GaAs.[15] Auger recombination, on the other hand, requires a high density of carriers and therefore dominates under high doping and high injection conditions.[16] When the process in which carriers are being generated instead of being destroyed is discussed, the *generation lifetime* τ_g is pertinent as defined by

$$\tau_g = \Delta n/R_g \tag{6.12}$$

where R_g is the generation rate of electron–hole pairs. Although the measured lifetime is usually referred to as *minority carrier lifetime* in practice, it is important to understand the concept of the two lifetimes τ_r and τ_g, which not only have different physical origins but also can be very different in magnitude, for example, $\tau_g \simeq (25–30)\tau_r$.[14]

Minority carrier lifetime has been one of the parameters to characterize the crystal perfection, since the lifetime depends strongly on the crystal purity mainly with respect to metallic impurities—that is, so-called *lifetime killers*, which act as trap centers.[17] From the device operational point of view, although MOS devices are majority carrier operated, the minority carrier lifetime of the base material as an important parameter controlling leakage currents of the channel, source, and drain. Moreover, a high minority carrier lifetime is desired to reduce the refresh time in RAM circuits, and to perform efficient transfer in charge-coupled devices (CCDs). However, in some devices such as fast switching devices, the lifetime is intentionally minimized by doping with a metallic impurity such as Au. It also should be noted that high-lifetime material is susceptible to discharge of the DRAM stored logic state due to transient electronic phenomena. For example, minority carrier currents induced by alpha particles result in the loss of information. Consequently, it is necessary to optimize the lifetime according to the device structure.

Lifetime Measurement Methods   Various experimental methods for lifetime measurements have been established.[4] An experimental method to measure the lifetime requires injecting carriers into the sample by means of a flash of light or voltage pulses, and observing their decay by any one of the following phenomena: (1) change in the potential of a collector point as a function of time, (2) reverse characteristics of p–n junctions, (3) conductivity change, (4) surface photovoltage effect, (5) microwave or infrared absorption, (6) diffusion, (7) photoelectromagnetic (PEM) effect, (8) steady-state photoconductivity, (9) shot noise, and (10) magnetoresistance effect. Experimental methods based on these phenomena can in general be classified into two categories as summarized in Fig. 6.6.[18] The classification depends on whether the lifetime is measured directly as a time interval or indirectly through a related parameter like the diffusion length. The methods can also be separated into two broad categories. The first uses a sample of the material as is, and allows the lifetime measurement nondestructively in which the samples are free of any mechanical or electrode contacts. The second makes use of the properties of a device in its final state, which requires electrode contacts using p-n junctions or MOS capacitors. When lifetime measurements are to be used to evaluate virgin material and the processes, those methods that require no additional sample preparation step are desirable. However, when lifetime measurement is for a sample in the form of a finished device, considerable caution must be exercised in interpreting the material quality since the lifetime in finished devices is very sensitive to process conditions, which may themselves introduce inpurities and physical defects. The lifetime measurement techniques most commonly used for silicon are reviewed next.

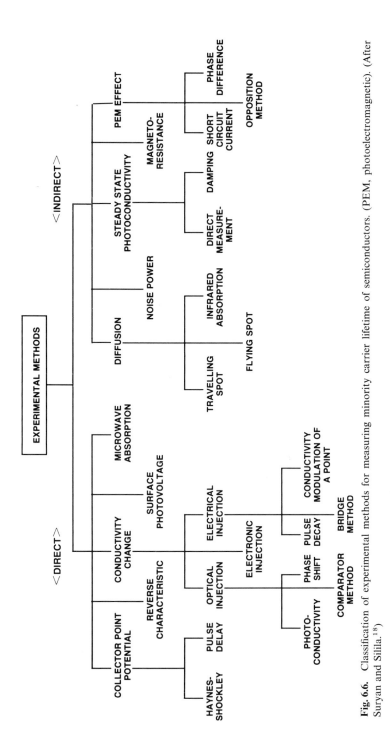

Fig. 6.6. Classification of experimental methods for measuring minority carrier lifetime of semiconductors. (PEM, photoelectromagnetic). (After Suryan and Silila.[18])

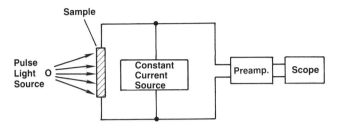

Fig. 6.7. Experimental arrangement for lifetime measurement using ohmic contact.

Photoconductivity Decay Method In this technique, the excess carriers are generated by illuminating the sample with a short pulse of penetrating light. The conductivity σ of the sample is directly proportional to the number of carriers, and the conductivity change is given by

$$\Delta\sigma = \Delta n \exp(-t/\tau_r) \qquad (6.13)$$

Thus the minority carrier lifetime can be determined by monitoring the conductivity after illumination is removed. The photoconductive decay can be detected directly using ohmic contacts[19] or indirectly by using microwave[20,21] or infrared[22,23] absorption techniques, measuring eddy-current losses,[24] or using capacitive coupling.[25] Figure 6.7 shows an experimental arrangement for the lifetime measurement using ohmic contacts. If the time dependence of the excess carrier density is obtained from the display of the voltage change on an oscilloscope, then the analysis of the oscilloscope trace yields the sample minority carrier lifetime.[19]

Generally in photoconductive decay methods, surface recombination due to carrier trapping can seriously affect the measured decay curves. The temporary trapping of carriers[26] superimposes a long lifetime constant tail on the exponential decay, which results in a longer lifetime than the actual one. For ordinary lifetime measurements, this effect can be minimized by using light with wavelengths near the absorption band edge, that is, with relatively low absorption coefficients, since the carriers are generated in the bulk of the sample and not at all at the surface. The use of a filter made of the same material as the sample, interposed between the light source and the sample, is also useful to minimize the surface effect. The filter eliminates the short-wavelength nonpenetrating photons, but passes a fair proportion of those wavelengths near the absorption edge, which can be absorbed reasonably uniformly throughout the thickness of the sample. Moreover, the bulk lifetime can be calculated by subtracting the surface recombination velocity s from the lifetime observed. If the sample is rectangular and is sandblasted so that the surface recombination velocity s is very large,[27] then

$$\mathbf{s} = \pi^2 D_a(1/a_2 + 1/b_2 + 1/c_2) \qquad (6.14)$$

where a, b, and c are the dimensions of the sample, and D_a is the ambipolar diffusion coefficient, defined by

$$D_a = (n_e + n_h)/\{(n_e/D_h) + (n_h/D_e)\} \qquad (6.15)$$

where n_e is the electron density, n_h is the hole density, and D_h and D_e are the diffusion coefficients of holes and electrons, respectively. For heavily doped silicon, D_a reduces to D_e for p-type, and D_h for n-type. Once the surface recombination velocity **s** has been calculated, the bulk lifetime τ_B is obtained by the following correction equation:

$$\tau_B = (1/\tau_{obs} - s)^{-1} \qquad (6.16)$$

where τ_{obs} is the observed lifetime. In the case when the surface recombination is not extremely large, the correction equations are much more complex.[19,28,29]

MOS Capacitor Method   Minority carrier lifetimes can be determined by observing the transient capacitance response of the metal insulator semiconductor (MIS) junction after the application of a large reverse-bias step. This lifetime measurement technique, which is commonly called the MOS capacitance method, is very useful for silicon wafers since a thin SiO_2 insulating layer can easily be grown on the surface. In physical terms, this method consists of applying a step voltage to the gate of an MOS capacitor in order to deplete majority carriers from the semiconductor surface. Initially, a large depletion layer is formed. When the voltage is kept constant, electron–hole pairs are generated in the depletion region. The generation rate R_g, assuming most of the generation centers are located near the center of the forbidden gap, is approximately given by[30]

$$R_g = n_i/2\tau_g \qquad (6.17)$$

In a time t_r, enough carriers will be generated to neutralize the entire depletion layer width W:

$$R_g t_r W = [N]W \qquad (6.18)$$

$$t_r = [N]/R_g = 2\tau_g([N]/n_i) \qquad (6.19)$$

where $[N]$ is the doping concentration and $n_i = 1.4 \times 10^{10}$ cm^{-3} for silicon at room temperature.[31] As minority carriers are generated in the depletion region, they are swept to the Si/SiO_2 interface, where they accumulate in an inversion layer. The majority carriers generated flow to the edge of the depletion region, where some of them neutralize the ionized impurity sites, reducing the width of the depletion region. The width of the depletion region

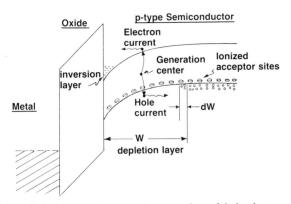

Fig. 6.8. Relaxation of depletion region due to generation of hole–electron pairs in the depletion region of MOS capacitor. (After Heiman.[31] © 1967 IEEE.)

thus decays as the inversion layer forms until equilibrium is reached. This phenomenon for *p*-type silicon is schematically shown in Fig. 6.8.[31] Thus the MOS capacitor lifetime measurement method can be explained by the following procedure: (1) the gate bias is pulsed to a new value, (2) the depletion layer width expands, and then (3) the decay of depletion width toward its equilibrium value is measured.[32] Accordingly, this method consists of measuring either the MOS capacitance or the external current as a function of time. The procedure used practically to determine lifetime, which was developed by Heiman,[31] uses the recording of the normalized MOS capacitance versus time after the application of a large depleting voltage step, for example, positive for *p*-type semiconductor. Consequently, the minority carrier generation lifetime is calculated by Eq. (6.19) and the following equation:

$$\ln[(C_f/C - 1)/(C_f/C_0 - 1)] + C_f/C - C_f/C_0$$
$$= -(C_f/C_{ox})(t/t_r) \tag{6.20}$$

where C_0 is the initial capacitance, C_f final capacitance, C transient capacitance, C_{ox} oxide capacitance, and t time. In practice, the lifetime is extracted from the slope of the C/C_{ox} versus t curve by writing

$$dC/dt = C^2/t_r[1 - (C/C_f)] \tag{6.21}$$

According to Eq. (6.21), measurement of the slope at any value of C will yield the time constant t_r, which is related to τ_g through Eq. (6.19). It has also been shown by Zerbst[33] that a plot of $-d(C_{ox}/C)^2/dt$ versus $(C_f/C - 1)$, as in Fig. 6.9, yields a straight line whose slope is related to τ_g, and the intercept of this line is proportional to the surface recombination velocity **s**.

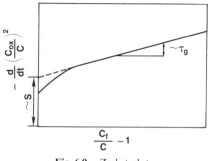

Fig. 6.9. Zerbst plot.

6.2 Chemical Characterization

6.2.1 Impurities in Silicon

The chemical properties that are of most exclusive concern are related to adventitiously induced impurities in processed silicon wafers as well as impurities incorporated into the starting silicon. Many of the impurities tend to be introduced at high temperatures where solubilities are high. On cooling or heat treatment at lower temperatures, precipitation occurs because of decreased solubility. It is very important, as already discussed repeatedly, to characterize silicon wafers by measuring impurity precipitation as well as the concentration. Numerous kinds of diagnostic techniques have been utilized for the measurement of impurities in silicon.[1-3] In this section, some techniques that are most commonly used and most useful for silicon characterization are demonstrated.

6.2.2 Radioactivation Analysis

Principle and Characteristics Radioactivation analysis[34] is a method considered for the analysis of trace impurities in semiconductor materials in which impurity concentrations in the parts per million to parts per billion range are becoming more and more important. If an analytical sample is exposed to the effect of neutrons or electrically charged particles of suitable energy, nuclear reactions occur in atoms. The products are usually radioactive and the amount is related to the amount of the desired element present in the sample. Activation analysis thus consists of two major processes: (1) the production of a radioactive nuclide from the desired element by some nuclear reaction, and (2) the measurement of the amount of the product. Typically, a relative method is taken for the analysis. This involves activating a standard containing a known amount of the element m_s, as well as the sample containing an unknown amount of the element m_x. The fundamental relation between the activities formed by activation of the given element in the sample and in the standard, A_x and A_s, respectively, is given by

$$A_x/A_s = m_x/m_s \qquad (6.22)$$

The activity A of nuclide i is commonly given by

$$A_i = \Phi\sigma_i N_i[1 - \exp(-\lambda_i t)] \qquad (6.23)$$

where Φ is the irradiation particle flux, σ_i the cross section for the reaction producing the nuclide i, N_i the number of target nuclei in the sample available for the reaction producing the nuclide i, λ_i the decay constant for the nuclide i, and t the duration of the irradiation.[34] In the course of activation, nuclides are formed not only from the element to be determined but also from most of the other elements present in the sample to be analyzed. The main problem is thus to distinguish radiation of the given nuclide from that of other nuclides. This differentiation can be carried out by either chemical or physical means or by some combination of the two.

Irradiation Sources   The choice of an irradiation source implies the choice of the nuclear reaction for producing the desired nuclide. The irradiation sources most commonly used for the analysis of trace impurities in silicon are neutrons and charged particles. Neutrons produced by nuclear reactions have in common the ability to be thermalized, in which energy state they have large probabilities for capture by most elements to form measurable radio-active nuclides as a result of the (n, γ) nuclear reaction. On the other hand, charged particles are very useful for light elements that do not yield convenient neutron reactions or that require neutron energies not attainable in nuclear reactions or neutron generators.

Neutron Activation Analysis   Neutron activation analysis (NAA) meets most of the requirements for the study of trace impurities in silicon.[35,36] Gamma-ray (γ-ray) spectroscopy, in which the (n, γ) reaction is used, has a very high sensitivity for most elements of the periodic table beyond neon. Exceptions are light elements (i.e., B, P, and S) and heavy elements (i.e., Tl, Pb, and Bi) that do not become γ-ray emitters but only beta (β) emitters. Therefore, the analysis of P, one of the common dopants in silicon, has been performed with β-ray spectroscopy. As for B, another common dopant in silicon, the charged-particle activation analysis (CPAA) discussed next has been used. By means of NAA, a systematic survey on the transition-metal group impurities, which reside in starting silicon and are introduced during the device manufacturing processes, has been conducted by Schmidt and Pearce.[37]

Table 6.2 Detection Limits for CPAA[a]

Detection limit	Elements
< 1 ppba	B,[b] C,[b] N,[b] O,[b] Ca, Y, Pr, Gd
1–10 ppba	Li, Ti, Cr, Co, Ni, Cu, Zn, Ga, Ge, Br, Rb, Zr, Ru, Cd, Sn, Te, Ba, Hf, Os, Ir, Pt, Tl, Pb, Bi, La, Ce, Eu, Ho, Er, Yb, Lu
10–100 ppba	H, He, Na, Mg, Al, Si, P, S, V, Fe, As, Se, Sr, Nb, Rh, Pd, Ag, Sb, Cs, Au, Hg
0.1–1 ppma	Cl, Mn, In, Re
1–10 ppma	K, Ta

[a] After Nakajima and Kohara.[40]

[b] The detection limits for these elements in silicon have been reported as: B, 10 ppba[35]; C, 20 ppba[41]; N, 2 ppba[42]; and O, 10 ppba.[42]

Charged-Particle Activation Analysis Charged-particle activation analysis (CPAA) is well suited to determining light elements (e.g., B, C, O, N) at the parts per billion level in silicon.[38,39] The particles used in CPAA include protons ($^1H^+$), deutrons ($^2H^+$), tritrons ($^3H^+$), helium-3 ions ($^3He^+$), and helium-4 ions or alpha particles. ($^4He^+$ or α). A major feature of CPAA is that different nuclides can be obtained from a given target isotope, depending on the nature and energy of the irradiating particle. The nuclear reactions commonly adopted for the analysis of B, C, N, and O are $^{11}B(p, n)^{11}C$, $^{12}C(^3He, ^4He)^{11}C$, $^{14}N(p, ^4He)^{11}C$, and $(^3He, p)^{18}F$, respectively.[38,40]

Although the major thrust of CPAA remains in the field of light-element analysis, highly sensitive CPAA procedures have been devised for most stable elements. The detection limits reported for 72 elements are given in Table 6.2.[40–42]

6.2.3 Infrared Spectroscopy

Principle and Characteristics Atoms or atomic groups in molecules are in continuous motion with respect to each other. The possible vibrational modes in a polyatomic molecule, acetaldehyde for example, can be visualized from mechanical models of the system as illustrated in Fig. 6.10.[43] Atomic masses and bonding forces of the chemical links are represented by balls and by springs that connect and keep the balls in positions of balance. Their motion can be regarded as being composed of two components, the stretching and bending vibrations. If the vibrating mode is observed with a stroboscopic light of variable frequency, certain light frequencies will find the balls appear to remain stationary. These represent the specific vibrational frequencies.[43] The frequencies depend not only on the nature of the particular bonds themselves, but also on the entire molecule and its environment.

As shown in Fig. 3.13, the infrared (IR) covers the region from 0.7 to 500 μm in wavelengths or from 14,000 to 20 cm^{-1} in wave numbers. The

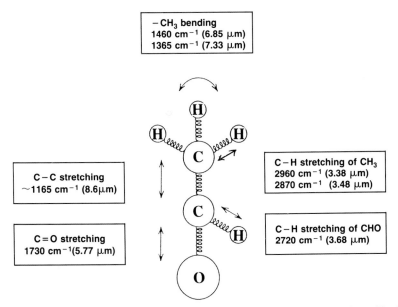

Fig. 6.10. Vibrational models and characteristic frequencies of acetaldehyde. (After Willard *et al.*[43] Reproduced with the permission of Wadsworth Publishing Company.)

range of special interest is between 400 and 2000 cm^{-1}, which covers almost all absorption due to impurities of interest in silicon. When the sample is irradiated by an IR beam whose frequency is changed continuously, the molecule will absorb certain frequencies as the energy is consumed in stretching or bending different bonds. The transmitted beam corresponding to the region of absorption will naturally be weakened, and thus recording of the intensity of the transmitted beam versus wave numbers or wavelength will give spectra that show characteristic absorption bands. The absorption bands are a direct reflection of the state of molecular bonds, and thus this aspect leads to the qualitative analysis. Moreover, the absorption intensity is proportional to the amount of the specific element present in the sample, and this leads to quantitative analysis by converting into concentration using the calibration curve obtained with another direct measuring technique such as CPAA. Infrared spectroscopy has been used for the analysis of various impurities such as dopants,[44] oxygen, carbon, and nitrogen in silicon. The great advantage of IR spectroscopy is that both qualitative and quantitative analyses can be performed nondestructively, quickly, and less expensively. Analytical sensitivity strongly depends on sample thickness. For silicon, routine analyses can be performed for common silicon wafers of several hundred micrometers to 1 mm in thickness; however, thicker samples (e.g., 2–10 mm) are recommended for precise analyses. Moreover, IR spectroscopic

Table 6.3 Detection Limits of Impurities for IR Spectrophotometry at ~ 10 K and 300 K[a]

Element	Limits (atoms/cm³)	
	~ 10 K	300 K
Oxygen	3×10^{13}	2.5×10^{15}
Carbon	—	5×10^{15}
Nitrogen	—	2×10^{15}
Boron	3.1×10^{11}	—
Aluminum	2.1×10^{12}	—
Gallium	2.7×10^{12}	—
Indium	1.6×10^{13}	—
Phosphorus	1.5×10^{11}	—
Arsenic	2.7×10^{11}	—
Antimony	3.1×10^{11}	—

[a] After Keenan and Larrabee,[6] Pajot,[45] ASTM,[46] and Ito et al.[47]

analysis at low temperatures results in higher sensitivity for impurities in a sample, since the ionized carriers in silicon begin to "freeze out"—that is, the absorption due the free carriers becomes negligibly small—at low temperatures.[44] Table 6.3 lists the detection limits of impurities in silicon for IR spectroscopy at ~ 10 K and 300 K.[6,45–47] It should be noted that IR spectroscopy cannot be used for heavily doped ($> 5 \times 10^{17}$ atoms/cm³ or $< 0.1\ \Omega$ cm) n^+ and p^+ silicon substrates due to free carrier absorption interferences,[48] unless the sample is irradiated with electrons that trap the free carriers.[49]

Spectrophotometers   Most IR spectrophotometers are double-beam instruments in which two equivalent beams of radiant energy are taken from the source.[43] The source beam strikes alternately the reference and sample paths by means of a combined rotating mirror and light interrupter. In order to disperse the radiation and focus successive wavelengths on a detector, a conventional spectrophotometer uses a prism or grating.

An alternative spectrophotometer is a Fourier-transform IR (FTIR) spectrophotometer,[6,43] which has been widely used in the silicon industry. An FTIR spectrophotometer basically consists of two parts: (1) an optical system that uses an interferometer, and (2) a dedicated computer that controls optical components, collects and stores data, performs computations on data, and displays spectra. Although some negative aspects of FTIR spectroscopy have been pointed out,[50] FTIR generally offers potential advantages compared with conventional dispersion IR spectroscopy: (1)

higher signal-to-noise (SN) ratios for spectra obtained under conditions of equal measurement time, and (2) higher accuracy for spectra taken over a wide range of frequencies. The most important advantage of FTIR spectrophotometer results from the use of an interferometer instead of a prism or grating, which requires energy-wasting slits. As regards the detection limit and accuracy of impurity analysis, at least for oxygen in silicon, no essential difference has been found between a dispersive-type IR and an FTIR spectrophotometer.[51]

Oxygen Analysis Since oxygen was identified by IR absorption in CZ silicon crystals in 1956,[52] IR spectroscopy has been a routine technique to determine the oxygen concentration in silicon crystals.[45] Moreover, IR spectroscopy has been commonly used for the characterization of oxygen precipitation in silicon crystals.[53–56] An interstitial oxygen atom that bonds with two adjacent silicon atoms (Fig. 5.35) gives rise to three basic vibration modes, as shown in Fig. 6.11.[52,57] With Si–O distances of 1.6 Å and assuming the Si–Si distance to be essentially unchanged (2.34 Å), the bond angle Si–O–Si will be approximately 100° for a hypothetical nonlinear Si_2O molecule.[52] The characteristic frequencies of the three basic absorptions v_{01}, v_{02}, and v_{03} at room temperature are 1205 cm^{-1}, 515 cm^{-1}, and 1106 cm^{-1}, respectively.[57] Figure 6.12a shows IR absorption spectra obtained at room temperature for as-grown CZ silicon that is commercially available.[58] Absorption peaks due to interstitial oxygen are clearly observed at 515 cm^{-1} (v_{02}) and 1106 cm^{-1} (v_{03}); however, v_{01} absorption is too weak to be

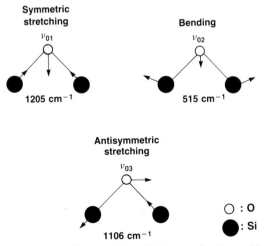

Fig. 6.11. Three basic vibrational modes of interstitial oxygen in silicon. (After Hrostowski and Kaiser.[57])

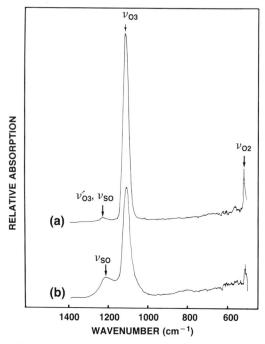

Fig. 6.12. Infrared absorption spectra for CZ silicon wafer: (a) as grown and (b) after heat treatment at 1000°C for 64 h. (After Shimura *et al.*[58])

observed at room temperature. The relative intensity of v_{02} and absorption to that of v_{03} is about 0.25. As for the weak absorption observed around 1225 cm^{-1}, it has been attributed to grown-in SiO_x precipitates, (i.e., v_{SO}),[54] which have absorption at 1224 cm^{-1},[53] and/or to the v'_{03} subpeak, which is expected at 1227 cm^{-1}.[59]

As mentioned briefly in the footnote on page 178, interstitial oxygen concentration $[O_i]$ is calculated using the absorption coefficient of v_{03} at 1106 cm^{-1} (i.e., $\lambda = 9.04$ μm) as follows:

$$[O_i] = f_c\alpha_{03} \quad (\times 10^{17} \text{ atoms/cm}^3)$$
$$= 2f_c\alpha_{03} \quad (\text{ppma}) \tag{6.24}$$

where f_c is the conversion factor and α_{03} is given by

$$\alpha_{03} = \ln(I_0/I)/x \tag{6.25}$$

where I is transmitted intensity at the absorption peak, I_0 the baseline intensity, and x the specimen thickness (cm). The detailed procedure for $[O_i]$ measurement is given in Ref. 46. The trouble here is that three different conversion factors are in common use in the silicon industry today for obtaining $[O_i]$ by IR absorption measurement at room temperature. The factor $f_c = 2.45$ is commonly referred to as "DIN"[60] or "New ASTM,"[46] while $f_c = 4.81$ is "Old ASTM."[61] In addition, $f_c = 3.03$, which is commonly referred to as "JEIDA,"[51] has been recently reported. Because of the diversity of these values, it is necessary to identify the conversion factor used when $[O_i]$ in silicon is discussed. In this book, the conversion factor $f_c = 2.45$ is used. However, it should be noted that JEIDA value ($f_c = 3.03$) is expected to be universally accepted in the near future.[62] Heat treatment leads to the precipitation of supersaturated interstitial oxygen in CZ silicon. For example, Fig. 6.12b shows the IR absorption spectra of CZ silicon subjected to a heat treatment at 1000°C for 64 hr.[58] Comparing the IR spectra for the as-grown sample shown in Fig. 6.12a, one clearly observes that the absorption v_{03} due to interstitial oxygen decreases, and inversely the absorption v_{SO} due to oxygen precipitates (SiO_x) increases. If the initial interstitial oxygen concentration is denoted as $[O_i]_0$ and the concentration after a heat treatment as $[O_i]_f$, the precipitated oxygen concentration $\Delta[O_i]$ is simply given as the difference:

$$\Delta[O_i] = [O_i]_0 - [O_i]_f$$

The size of measurement area, namely an IR beam size, depends on the respective spectrometer. Typical sizes are 5 mm (width) × 15 mm (height) and 15 mm in diameter for the most commercial IR spectrometers.[51] In order to obtain the microprofiles of oxygen concentration or oxygen precipitation, a scanning IR absorption technique using a collimated laser beam, for example, a double-heterostructure $PbTe/Pb_{0.82}Sn_{1.82}Te$ laser ($\lambda = 9.04\ \mu m$, spatial resolution $\approx 200\ \mu m$ diameter),[63] and a CO_2 laser ($\lambda = 9.17\ \mu m$, spatial resolution $\approx 30\ \mu m$ diameter)[64] has been developed. By applying this technique, oxygen striations in CZ silicon crystals have been observed and correlated with thermally induced microdefects.[65]

It should be noted that the absorption v_{03} involves only interstitial oxygen, but not all oxygen, although almost all oxygen atoms occupy interstitial sites in silicon.[66] CPAA, instead of IR spectroscopy, is adequate for the measurement of the absolute concentration of oxygen in silicon.

Carbon Analysis Substitutional carbon in silicon gives rise to the IR absorption at 607 cm^{-1}, v_c, and this peak has been used for the determination of carbon concentration $[C_s]$ in silicon crystals.[67] As shown in Fig. 6.12a, the absorption due to carbon is usually not observed in common CZ silicon crystals, which in turn implies that the concentration of carbon in common CZ silicon is less than 5×10^{15} atoms/cm^3. Figure 6.13 shows the IR absorption spectra for a CZ silicon crystal in which carbon is intentionally

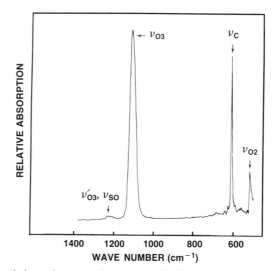

Fig. 6.13. Infrared absorption specta for as-grown CZ silicon doped with carbon at ∼6 ppma. (After Shimura *et al.*[58])

doped.[58] The concentration of substitutional carbon $[C_s]$ is calculated with the following equation according to the standard procedure given in Ref. 68:

$$[C_s] = \alpha_c \quad (\times 10^{17} \text{ atoms/cm}^3)$$

$$= 2\alpha_c \quad \text{(ppma)} \tag{6.26}$$

where α_c is the absorption coefficient for ν_c at 607 cm^{-1}. In contrast to the conversion factor for interstitial oxygen, the factor of unity has been universally used in the silicon industry since 1965. However, detailed analyses have recently been performed for the concentration of carbon in silicon by means of both IR and CPAA techniques, and the following equation has been proposed[69]:

$$[C_s] = (0.85 \pm 0.09)\alpha_c \times 10^{17} \quad \text{atoms/cm}^3 \tag{6.27}$$

A detailed examination of the calibration of carbon concentration is required in order to meet an increasing interest in carbon in silicon crystals. Again, it should be noted that the absorption ν_c involves only substitutional carbon, but not all carbon, in silicon.

6.2.4   Secondary Ion Mass Spectroscopy

Principle and Characteristics   When an energetic primary ion beam with an energy typically between 5 keV and 15 keV impinges on the surface of a sample, the material is sputtered off in the form of atoms and ions, and ejected from the surface. A small fraction of these atoms is ionized either positively or negatively, and the sputtered secondary ions are mass analyzed with a magnetic sector or quadrupole analyzer.[70] The secondary ion signal can be recorded in four ways[71,72]: (1) by a value for the ion current of the emitted ion selected, (2) by an image displayed on a fluorescent screen in the form of the distribution of an element on the sample surface, (3) by an image formed on an electron-sensitive photographic film, and (4) by an image on a resistive anode encoder (RAE). The mass analyzed is directly related to species, and the intensity of the detected signal is proportional to the concentration of the element in the sample.

The diameter of an analyzed surface is usually a few hundred micrometers, but information may be obtained from a surface of $1-25$ μm in diameter by using a focusing system.[73] The sputtering rate for silicon is usually a few angstroms per second, but can be as high as 1000 Å per second for deep depth profiles. Thus, the secondary ion mass spectroscopy (SIMS), which is alternatively called *ion microprobe mass analysis* (IMMA or IMA), leads to a localized qualitative and quantitative analysis for impurities contained in a small selected volume of the sample surface. This aspect characteristically distinguishes SIMS from the other above-mentioned analytical techniques, which offer the bulk concentrations of impurities in a sample. This technique thus provides in-depth information on atomic constituents by monitoring one or more secondary ion signals with time as the sputtering process removes the surface atoms of a sample. From the electronic device point of view, this depth-profiling aspect has become the most important use of SIMS.[74]

Moreover, SIMS has two major capabilities that make it highly useful for the characterization of semiconductor materials used for electronic device fabrication: (1) it is capable of detecting all the elements including H and He, and (2) it can quantify these elements at parts per million atomic to parts per billion atomic orders. Table 6.4 lists the SIMS bulk detection limits of "order of magnitude" in silicon.[75] These are the optimum detection limits obtained by state-of-the-art SIMS technology, and apply to bulk measurements. If SIMS is used to profile or image an element, the detection limit is expected to be worse. It should be noted also that the detection limit strongly depends on the SIMS system used, particularly on the species of primary ion beams and sputtering rate, and because of the complexity of SIMS, process standards are required to apply SIMS to a problem requiring quantitative analysis.

Depth Profiling of Impurities For the semiconductor device technology, it is very important to measure impurity distribution, particularly the distribution of dopants, in the device active regions. Moreover, the analysis of the diffusive and reactive processes between metal, dielectric films, and substrates

Table 6.4 Bulk Detection Limits in Silicon by SIMS[a]

Element	Detection limit (atoms/cm^3)	Element	Detection limit (atoms/cm^3)
H	10^{17}	Mn	10^{14}
Li	10^{13}	Fe	10^{15}
B	mid 10^{12}	Co	10^{15}
C	mid 10^{15}	Ni	10^{15}
N	mid 10^{14}	Cu	10^{16}
O	low to mid 10^{16}	Zn	10^{16}
F	10^{14}	Ga	10^{14}
Na	10^{14}	As	10^{15}
Mg	10^{14}	Mo	10^{15}
Al	10^{15}	Ag	10^{14} (?)
P	10^{15}	Sn	10^{15}
Cl	10^{14-15}	Sb	10^{15}
K	10^{14}	Ta	10^{15} (?)
Ca	10^{15}	W	10^{15} (?)
Ti	10^{14}	Au	$< 10^{14}$
V	10^{14} (?)	Pb	10^{16}
Cr	10^{14}		

[a] Courtesy of P. Chu and R. S. Hockett, Charles Evans & Associates.[75]

is indispensable to the design of various semiconductor devices. The analytical requirements for depth profiling include good spatial resolution as well as elemental sensitivity and specificity.[76] It has been widely accepted that SIMS is an analytical technique well suited to impurity depth profiling in semiconductor materials because of its high sensitivity and good depth and lateral resolution.[77] Although the lateral resolution and depth resolution depend on many factors, SIMS may achieve about 0.1 μm and 10 Å, respectively, at optimum operational conditions.[3] Figure 6.14 shows a depth profile of the boron concentration in a silicon wafer that was implanted with B$^+$ ions at 10^{15} ions/cm^{-2} and then annealed at 950°C for 40 min.[78] The SIMS profile is compared with the charge-carrier distribution measured with Hall-effect measurements combined with later stripping.

Gettering efficiency (see Section 7.4) can be monitored chemically by means of SIMS. Figure 6.15 shows the SIMS concentration profile of Ge and Au that was deposited on the surface of an Si-epi/Ge-doped Si-epi/Si-substrate structure.[79] The Ge-doped silicon epitaxial layer on the silicon substrate was deposited in order to induce misfit dislocations as gettering sites. The sample was heat-treated at 900°C for 15 min after a 20 Å-thick Au deposition on the surface. At the 6-μm-deep interface with low misfit

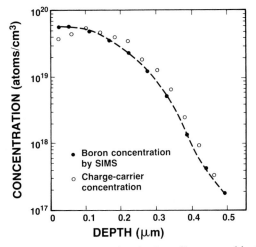

Fig. 6.14. Comparison of boron concentration depth profile measured by SIMS with charge-carrier distribution of annealed silicon implanted with B^+ (10^{15} ions/cm^2, 25 keV). After Hofker.[78])

dislocation density, a small Au peak is observed, whereas a strong peak exists at 9 μm where the misfit dislocation density is largest. These results clearly demonstrate the expected gettering by the misfit dislocations, which were intentionally induced at the interface of the substrate and epitaxial layer by doping Ge in the epitaxial layer.

Oxygen and Carbon Characterization   The outdiffusion and precipitation of oxygen in CZ silicon wafers during device fabrication processing can dramatically affect the device performance. SIMS has been used to characterize the diffusion[80,81] and precipitation[82–85] of oxygen in silicon crystals after thermal processing. SIMS oxygen concentration depth profiles versus the precipitated oxygen $\Delta[O_i]$ measured by FTIR after thermal processes are shown in Fig. 6.16.[85] In addition to the oxygen outdiffusion profiles, fluctuations or spikes in the oxygen concentration that are correlated to oxygen precipitates are clearly observed.

As discussed in Section 7.2.4, carbon enhances oxygen precipitation in silicon crystals over a wide range of temperature.[86] Direct evidence for the coaggregation of carbon and oxygen in CZ silicon heat-treated at 1000°C for 64 hr has been provided by measuring SIMS depth profiles and SIMS two-dimensional images of these impurities.[83] Figure 6.17 shows the carbon and oxygen SIMS profiles taken with a reduced sputtering rate to increase spatial sensitivity to the carbon fluctuations. Although the fluctuations in the carbon profile are less pronounced than those in the oxygen profile, a definitive

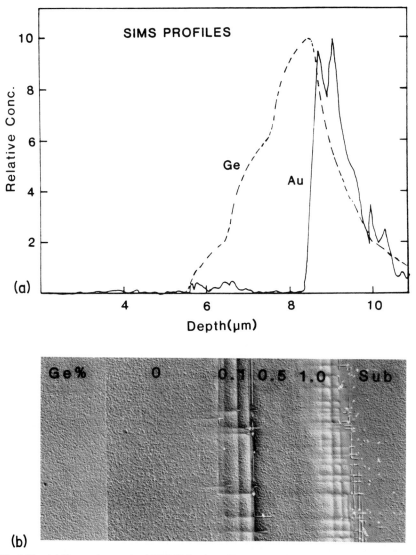

Fig. 6.15. (a) Germanium and gold SIMS depth profiles and (b) chemically etched cross section of Si-epi/Ge-doped Si-epi/Si-substrate structure. (After Salih *et al.*[79])

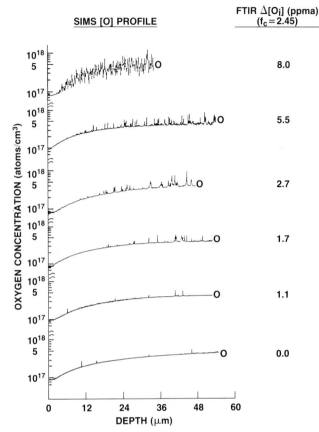

Fig. 6.16. SIMS depth profiles of oxygen in thermally processed CZ silicon versus the precipitated oxygen concentration $[O_i]$ measured by FTIR. (After Hockett *et al.*[85])

correlation in the carbon and oxygen fluctuations can be seen at the depths denoted with arrows. The two-dimensional SIMS resistive anode encoder (RAE) images[72] of oxygen and carbon are shown in Fig 6.18a and b, respectively. The carbon image was obtained from the region about 100 Å deeper than that for the oxygen image. A schematic map of superimposed images from Fig. 6.18a and b is shown in Fig. 6.19. It is evident that the eight carbon image spots of intensity above the background noise coincide exactly with some oxygen image spots. Accordingly, the SIMS results shown in Figs. 6.17–6.19 clearly imply direct evidence for the coaggregation of oxygen and carbon in a heat-treated CZ silicon wafer.

As described above, SIMS is an extremely powerful tool to analyze impurity behavior in the region close to wafer surfaces, which are very

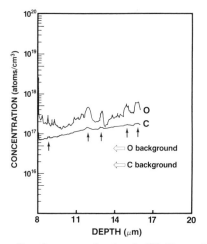

Fig. 6.17. SIMS depth profiles of oxygen and carbon in CZ silicon subjected to heat treatment at 1000°C for 64 h. (After Shimura et al.[83])

important from the device operation point of view. Moreover, SIMS has become a very versatile analytical technique, as mentioned in Section 6.2.3, for oxygen in heavily doped ($>5 \times 10^{17}$ atoms/cm^3 or <0.1 Ω cm) n^+ and p^+ silicon substrates for which IR spectroscopy cannot be used to measure the oxygen concentration. Figure 6.20 shows the SIMS depth profiles of oxygen and antimony for a heavily Sb-doped CZ silicon (0.02 Ω cm) that was subjected to three different heat treatments.[82] Oxygen outdiffusion and oxygen precipitation (heat treatment C) are observed, while antimony outdiffusion is not observed or is negligible because of its low diffusivity.

Surface Cleanliness Analysis   The surface cleanliness requirements for silicon substrates have become increasingly more stringent as the size of the active elements has decreased. Of particular concern in the device manufacturing process is the introduction of contamination into the surface of semiconductor substrate. The surface has active dangling bonds, and has the potential to be contaminated at anytime. The levels of contamination affecting electrical characteristics have been found to reach the order of parts per billion atomic concentrations.[87] The traditional surface analysis techniques such as Auger electron spectroscopy (AES) and electron spectroscopy for chemical analysis (ESCA) no longer have sufficient sensitivity to detect and quantitate the surface contamination levels.[88] By using SIMS for silicon wafers intentionally contaminated and cleaned with several different procedures, it has been found that all of the cleaning procedures removed some of each contaminant; however, there are large differences in the amounts removed.[88] Focusing on aluminum contamination, commercial CZ silicon

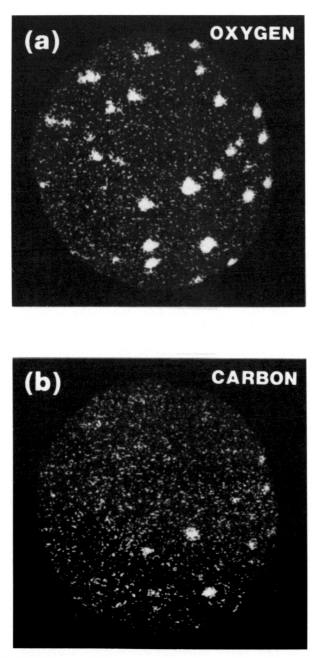

Fig. 6.18.   SIMS-RAE images of (a) oxygen and (b) carbon for 150-μm-diameter field of view in CZ silicon subjected to heat treatment at 1000°C for 64 h. (After Shimura *et al.*[83])

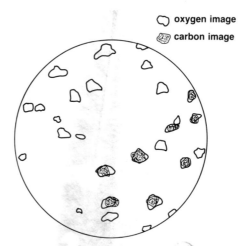

Fig. 6.19. Schematic map of superimposed images of oxygen and carbon obtained from Fig. 6.18 (After Shimura *et al.*[83])

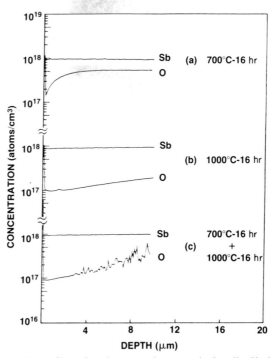

Fig. 6.20. SIMS depth profiles of antimony and oxygen in heavily Sb-doped CZ silicon subjected to different heat treatments. (After Shimura *et al.*[82] Reproduced with the permission of The Electrochemical Society.)

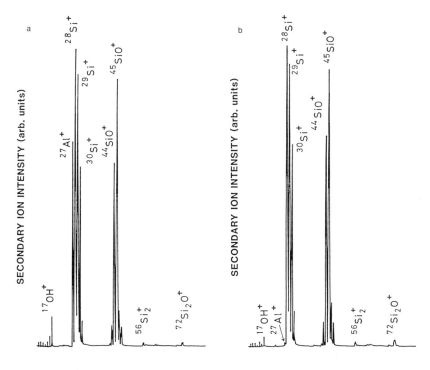

Fig. 6.21. SIMS mass spectra for (a) an "as-received" silicon wafer and (b) after chemical etching with HF–HCl–H$_2$O solution for 15 sec. (After Kawado *et al.*[89] Reprinted with permission of The Electrochemical Society, Inc. Courtesy of S. Kawado, Sony Corporation.)

wafers have also been analyzed by SIMS.[89] Figure 6.21a shows an example of the SIMS mass spectra obtained for an as-received silicon wafer. A strong aluminum peak is observed, together with several peaks originating from the substrate silicon. As shown in Fig. 6.21b, a remarkable reduction of aluminum intensity is observed on the wafer surface that was chemically etched with an HF–HCl–H$_2$O solution for 15 sec. From the SIMS depth profile of aluminum, the thickness of Al-contaminated layer was estimated to be 30–50 Å.

6.3 Physical Characterization

6.3.1 Defects and Diagnostic Techniques

Physical characterization of semiconductor crystals involves the analysis of crystallographic defects such as shown in Table 3.8 and Fig. 3.27. Basically, physical defect characterization measures the deviation from perfection of the

Table 6.5   Comparison of Techniques Commonly Used to Study Crystal Defects in Silicon[a]

	Chemical etching	XRT	TEM
Principle	Reaction of solution	Diffraction	Diffraction
Sample preparation	◎	◎	●
Nondestructive observation	◐	◎	●
Specimen diameter	◎	◎	●
Specimen thickness	◎	◎	●
Defect size	○	◐	◎
Density measurement	○	●	○
Defect mapping	○	◎	●
In-depth defect profile	○	◐	○
Defect identification	◐	○	○
Defect structural analysis	●	○	○
Defect composition	●	●	○
Ease and cost	◎	◐	●
Remarks	Most convenient for defect distribution observation	Very sensitive to lattice distortion; truely nondestructive	Only technique to observe microdefects with analytical methods

[a] ◎, excellent; ○, good; ◐, acceptable; ●, poor.

crystal lattice by means of direct or indirect observation methods. Defects in silicon have been characterized by various techniques. This section reviews the defect observation techniques most commonly used for the characterization of silicon crystals: (1) preferential chemical etching, (2) X-ray diffractometry, which consists of X-ray topography (XRT) and X-ray goniometry such as X-ray rocking curve (XRC), and (3) transmission electron microscopy (TEM). When discussing analytical techniques, it is appropriate to judge both the quality of the information obtained, and the ease and cost of obtaining results. Table 6.5 compares those techniques commonly used to study crystal defects in silicon. It is clear that the successful characterization of silicon crystals can only be achieved by complementary use of several existing techniques.

6.3.2   Chemical Etching

Principle and Characteristics   The rate of reaction of a solution, the *etchant*, with a solid surface depends distinctly on the crystallographic orientation. Moreover, the rate is significantly affected by local stress caused by defects. In regions near dislocations or any other physical defects, chemical etching proceeds more rapidly compared with perfect regions. As a result, etch pits that often have crystallographic symmetry are formed on the surface of sample. Although not as common as etch pits, etch hillocks may also form and can often be confused with pits in the optical microscope. Hillocks are normally caused by precipitates whose etch rate is smaller than the matrix crystal, or by small undissolved oxide particles that mask the surface.

Etch pits are commonly observed with a Nomarski microscope, which gives rise to the differential interference contrast due to etched features on the sample surface.[90] This method, in conjunction with preferential chemical etching and optical microscopy, leads to the easiest and least expensive technique for the analysis of crystal defects in silicon crystals. This technique allows for excellent definition of the defect density and distribution of defects either across the polished surface of the wafer or vertically along a cleaved face perpendicular to the wafer surface. However, one of the major limitations of chemical etching technique is that the technology is similar to archeology: the object of interest has already been consumed before the optical analysis is performed. Therefore, it should be noted that preferential chemical etching is a destructive process and that those etch pits observed are the cast-offs, but not the substance, of defects.

Etching Solutions   Etching solutions or etchants for semiconductor materials can be classified into three categories according to the purpose of etching: (1) delineation of defects (preferential or selective etching), (2) polishing, and (3) staining. The action of both preferential and polishing etchants in removing material is essentially the same, and those based on acid agents usually consist of three components. One component oxidizes the material, while another dissolves the oxide into the etching solution. A third, such as water and acetic acid, is normally added as a diluent, which controls the etching rate. A number of preferential etchants have been reported for the delineation of defects in silicon crystals.[3] Table 6.6[91-93] summarizes those etchants most commonly used for silicon.

Defect Observation   The following demonstrates defects delineated by preferential chemical etching in conjunction with an optical microscope or a scanning electron microscope (SEM) for CZ silicon wafers. Figure 6.22a, b, and c shows Wright-etched figures that correspond to oxidation-induced stacking faults (OSFs) (see Section 7.2.2) on the surface of (100), (111), and (511) wafers, respectively.* As schematically illustrated in Figs. 3.46 and 3.47, stacking faults are two-dimensional defects that lie on $\{111\}$ planes, and they intersect the surface of (100) and (111) wafers along the $\langle 110 \rangle$ direction. As a consequence, referring to Table 3.2, in a (100) wafer, the faults lie on $\{111\}$ planes inclined at angles of 54.74° to the surface and result in linear etch pits due to stacking faults along orthogonal $\langle 110 \rangle$ directions, since the $\{111\}$

---

* All the waters were oxidized at 1100°C for 80 min in wet O_2; however, one may find the difference in the length of OSFs (see Section 7.2.2).

Table 6.6 Preferential Etching Solutions for Delineation of Defects in Silicon

Etchant	Formula	Etching rate[a]	Comment
Sirtl and Adler[91]	HF, 100 cm³	~ 3.5 μm/min	Best applicable to (111)
	CrO₃ (5 M) 100 cm³		
	H₂O, 100 cc		
Secco d'Aragona[92]	HF, 100 cm³	~ 1.5 μm/min	Best applicable to (100)
	K₂Cr₂O₇ (0.15 M), 50 cm³		Ultrasonic agitation recommended
Wright[93] (Jenkins)	HF, 30 cm³	~ 1.0 μm/min	Widely applicable for all
	HNO₃, 15 cm³		
	CrO₃ (5 M), 15 cm³		
	Cu(NO₃)₂, 1 g		
	H₂O, 30 cm³		
	CH₃COOH, 30 cm³		

[a] Etching rate is strongly affected by several factors, such as the freshness and temperature of the etchant and the agitating conditions.

planes intersecting the (100) surface are orthogonal to each other. On the other hand, in a (111) wafer, the fault planes lie at angles of 70.53° to the surface, and the ⟨110⟩ directions are at 60° to each other. Consequently, linear etch pits along three different ⟨110⟩ directions appear on the (111) surface.

Figure 6.23 compares the etched figures delineated by Secoo and Wright etching for the matched (110) cleaved faces of a (100) CZ silicon wafer subjected to a two-step heat treatment (i.e., 720°C for 3 hr + 1020°C for

(a)   (b)   (c)   10μm

Fig. 6.22. Etched figures corresponding to oxidation-induced stacking faults: (a) (100) silicon wafer, (b) (111) silicon wafer, and (c) (511) silicon wafer. All the wafers were oxidized at 1100°C for 80 min in wet O₂. (Courtesy of A. Yamaguchi, Monsanto Japan, Limited.)

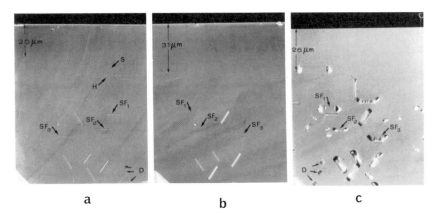

Fig. 6.23. Optical micrographs for matched cleaved faces of (100) CZ silicon wafer subjected to two-step heat treatment (720°C/3 h + 1020°C/60 h): (a) 15-sec Secco etching, (b) 22.5-sec Wright etching, and (c) 5-min Wright etching. (Courtesy of M. K. El Ghor.[94])

60 hr).[94] In Fig. 6.23a the cleaved face was etched with Secco etchant for 15 sec, while the matched face in Fig. 6.23b was etched with Wright etchant for 22.5 sec, and Fig. 6.23c shows the same area as shown in Fig. 6.23b but further etched with Wright etchant for 5 min. For the matched faces shown in Fig. 6.23a and b, the etching time was adjusted to remove the same amount of material from each half, since the two solutions have different etching rates as shown in Table 6.6. The Secco etching shows a much higher sensitivity to microdefects by showing up more small pits (S) and hillocks (H) compared with the Wright etching, which produces a much smoother background free of any traces of microdefects all over the sample face. The one-to-one correspondence still prevails between stacking faults SF_1, SF_2, and SF_3, in addition to the three dislocations (D) at the lower right corner in Fig. 6.23a and better shown in Fig. 6.23c, after a longer etching time in Wright etchant. The comparison shown in Fig. 6.23 implies that such delineation is crucial in determining the actual denuded zone of the sample: Fig. 6.23a gives 20 μm, while Fig. 6.23b gives 31 μm.

Figure 6.24 shows dense saucer pits (S-pits), which are commonly revealed by preferential etching for the surface of a contaminated silicon wafer subjected to a heat treatment at temperatures higher than 1100°C.[95] These surface microdefects that originate S-pits have been characterized by electron microscopy to be transition-metal clusters or tiny stacking faults generated by those clusters.[95,96]

As shown in Fig. 6.23, in-depth observation of thermally induced defects is easily performed by preferential etching for the cleaved face of a silicon wafer. When the two-dimensional distribution of interior defects in the plane

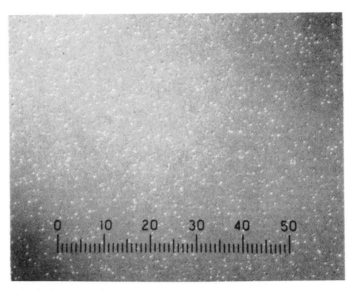

Fig. 6.24. Optical micrograph of S-pits caused by surface microdefects in a CZ silicon wafer subjected to heat treatment at 1150°C for 2 h. One division is 2.5 μm.

parallel to the wafer surface is to be observed, the wafer is polished or chemically thinned to the desired depth from the surface, and then the plane obtained is preferentially etched. Figure 6.25a shows the Wright-etched octants that were cut from the same silicon wafer and then were subjected to a heat treatment for 64 hr at the temperature noted. Material to about 100 μm depth from the surface of each octant was etched off before Wright etching for 4 min. Concentric striations (i.e., swirl) due to thermally induced interior defects are observed in the octants except for three (as-grown, 600°C, and 1150°C). Optical micrographs of those etch pits in octants are shown in Fig. 6.25b. Although no etch pits are observed in the as-grown and 600°C octants, etch pits with different shapes and densities are found in other heat-treated octants. The etch pit density can be obtained easily by counting the number of etch pits in a specific area; however, the characterization of the defects that result in the etch pits must depend on transmission electron microscopy. Thus another major limitation of chemical etching technique is that most of defects resulting in etch pits can not be identified directly by this method.

As mentioned above, etch pits have been observed almost exclusively by an optical microscope. Although SEM has been most powerfully used for the observation of surface morphology, and of electrical effects at crystallographic defects and impurities in the semiconductor materials, etch pits and etch hillocks can also be examined by SEM with much higher magnifications and higher resolution than with an optical microscope. Since the information signal obtained by SEM is electronic in nature, a variety of signal-processing techniques can be utilized to enhance the display. Figure 6.26a shows striations delineated by Wright etching for a (100) rectangular plate of CZ silicon cut along the growth direction $\langle 111 \rangle$. The sample was subjected to a heat treatment at 1000°C for 64 hr, and corresponds to the 1000°C octant in Fig. 6.25; thus the striations are due to thermally induced defects, which are attributed to oxygen precipitation as discussed in Section 7.2. The optical micrograph of etch pits that compose the striations is shown in Fig. 6.26b, while the SEM image of *a pit* observed by an optical microscope is given in Fig. 6.26c with high magnification. Although optical microscopy limits its magnification to $\times 1000$, SEM can allow observation with much higher magnification up to the order of $\times 10^5$.

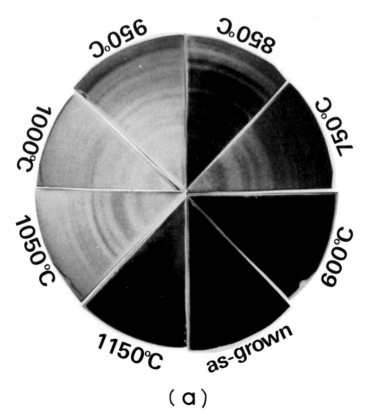

(a)

Fig. 6.25. Wright-etched figures for CZ silicon octant samples subjected to heat treatment for 64 h at temperature shown: (a) optical photograph of octants.

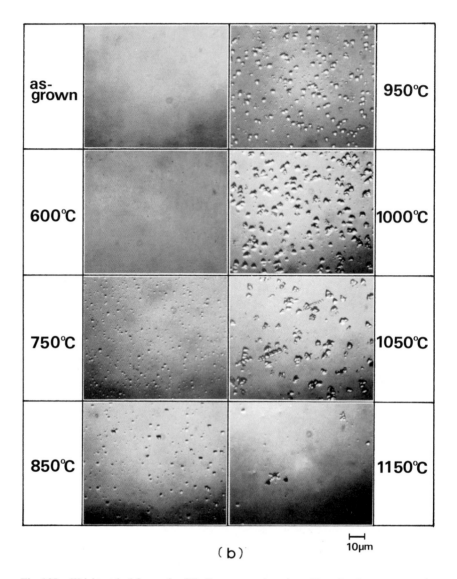

(b)   10μm

Fig. 6.25.   Wright-etched figures for CZ silicon octant samples subjected to heat treatment for 64 h at temperature shown: (b) optical micrograph of etch pits in each octant.

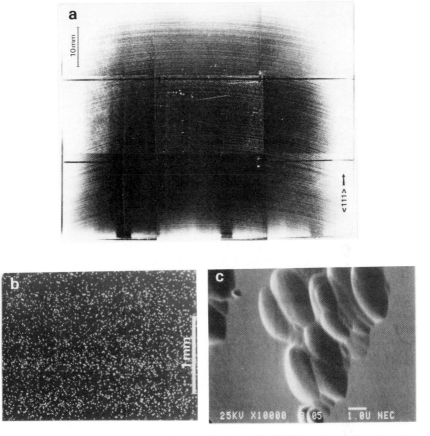

Fig. 6.26. Striations delineated by Wright etching for (110) CZ silicon plate subjected to heat treatment at 1000°C for 64 h: (a) optical photograph of striations, (b) optical micrograph of etch pits, and (c) SEM image of an etch pit.

6.3.3 X-Ray Diffraction Methods

Principle and Characteristics X-Ray diffraction methods for the characterization of semiconductor crystals are based on the interaction of incident X-rays with crystal lattices, as discussed in Section 3.2. The X-ray diffraction and absorption phenomena are *disturbed* at imperfections or defects in the crystal lattice. The disturbance can be quantitatively or qualitatively analyzed for the characterization. That is, the intensity of the X-rays diffracted by deformed planes differs from that diffracted by a perfect crystal. Primarily, the intensity of diffracted X-rays can be implied with the following Bragg equation, which was discussed in Section 3.2.4:

$$\lambda = 2d \sin \theta_{\mathrm{B}} \tag{3.33}$$

It is obvious that the Bragg condition will not apply simultaneously to the perfect region and to the distorted region in which the lattice spacing or lattice plane orientation varies locally due to crystal defects. Consequently, a difference in the diffracted X-ray intensity at the Bragg angle arises, or some intensity can be obtained at diffraction angles that deviate from the Bragg angle. The intensity of diffracted X-rays is characteristic for a specific diffraction plane, that is, diffraction vector $\mathbf{g}_{hkl}$; therefore, the crystallographic nature of defects can be characterized by applying several different diffraction vectors similar to that performed by transmission electron microscopy (TEM).

For precise treatment of these changes in intensity, the theory of X-ray diffraction in solids must be considered. The kinematical diffraction theory can be employed when it is assumed that the intensities of the scattered waves are always negligibly small compared with the incident wave intensity in such a case as "nearly perfect" crystals. It is found that the integrated diffraction from an "ideally perfect" crystal is usually one to two orders of magnitude smaller than that from an "ideally imperfect" one, and also that the angular range of diffraction from a perfect crystal is limited to a few seconds of arc.[97] However, in highly perfect crystals, the intensity of a diffracted wave becomes comparable with that of the incident beam, which results in interchanges of energy between incident and diffracted beams occurring as a result of multiple reflections at the net planes. For such processes in which the diffracted wave must be considered, the dynamical diffraction theory is required. Readers interested in these diffraction theories may study the related literature in Ref. 98.

As regards the source of X-rays, in addition to the conventional X-rays discussed in Section 3.2.2, synchrotron radiation (SR) has been used as an X-ray source for topography since 1974.[98] Since then, it has become more and more apparent that SR is very well suited for X-ray diffraction experiments because of its unique properties, namely, continuous spectrum, extremely high power, extreme collimation, and defined polarization states.

Various X-ray diffraction methods for crystal defect analysis have been used according to the objective. These X-ray diffraction methods can be classified into two major categories: *goniometry* and *topography*.[99] In goniometry, the X-ray diffraction intensity as a function of angular position of the crystal is quantitatively analyzed. On the other hand, in topography, the diffraction of X-ray beams is recorded on a photographic plate as a function of position in the crystal—in other words, a photographic image of defects is produced.[100] The term "topography" is occasionally misleading. The topography is not necessarily of the exterior features, but is of the diffracting planes in the crystal, although the contours of the crystal surfaces are important in determining the contrast on XRT. Since the photograph obtained by XRT is a one-to-one reproduction of the sample, any magnification of the image must be achieved by subsequent enlargement of the film and is restricted to magnification of a few hundred because of the film grain size. For this reason, XRT is not useful for the detection of small defects of less than a few micrometers in size. Under this circumstance, it is desirable to use XRT complementarily with goniometry or other characterization techniques such as TEM. One of the greatest advantages of X-ray diffraction methods over other characterization methods is its capability of characterizing the sample of interest truly nondestructively.

Transmission X-Ray Topography The most popular and most widely used XRT in the semiconductor industry is the *Lang method* for the transmission mode.[97] The fundamental diffraction image that leads to the Lang XRT image can be obtained with *section topography*.[101] The principle of the section topography is schematically illustrated in Fig. 6.27. Incident X-rays strike the crystal under investigation. The incident X-rays typically consist of characteristic lines, mostly $K\alpha_1$, whose horizontal divergence is limited to about 4 min of arc by a slit at the end of the collimating tube. The crystal sample is usually oriented so that the lattice planes (*hkl*) normal to the sample front and back faces will reflect with the Bragg condition. Thus the Bragg-reflected rays are transmitted through the crystal and produce a section topograph image on a photographic plate as shown. In the Lang method, the crystal sample and the photographic plate are synchronously traversed in order to obtain the transmission XRT of the whole sample, which can be obtained as shown in Fig. 6.28. Other diffracting planes can easily be chosen by rotating the crystal sample.

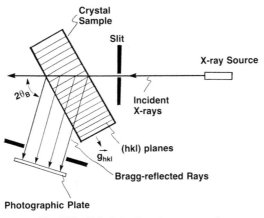

Fig. 6.27. Principle of section topography.

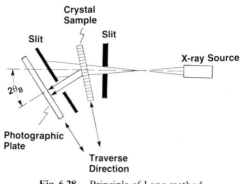

Fig. 6.28.   Principle of Lang method.

As discussed in Section 3.2.3, as X-rays pass through the sample, their intensity is reduced exponentially with thickness:

$$I = I_0 \exp(-\mu_a t) \tag{3.26}$$

When $\mu_a t \lesssim 1$, the kinematical image contributes dominantly to the XRT of a crystal sample. That is, any crystallographic imperfection that disturbs the regular lattice periodicity will manifest itself by an increase in diffracting power in the vicinity of the imperfection.[97] On the other hand, when $\mu_a t \gtrsim 10$, diffracted beams anomalously transmit with low absorption at perfect crystal regions because of the *Borrmann effect*.[102] Hence, crystal imperfections give a weaker image (i.e., dynamical image) than does the perfect matrix. The dynamical image results in high sensitivity to the imperfection, although it takes more exposure time because of the weak intensity of the diffracted beam. The thickness t_1 ($= 1/\mu_a$) of single-crystalline silicon for several characteristic X-ray radiations is given in Table 6.7. Since the thickness of standard silicon wafers used for the fabrication of electronic devices ranges between 350 and 700 μm, kinematical images and dynamical images can be obtained by chosing $AgK\alpha_1$ or $MoK\alpha_1$ and $CuK\alpha_1$ or $CrK\alpha_1$

Table 6.7   Thickness t_1 ($= 1/\mu_a$) of Single-Crystalline Silicon for Different X-Ray Radiations

X-ray	Wavelength (Å)	t_1 (mm)
$CrK\alpha_1$	2.290	0.022
$CuK\alpha_1$	1.541	0.068
$MoK\alpha_1$	0.7093	0.642
$AgK\alpha_1$	0.5594	1.26

X-rays, respectively. The Lang XRT method has been extensively used for the investigation of thermally induced defects in silicon crystals,[103] as well as for the incoming inspection of silicon wafers, yield analysis, and problem solving in device manufacturing processes.[104]

Silicon wafers, particularly large-diameter wafers, usually have some amount of lattice strain induced by several possible causes, such as mechanical damage induced during wafer shaping processes, grown-in or thermally induced defects, and warp or bow induced during device manufacturing processes. In such a case, with the Lang XRT technique, it is difficult to obtain large-area topographs of silicon wafers principally because of local variations in crystal perfection, which result in the deviation from the Bragg condition. A novel XRT technique that is capable of recording large-area transmission topographs of crystal wafers has been developed by Schwuttke.[105] This technique is called the *scanning oscillator technique*

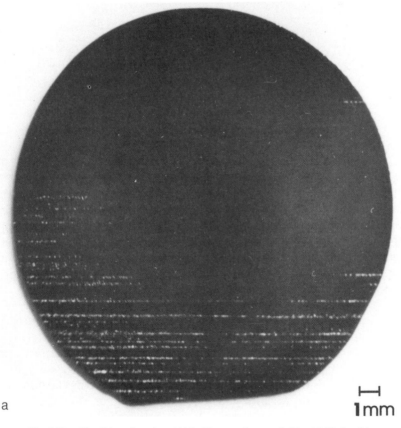

a
⊢—⊣
1mm

Fig. 6.29.   Slip dislocations in a (111) silicon wafer revealed by (a) Sirtl etching.

b

Fig. 6.29. Slip dislocations in a (111) silicon wafer revealed by (b) Lang X-ray topograph with
$AgK\alpha_1/g_{2\bar{2}0}$ diffraction. (Courtesy of J. Matsui, NEC Corporation.)

(SOT), where, while the sample is being scanned, the crystal sample and
photograpic plate are also oscillated synchronously around the normal to the
plane containing incident and reflected beams. In this way, crystallographic
defects are visible at any region with different contrast in the topograph
recorded by different Bragg reflections.

The following are some examples of Lang XRT images obtained for
thermally processed CZ silicon wafers. Figure 6.29 compares the Sirtl-etched
figure (a) and the Lang XRT image (b) for the slip dislocations in a (111)
silicon wafer subjected to a heat treatment. Slip bands which intersect the
wafer surface are revealed by Sirtl etching as etch pit lines parallel to a $\langle 110 \rangle$
direction. The transmission XRT, however, delineates expanding slip dislocations on the interior declined $\{111\}$ planes as well as the slip bands
intersecting the (111) wafer surface. Figure 6.30a shows a (111) wafer
subjected to heat treatment at 1000°C for 1 hr and cooled down rapidly. Slip

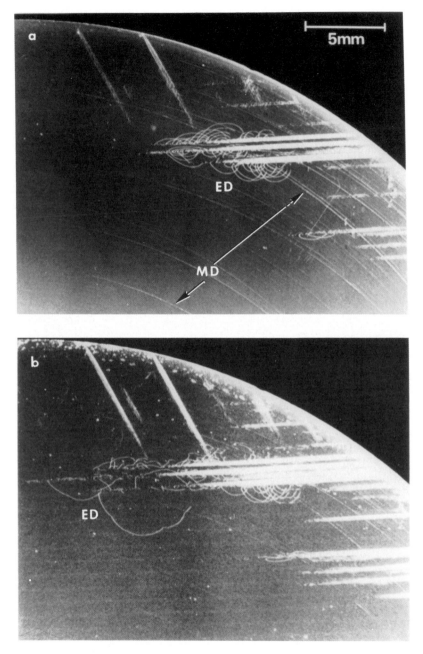

Fig. 6.30. Lang X-ray topograph with $MoK\alpha_1/g_{\bar{2}20}$ diffraction for CZ silicon wafer subjected to heat treatment at $1000°C$ for 1 h: (a) after first-step heat treatment and (b) after second-step heat treatment. ED and MD refer to expanding dislocations and mechanical damage, respectively.

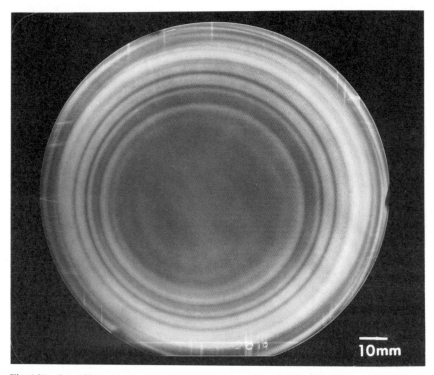

Fig. 6.31.   Lang X-ray topography with $MoK\alpha_1/g_{\bar{2}20}$ diffraction for CZ silicon wafer subjected to heat treatment at 1000°C for 64 h.

dislocations toward $\langle 110 \rangle$ directions and expanding dislocations (ED) are observed, in addition to the backside mechanical damage (MD) for extrinsic gettering. After an additional heat treatment (Fig. 6.30b), the movement of expanding dislocations is clearly observed. Swirl defects caused by oxygen precipitation in a (100) wafer subjected to a two-step heat treatment are shown in Fig. 6.31. Figure 6.32 correlates the white spot defects, which have swirl-like distribution in CCD images obtained from several chips fabricated in one wafer. It has been known that these white spot CCD defects are due to leak currents at the *p-n* junction. After removing electrodes, XRT was taken for the lower right quadrant of the wafer shown in Fig. 6.32a. The XRT image clearly reveals dislocations generated at the pattern edge with swirl-like distribution (Fig. 6.32b and c), which corresponds to the distribution of white spot defects in Fig. 6.32a.

X-Ray films and nuclear emulsions have been most commonly used for photographic recording of XRT images[106]; thus it takes some time to develop the images through common photographic procedures. Recently,

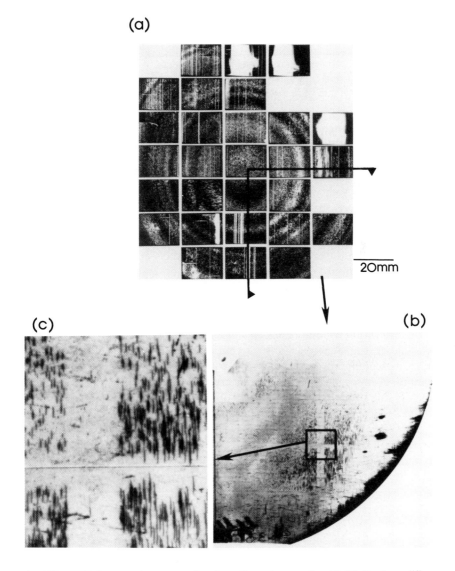

Fig. 6.32. CCD image and corresponding Lang X-ray topographs with $MoK\alpha_1/g_{\bar{2}20}$ diffraction: (a) CCD images obtained by several chips frabricated in a CZ silicon wafer, (b) Lang X-ray topograph for lower right quadrant of the wafer, and (c) enlarged X-ray topograph shown in (b).

real-time or *in situ* XRT has been developed by the combination of high-power X-ray generation and a television unit with an X-ray–sensing vidicon camera tube.[107,108] This technique has been a powerful tool to study directly the dynamic behavior of strain fields and defects in crystals under various influences. However, this technique is presently limited to a spatial resolution of 20–25 μm,[109] which is about one order of magnitude poorer than the spatial resolution obtained with conventional emulsions. More recently, a multiple-stage imaging technique that enables real-time observation with higher spatial resolution (~ 10 μm) has been developed.[110] The X-ray image is first converted into a visible pattern by a fluorescent screen; then the pattern is optically coupled, either by a lens or by a fiber-optic plate, to the input photocathode of a light-sensitive electro-optical device, and finally the output image is displayed on a television monitor.

Decoration Technique   Crystallographic imperfections that are transparent to visible and infrared radiation can be selectively *decorated* by precipitation of a suitable impurity, and thus can be made observable with an optical or infrared microscope. A valuable decoration technique for silicon has been developed using copper as precipitating (decorating) impurity in order to enhance the visibility of dislocations by optical or infrared microscopy.[111] This decoration technique has been applied also to the observation of lattice defects by transmission X-ray topography.[112] Although other impurities such as aluminum, nickel, and iron have also been used, copper is almost exclusively used in investigation of crystal imperfections in silicon. This is mainly due to the relative simplicity of the technique. Copper decoration by diffusion is usually carried out by heating a silicon sample covered with $Cu(NO_3)_2$.[111] The sample is heated for about 30 min at 950°C or a higher temperature in an argon atmosphere, and after the heat treatment the sample is quenched to room temperature within a few seconds in air in order to achieve clear precipitation at crystal imperfections. This copper-decoration technique in conjunction with the Lang XRT method has been proved to be very powerful in the investigation of microdefects such as vacancy clusters in silicon.[113] The major disadvantage of this technique, however, is that additional defects can be created, and the existing defects can be modified, during the copper diffusion process at 950°C or a higher temperature. In addition, the effect of thermal shock, induced into the sample due to quenching from such a high temperature, on crystal imperfections of interest may not be ignored. Under the circumstances, lithium decoration is preferable to copper decoration because lithium can decorate crystal imperfections at a low temperature such as 400°C. Furthermore, defects that do not act as nucleation sites for copper may do so for lithium because of its small size and high solubility in silicon.[112]

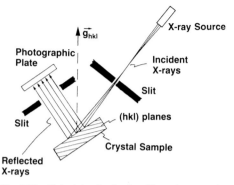

Fig. 6.33. Principle of reflection X-ray topography.

Reflection X-ray Topography Defects in the surface region of a crystal sample can be observed by means of reflection XRT, whose principle is illustrated in Fig. 6.33. This method consists of illuminating the surface of a crystal sample with X-rays coming from a focus in the form of a line that satisfies the Bragg condition for the (*hkl*) reflection plane. Based on the primary extinction effect of X-rays, the reflection XRT image of surface regions of a few micrometers depth can be obtained on the photographic plate. This arrangement is commonly referred to as the *Berg–Barrett technique*.[114,115] Similar to the Lang technique, the reflection XRT of the whole sample area can be obtained by synchronously traversing the crystal sample and the photographic plate. When soft X-ray radiation (e.g., $CrK\alpha_1$) is used, the X-rays penetrate only a very small distance into the crystal, and then a very thin layer of the crystal close to the surface can be examined. Thus reflection XRT is a very useful technique to investigate epitaxial layers or superlattice structures on semiconductor substrates, as well as device active regions that are fabricated in the surface regions of substrates.

Multiple Crystal Arrangement By using high-order Bragg reflections, a beam divergence of a few tenths of a second of arc is obtained. A double-crystal diffraction technique[98,100,116] utilizes two successive Bragg reflections, as schematically shown in Fig. 6.34. In order to obtain X-ray beams of extremely narrow angular divergence, a triple-crystal diffractometer has also been developed.[117] The angular width of 0.01 sec has been obtained with a monochromator system consisting of three crystals prepared from a block of silicon single crystal, in which asymmetric (422) reflections of $CuK\alpha_1$ are repeated.[118] Thus these double- or triple-crystal diffraction techniques are extremely sensitive to lattice distortion or misorientation, and can detect very small lattice stain such as $\Delta d/d = 10^{-8}$. However, the major difficulties

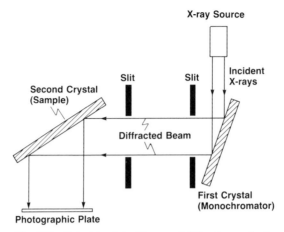

Fig. 6.34.   Principle of double-crystal diffraction method.

associated with multiple-crystal diffractometry are in set-up time and exposure time, because of its very weak diffracted beam. In order to meet the
requirement of obtaining accurate and detailed information on crystal defects
or imperfections, various arrangements of multiple-crystal diffractometer
have been developed.[100,119] This multiple-crystal diffractometry, of course,
can be applied to X-ray goniometry as well.

X-ray Goniometry   As described above, X-ray goniometry quantitatively
analyzes the X-ray diffraction intensity as a function of angular position,
which in turn is a function of lattice constant as was given by Eq. (3.33).
Therefore, the displacement of diffraction angle from the Bragg angle can be
directly correlated to the distortion of lattice spacing due to crystallographic
imperfections. Moreover, it is then interpreted that the width of the diffraction intensity profile as a function of angular position—that is, the X-ray
rocking curve—can be a parameter of crystal perfection. Usually the full
width at half value (FWHV) of the X-ray rocking curve is discussed in
evaluating the crystal perfection. It is obvious that a perfect crystal gives rise
to a small FWHV, while a crystal that contains a large amount of imperfections results in a large FWHV.

Double-crystal X-ray goniometry has been used to investigate oxygen
diffusion into silicon crystals.[120] Measuring the diffraction peak angular
position, which displaces by $\Delta\theta$ from the Bragg angle θ_B, one finds the lattice
strain caused by the oxygen atoms. The strain is estimated according to the
following relation:

$$\Delta d/d = -\Delta\theta \cot \theta_B \qquad (6.28)$$

As a result, it has been confirmed that diffused oxygen atoms expand the silicon lattice, and the strain caused by oxygen is proportional to the concentration. The expansion coefficient of oxygen is calculated as 4.5×10^{-24}/atom.

Triple-crystal X-ray diffractometry has also been used to study defects induced by the diffusion of metallic impurities.[121] The incident X-ray beam of 6 sec of arc, which was obtained by successive reflections from a grooved crystal (first crystal), struck on the sample crystal (second crystal). The angular distribution of diffracted X-rays from the sample set various angular positions around the exact Bragg angle was measured by rotating the analyzer crystal (third crystal). From the intensity curves of diffracted beam, the relative defect volume in the silicon crystal was estimated.

Bond Method   High precision and high accuracy in lattice-constant determination can be achieved by the *Bond method*.[122] In this method, the crystal sample is first placed for the (*hkl*) reflection at $+2\theta_B$ as shown in Fig. 6.35. The diffracted beam is detected by detector 1. If the sample is then rotated to the position for ($\bar{h}k\bar{l}$) reflection at $-2\theta_B$, the beam is diffracted into detector 2. Whether measuring the angle between the two detector positions or the angle between the two positions of the sample, one finds the eccentric errors can be very effectively removed. In this way, a lattice constant can be determined with precision on the order of 1 ppm under the carefully considered alignment of the instrument. By means of this method, the lattice constants of 5.431062 Å ± 4 ppm (at 25°C) and 5.431073 Å ± 0.08 ppm (at 25°C) have been reported for FZ and CZ silicon crystals, respectively.[123] Although the Bond method has not been extensively used in the semiconductor materials field, it is potentially a powerful tool to study point defects and impurities in silicon crystals. In particular, it is expected that the behavior of self-interstitials and vacancies, which play a very important role in the generation

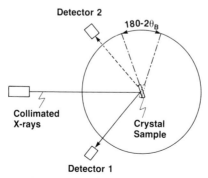

Fig. 6.35.   Principle of Bond method.

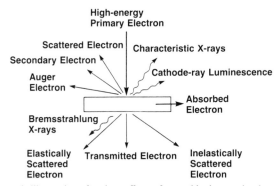

Fig. 6.36. Schematic illustration of various effects of caused by interaction between high-energy electron and thin specimen.

of lattice defects, can be investigated successfully by measuring precisely the change in lattice constant.

6.3.4   Electron Microscopy

Principle and Characteristics   As described in Section 3.2.5, an electron beam interacts with a substance in many ways. By analyzing the interaction, the physical and chemical nature of the substance can be characterized. The wavelength of an electron beam is very short compared with that of an X-ray, which results in various characteristic features of an electron beam from the analytical point of view. When a high-energy electron beam strikes a thin sample, various kinds of effects occur, as schematically shown in Fig. 6.36. In theory, all the information can be useful for characterizing the material.[124,125] However, transmission electron microscopy,[126,127] which is one of the most powerful tool to characterize silicon crystals, will be focused on in this section.

Transmission Electron Microscopy   Transmission electron microscopy (TEM) is based on the fact that electron beams can be focused by an electrostatic or magnetic field as described in Section 3.2.5. Eventually, extremely high magnifications up to $\times 1,000,000$, which allow spatial resolutions in the range of 2–5 Å, are readily attainable. It should be noted, however, that this spatial resolution does not necessarily mean the size of a defect that can be characterized. It is true that TEM attains the highest resolution for the investigation of defects in silicon; however, it may be still difficult to characterize a lattice defect or precipitate of less than 50 Å in size.

The incident electron beam, usually accelerated to 100–1000 keV and focused electromagnetically, diffracts, and two beams (i.e., the *direct beam* and the *diffracted beam*) emerge from the sample. By using a suitable aperture

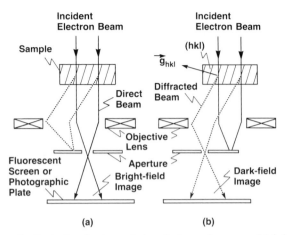

Fig. 6.37. Schematic illustration for transmission electron microscopy: (a) bright-field image formed with direct beam and (b) dark-field image formed with diffracted beam.

and focusing lenses, either the direct beam or the diffracted beam can be selected to form an image on a fluorescent screen or a photographic plate, as schematically shown in Fig. 6.37. The TEM images formed by direct beams and diffracted beams are called the *bright-field image* and the *dark-field image*, respectively. The contrast in the image depends on the intensities of the electron beam exiting from the sample. The image contrast can be predicted by the kinematical and the dynamical theories of electron diffraction.[126,127] The bright-field image is obtained by subtracting the intensities absorbed and scattered into the Bragg reflection for the particular g_{hkl} from that of the incident beam. The contrast formed by the diffracted beam—that is, the dark-field image—is sensitive to any lattice distortion caused by defects disturbing the local Bragg condition. Since the contrast effects in TEM images are produced by the displacement of atoms from their ideal positions when it is normal to the diffraction vector g, the nature of the defect (e.g., the Burgers vector) can be characterized by observing the images with different diffraction conditions. In a bright-field image, dislocations appear as dark lines and stacking faults give rise to interference fringes when they are observed with a certain Bragg condition, while defects that disturb the lattice spacing appear as white lines in a dark-field image.

The following are examples of TEM images of thermally induced defects in CZ-silicon crystals. The observations of these defects by preferential chemical etching for the corresponding samples has been shown in Fig. 6.25. Although chemical etching does not reveal defects in the sample subjected to heat treatment at 600°C for 64 h, the TEM dark-field *weak beam image*[127] indicates small precipitates at a density of $\sim 10^{14}/cm^3$ as shown

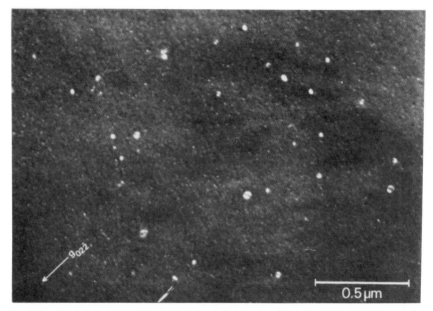

Fig. 6.38. Weak-beam dark-field TEM image of microprecipitate in (111) CZ silicon wafer subjected to heat treatment at 600°C for 64 h. (After Shimura *et al.*[128] Reprinted with the permission of The Electrochemical Society, Inc.)

in Fig. 6.38.[128] Dominant precipitates observed are around 60 Å in size, but are easily recognized by the strong compressive stress field in the surrounding matrix. The largest precipitate observed is about 350 Å in size and shows platelike morphology with a thickness of about 50 Å. This TEM observation result implies that a chemical etching technique may not be suitable for the investigation of defects that are smaller than 350 Å in size. Figure 6.39 shows platelike SiO_2 precipitates and dislocations generated by these precipitates, that is, *precipitate-dislocation complexes* (PDC), in the sample subjected to heat treatment at 1000°C for 64 h.[129] These PDC defects give rise to etch pits such as shown in Fig. 6.25 and Fig. 6.26. Figure 6.40 shows three TEM images taken with different diffraction conditions for the same area of CZ silicon subjected to heat treatment at 950°C for 64 h.[128] Various kinds of microdefects, such as precipitates, perfect dislocation loops, and stacking faults, are observed. By observing the images of four stacking faults, denoted A, B, C, and D, under different diffraction conditions, one can characterize their fault planes as $(11\bar{1})$, $(\bar{1}11)$, (111), and (111), respectively.

A diffraction pattern can be also magnified and displayed by selecting a suitable aperture and a lens system. The diffraction pattern is indispensable to ensure the diffraction conditions for the TEM image obtained. Moreover,

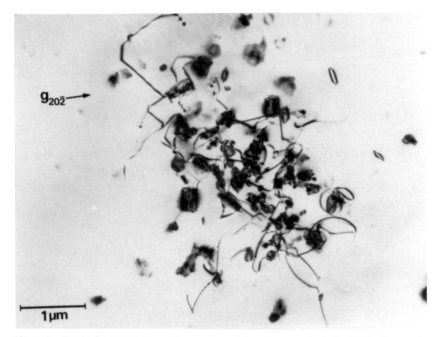

Fig. 6.39. Bright-field TEM image of precipitate–dislocation complex in (111) CZ silicon wafer subjected to heat treatment at 1000°C for 64 h. (After Shimura and Tsuya.[129] Reprinted with the permission of The Electrochemical Society, Inc.)

it can be used to identify the structure, and hence the chemical nature, of the second phase in silicon such as SiO_2 and metallic impurity precipitates.

Sample Preparation   As described in Section 3.2.5, electrons interact with substances much more than do X-rays. Therefore, electrons penetrate much less into material than do X-rays, and are very easily absorbed even by air. Since TEM involves transmitting electrons through the sample substance, the sample must be thin enough for high electron transparency. For example, a silicon sample should be thinner than 1 μm and 2 μm for the 100-keV and 200-keV electron accelerating voltages, respectively, for best defect characterization results. As a consequence, silicon wafers and crystals must be appropriately thinned for the TEM observation. The preparation of a thin specimen is cumbersome and is one of disadvantages of TEM; however, it is true that the successful characterization of defects in silicon strongly depends on the sample preparation. Thin silicon foils of the required thickness can be prepared by mechanochemical polishing, chemical etching, and energetic ion beam sputtering, or their combinations.[130–132] In principle, a chemical

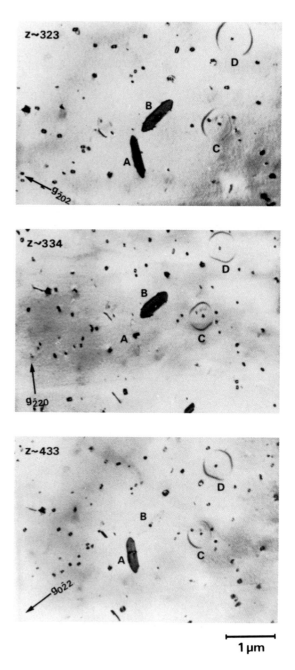

Fig. 6.40. Bright-field TEM image obtained with different diffraction conditions for various types of microdefects in (111) CZ silicon wafer subjected to heat treatment at 950°C for 64 h. Note the change in contrast of stacking faults denoted A, B, C, and D. (After Shimura *et al.*[128] Reprinted with the permission of The Electrochemical Society, Inc.)

thinning technique is preferred to other techniques because of its relative simplicity, and clean and damage-free nature.

High-voltage electron microscopy (HVEM), where accelerating voltages are 500–2000 keV, enables the observation of silicon foils with a thickness of 4–8 μm and expands the range of applicability of TEM.[133] Eventually, the whole active region of the device, for example, can be simultaneously observed. For MOS devices, in particular, HVEM contributes to obtaining useful TEM images since the higher-energy electron will not be influenced by the gate oxide, whereas a lower-energy electron will be scattered excessively in the layer, which results in a poor TEM image. In addition, the HVEM combined with an appropriate equipment will contribute highly to *in situ* observation studies on the dynamical behavior of dislocations that are representative of those in the bulk crystal.[134]

High-Resolution TEM   The transmission electron microscope produces images of thin specimens by an amplitude-contrast mechanism where the image resolution is determined by the extent of local lattice distortion caused by a defect. Alternatively, TEM allows image formation by a phase-contrast mechanism where phase variations between the chosen beams produce the direct image of crystal lattices. This requires more than one beam from diffracting planes of the specimen by selecting the proper objective lens systems, and this technique is commonly called *multibeam high-resolution TEM (HRTEM)*.[135] The direct lattice images of silicon observed as three kinds of projections onto the (001), (110), an (111) planes have been obtained by means of this technique.[136] The highest resolution of ~ 1.36 Å, which corresponds to the d_{400} spacing (1.357 Å), as the distance between observed spots corresponding to atomic position was obtained in the case of the (110) projection.[137] The HRTEM has recently been extensively used for the direct observation of lattice defects in silicon and for the investigation of the interface of silicon oxide and substrate silicon.[125,138] Figure 6.41 shows an example of the direct lattice image of SiO_2/Si interface in the (110) projection.

Scanning Transmission Electron Microscopy   The imaging process of scanning transmission electron microscopy (STEM) is explained by the reciprocal relationship to the essential elements for conventional TEM (CTEM).[139,140] As a consequence of the reciprocal relationship, it can be predicted that the contrast of the STEM image will be the same in principle as that of CTEM for identical electro-optical components and equivalent sources and detectors. The main differences in practice between the two forms of microscopy arise from the difference in the techniques used for the detection and recording of the images. The basic difference between CTEM and STEM lies in the fact that the whole area of interest is examined at one time in CTEM, whereas in

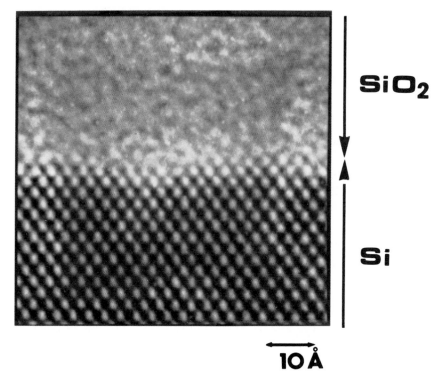

SiO₂

Si

10 Å

Fig. 6.41. High-resolution TEM image of SiO₂/Si interface in (110) projection. (Courtesy of P. Fraundorf, Monsanto Physical Science Center.)

STEM an electron probe far smaller (typically 100 Å) than that used in CTEM (typically 10 μm) is scanned across the area of interest to produce the image. As the beam scans the specimen, electrons scattered by the specimen are collected by an electron detector to produce an electrical image signal in serial form for display on a cathode ray tube (CRT), which is scanned synchronously with the electron beam scan on the specimen. The magnification of the image is given simply by the ratio of the length of the CRT scan to the length of the specimen scan. The ability of STEM to electrically process the information obtained from a specimen represents a major advantage of STEM over CTEM. In particular, STEM, which uses a very small electron probe, is a very powerful tool for producing selective crystallographic, chemical, and electronic information about materials with high spatial resolution comparable with that of a modern CTEM. In addition, the incoherent imaging mode of STEM allows the formation of images showing strong atomic number contrast. The images of Bi (0.2–1 at. %) and Sb (0.6–6 at. %) in ion-implanted silicon have been obtained by this technique.[141]

Analytical Electron Microscopy Analytical electron microscopy (AEM) provides chemical, physical and structural characterization of a thin specimen with high spatial resolution using spectroscopic and microdiffraction techniques.[124,142] The AEM is usually based on an STEM instrument because of its unique capabilities. When the electron probe is positioned on a point of interest in the specimen, the microdiffraction pattern can be obtained since the small electron probe defines the precise area from which the diffraction pattern is generated. If the microscope is fitted with appropriate X-ray analysis facilities, invariably using an energy-dispersive X-ray spectrometer (EDX), the characteristic X-rays from any small region can be detected and analyzed. Electron energy loss spectroscopy (EELS), which analyzes the energy distribution of electrons transmitted through a specimen, also provides a powerful microanalytical method giving detailed qualitative and quantitative information about the chemical and physical nature of the specimen.[143] The energy loss spectrum can be divided conveniently into three parts: (1) no loss, (2) low loss (< 50 eV), and (3) high loss (> 50 eV). However, EELS is mainly concerned with high-loss electrons, particularly those that loose more than a certain critical energy required to ionize an inner-shell (i.e., K, L, or M) electron. The EELS is particularly useful for the detection of light elements because of the increase in ionization cross section as atomic number decreases, whereas the characteristic X-ray production efficiency decreases. The EELS in conjunction with STEM has been successfully used for the analysis of oxygen and carbon related to precipitates in silicon crystals.[144,145] Although the electron beam induced current (EBIC) technique has been used in conjunction with scanning electron microscopy (SEM) for the detection of electrically active defects,[146] it can be complemented by observations in STEM for the identification of the chemical structural nature of microdefects in semiconductors.[147]

References

1. P. A. Barnes and G. A. Rozgonyi, eds., "Semiconductor Characterization Techniques." Electrochem. Soc., Princeton, New Jersey, 1978.
2. N. G. Einspruch and G. B. Larrabee, eds., "VLSI Electronics Microstructure Science," Vol. 6. Academic Press, New York, 1983.
3. F. Shimura and H. R. Huff, VLSI silicon material criteria. *In* "VLSI Handbook" (N. G. Einspruch, ed.), pp. 191–269. Academic Press, New York, 1985.
4. W. R. Runyan, "Semiconductor Measurements and Instrumentation." McGraw-Hill, New York, 1975.
5. R. Rymaszewski, Relationship between the correction factor of the four-point probe value and the selection of potential and current electrodes. *J. Sci. Instrum.* **2**, 170–174 (1969).
6. J. A. Keenan and G. B. Larrabee, Characterization of silicon materials for VLSI. *In* "VLSI Electronics Microstructure Science," Vol. 6, pp. 1–72. Academic Press, New York, 1983.

7. H. L. Libby, "Introduction to Electromagnetic Nondestructive Test Methods." Wiley (Interscience), New York, 1971.

8. G. L. Miller, D. A. H. Robinson, and J. D. Wiley, Contactless measurements of semiconductor conductivity by radio fraquency-free-carrier power absorption. *Rev. Sci. Instrum.* **47**, 799–805 (1976).

9. R. G. Mazur and D. H. Dickey, A spreading resistance technique for resistivity measurements on silicon. *J. Electrochem. Soc.* **113**, 255–259 (1966).

10. R. G. Mazur, Resistivity inhomogeneities in silicon crystals. *J. Electrochem. Soc.* **114**, 255–259 (1967).

11. P. J. Severin, Measurement of resistivity of silicon by the spreading resistance method. *Solid-State Electron.* **14**, 247–255 (1971).

12. F. L. Jones, "Physics of Electrical Contacts." Oxford Univ. Press, London and New York, 1957.

13. R. Brennan and D. Dickey, Determination of diffusion characteristics using two- and four-point probe measurements. *Solid State Technol.* Dec., pp. 125–132 (1984).

14. D. K. Shroder, The concept of generation and recombination lifetimes in semiconductors. *IEEE Trans. Electron Devices* **ED-29**, 1336–1338 (1982).

15. Y. P. Varshni, Band-to-band radiative recombination in groups IV, VI and III–V semiconductors. *Phys. Status Solidi* **19**, 459–514 (1967).

16. J. Dziewor and W. Schmid, Auger coefficients for highly doped and highly excited silicon. *Appl. Phys. Lett.* **31**, 346–348 (1977).

17. R. D. Westbrook, ed., "Lifetime Factors in Silicon." Am. Soc. Test. Mater., Philadelphia, Pennsylvania, 1980.

18. G. Suryan and G. Sisila, Lifetime of minority carriers in semiconductors: Experimental methods. *J. Sci. Ind. Res., Sect. A* **21**, 235–246 (1962).

19. D. T. Stevenson and R. J. Keyes, Measurement of carrier lifetimes in germanium and silicon. *J. Appl. Phys.* **26**, 190–195 (1955).

20. H. A. Atwater, Microwave measurement of semiconductor carrier lifetimes. *J. Appl. Phys.* **31**, 938–939 (1960).

21. R. D. Larrabee, Measurement of semiconductor properties through microwave absorption. *RCA Rev.* **21**, 124–129 (1960).

22. N. J. Harrick, Lifetime measurements of excess carriers in semiconductors. *J. Appl. Phys.* **27**, 1439–1442 (1956).

23. N. G. Nilsson, Determination of carrier lifetime diffusion length, and surface recombination velocity in semiconductors from photo-excited infrared absorption. *Solic-State Electron.* **7**, 455–463 (1964).

24. R. M. Lichtenstein and H. J. Willard, Jr., Simple contactless method for measuring decay time of photoconductivity in silicon. *Rev. Sci. Instrum.* **38**, 133–134 (1967).

25. I. R. Weingarten and M. Rothberg, Radio-frequency carrier and capacitive coupling procedures for resistivity and lifetime measurements on silicon. *J. Electrochem. Soc.* **108**, 167–171 (1961).

26. J. R. Hatnes and W. Shockley, The mobility and life of injected holes and electrons in germanium. *Phys. Rev.* **81**, 835–843 (1951).

27. Institute of Radio Engineers, Measurement of minority-carrier lifetime in germanium and silicon by the method of photoconductive decay. *Proc. IRE* **49**, 1292–1299 (1961).

28. J. P. McKelvey, Volume and surface recombination of injected carriers in cylindrical semiconductor ingots. *IRE Trans. Electron devices* **ED-5**, 260–264 (1958).

29. A. C. Sim, A Note on surface recombination velocity and photoconductive decays. *J. Electron. Control* **5**, 251–255 (1958).

30. C. T. Sah, R. N. Noyce, and W. Schokley, Carrier generation and recombination in P-N junctions and P-N junction characteristics. *Proc. IRE* **45**, 1228–1243 (1957).

31. F. P. Heiman, On the determination of minority carrier lifetime from the transient response of an MOS capacitor. *IEEE Trans. Electron Devices* **ED-14**, 781–784 (1967).
32. E. H. Nicollian and J. R. Brecos, "MOS (Metal Oxide Semiconductor) Physics and Technology." Wiley, New York, 1982.
33. M. Zerbst, Relaxationseffekte an Halbleiter-Isolator-Grenzflachen *Z. Angew., Phys.* **22**, 30–33 (1966) (in German).
34. P. Kruger, "Principle of Activation Analysis." Wiley, (Interscience), New York, 1971.
35. W. C. J. Gebauhr, Trace analysis in semiconductor silicon, a comparison of methods. *In* "Semiconductor Silicon/1969" (R. R. Haberecht and E. L. Kern, eds.), pp. 517–533. Electrochem. Soc., Princeton, New Jersey, 1969.
36. J. A. Martin, Neutron activation analysis and autoradiography of silicon. *In* "Semiconductor Silicon/1969" (R. R. Haberecht and E. L. Kern, eds.), pp. 547–557. Electrochem. Soc., Princeton, New Jersey, 1969.
37. P. F. Schmidt and C. W. Pearce, A neutron activation analysis study of the sources of transition group metals contamination in the silicon device manufacturing process. *J. Electrochem. Soc.* **128**, 630–637 (1981).
38. T. Nozaki, Y. Yatsurugi, and N. Akiyama, Charged particle activation analysis for carbon, nitrogen and oxygen in semiconductor silicon. *J. Radioanal. Chem.* **4**, 87–98 (1970).
39. E. A. Schweikert, Charged particle activation analysis. *Anal. Chem.* **52**, 827A–844A (1980).
40. S. Nakajima and R. Kohara, Chemical analysis of semiconductors. *Oyo Butsuri* **43**, 422–442 (1974) (in Japanese).
41. T. Nozaki, Y. Yatsurugi, and N. Akiyama, Concentration and behavior of carbon in semiconductor silicon. *J. Electrochem. Soc* **117**, 1566–1568 (1970).
42. Y. Yatsurugi, N. Akiyama, Y. Endo, and T. Nozaki, Concentration, solubility, and equilibrium distribution coefficient of nitrogen and oxygen in semiconductor silicon. *J. Electrochem. Soc.* **120**, 975–979 (1973).
43. H. H. Willard, L. L. Merritt, Jr., and J. A. Dean, "Instrumental Methods of Analysis," 5th ed. (Van Nostrand-Reinhold, Princeton, New Jersey, 1974.
44. S. C. Baber, Net and total shallow impurities analysis of silicon by low temperature Fourier transform infrared spectroscopy. *Thin Solid Films* **72**, 201–210 (1980).
45. B. Pajot, Characterization of oxygen in silicon by infrared absorption. *Analysis* **5**, 293–303 (1977).
46. ASTM F121, Interstitial atomic oxygen content of silicon by infrared absorption. *Annu. Book ASTM Stand.* pp. 240–242 (1984).
47. Y. Itoh, T. Nozaki, T. Masui, and T. Abe, Calibration curve for infrared spectrophotometry of nitrogen in silicon. *Appl. Phys. Lett.* **47**, 488–489 (1985).
48. R. J. Bleiler, R. S. Hockett, P. Chu, and E. Strathman, SIMS measurements of oxygen in heavily-doped silicon. *In* "Oxygen, Carbon, Hydrogen, and Nitrogen in Crystalline Silicon" (J. C. Mikkelsen, Jr., S. J. Pearton, J. W. Corbett, and S. J. Pennycock, eds.), pp. 73–79 Mater. Res. Soc. Pittsburgh, 1986.
49. H. Tsuya, M. Kanamori, M. Takeda, and K. Yasuda, Infrared optical measurement of interstitial oxygen content in heavily doped silicon crystals. *In* "VLSI Science and Technology/1985" (W. M. Bullis and S. Broydo, eds.), pp. 517–525. Electrochem. Soc., Princeton, New Jersey, 1985.
50. D. W. Green and G. T. Reedy, Matrix-isolation studies with Fourier transform infrared. *In* "Fourier Transform Infrared Spectroscopy" (J. R. Ferraro and L. J. Basile, eds.), Vol. 1, pp. 1–56. Academic Press, New York, 1978.
51. T. Iizuka, S. Takasu, M. Tajima, T. Arai, M. Nozaki, N. Inoue, and M. Watanabe, Conversion coefficient for IR measurement of oxygen in Si. *In* "Defects in Silicon" (W. M. Bullis and L. C. Kimerling, eds.), pp. 265–274. Electrochem. Soc., Princeton, New Jersey, 1983.

52. W. Kaiser, P. H. Keck, and L. F. Lange, Infrared absorption and oxygen content in silicon and germanium. *Phys. Rev.* **101**, 1264–1268 (1956).
53. K. Tempelhoff and F. Spiegelberg, Precipitation of oxygen in dislocation-free silicon. *In* "Semiconductor Silicon 1977" (H. R. Huff and E. Sirtl, eds.), pp. 585–595. Electrochem. Soc., Princeton, New Jersey, 1977.
54. F. Shimura, H. Tsuya, and T. Kawamura, Precipitation and redistribution of oxygen in Czochralski-grown silicon. *Appl. Phys. Lett.* **37**, 483–486 (1980).
55. S. M. Hu, Infrared absorption spectra of SiO_2 precipitates of various shapes in silicon: Calculated and experimental. *J. Appl. Phys.* **51**, 5945–5948 (1980).
56. P. Gaworzewski, E. Hild, F. G. Kirscht, and L. Vecserryes, Infrared spectroscopical and TEM investigations of oxygen precipitation in silicon crystals with medium and high oxygen concentrations. *Phys. Status Solidi A* **85**, 133–147 (1984).
57. H. J. Hrostowski and R. H. Kaiser, Infrared absorption of oxygen in silicon. *Phys. Rev.* **107**, 966–972 (1957).
58. F. Shimura, J. P. Baiardo, and F. Fraundorf, Infrared absorption study on carbon and oxygen behavior in Czochralski silicon. *Appl. Phys. Lett.* **46**, 941–943 (1985).
59. B. Pajot, H. J. Stein, B. Cales, and C. Naud, Quantitative spectroscopy of interstitial oxygen in silicon. *J. Electrochem. Soc.* **132**, 3034–3037 (1985).
60. Deutsche Normen DIN 50 438/1, "Bestimmung des Verunreinigungshalts in Silicium mittels infrarot-Absorption, Sauerstoff," pp. 1–5. Beuth Verlag GmbH, Berlin and Köln, 1978 (in German).
61. ASTM F121, Interstitial atomic oxygen content of silicon by infrared absorption. *Annu. Book ASTM Stand.* pp. 518–520 (1976).
62. W. M. Bullis, M. Watanabe, A. Baghdadi, Li, Y.-Z., R. I. Scace, R. W. Series, and P. Stallhofer, Calibration of infrared absorption measurements of interstitial oxygen concentration in silicon. *In* "Semiconductor Silicon 1986" (H. R. Huff, T. Abe, and B. O. Kolbesen, eds.), pp. 166–180. Electrochem. Soc., Princeton, New Jersey, 1986.
63. A. Ohsawa, K. Honda, S. Ohkawa, and R. Ueda, Determination of oxygen concentration profiles in silicon crystals observed by scanning IR absorption using semiconductor laser. *Appl. Phys. Lett.* **36**, 147–148 (1980).
64. P. Rava, H. C. Gatos, and J. Lagowski, Activation of the oxygen donor in Si on a microscale. *In* "Semiconductor Silicon 1981" (H. R. Huff, R. J. Kriegler, and Y. Takaishi, eds.), pp. 232–243. Electrochem. Soc., Princeton, New Jersey, 1981.
65. A. Ohsawa, K. Honda, S. Ohkawa, and K. Shinohara, Oxygen striation and thermally induced microdefects in Czochralski-grown silicon crystals. *Appl. Phys. Lett.* **37**, 157–159 (1980).
66. W. L. Bond and W. Kaiser, Interstitial versus substitutional oxygen in silicon. *J. Phys. Chem. Solids* **16**, 44–45 (1960).
67. R. C. Newman and J. B. Willis, Vibrational absorption of carbon in silicon. *J. Phys. Chem. Solids* **26**, 373–379 (1965).
68. ASTM F123, Substitutional atomic carbon content of silicon by infrared absorption. *Annu. Book ASTM Stand.* pp. 245–249 (1984).
69. T. Arai, N. Inoue, K. Endo, and T. Nozaki, Precise measurement of carbon concentration in silicon by IR spectroscopy and charged particle activation analysis. *Ext. Abstr., 33rd Spring Meet. Jpn. Soc. Appl. Phys. Relat. Soc.* p. 771 (1986) (in Japanese).
70. H. Liebl, Ion microscope mass analyzer. *J. Appl. Phys.* **38**, 5277–5283 (1967).
71. C. A. Anderson, Progress in analytic methods for the ion microprobe mass analyzer. *Int. J. Mass Spectrum. Ion Phys.* **2**, 61–74 (1969).
72. R. W. Odow, B. K. Furman, C. A. Evans, Jr., C. E. Bryson, W. A. Peterson, M. A. Kelly, and D. H. Wayne, Quantitative image acquisition system for ion microscopy based on the resistive anode encoder. *Anal. Chem.* **55**, 574–578 (1983).

73. C. W. Magee and R. W. Honig, Depth Profiling by SIMS—Depth resolution, dynamic range and sensitivity. *Surf. Interface Anal.* **4**, 35–41 (1982).
74. A. M. Huber and M. Moulin, Use of the ion microanalyzer for the characterization of bulk and epitaxial silicon and gallium arsenide. *J. Radioanal. Chem.* **12**, 75–83 (1972).
75. P. Chu and R. S. Hockett, private communication (1987).
76. E. Zinner, Sputter depth profiling of microelectronic structures. *J. Electrochem. Soc.* **130**, 199C–222C (1983).
77. P. R. Boudewijn and H. W. Werner, Quantitative SIMS depth profiling of semiconductor materials and devices. *In* "Secondary Ion Mass Spectroscopy SIMS-V" (A. Benninghoven, ed.), pp. 270–278. Springer-Verlag, Berlin and New York, 1985.
78. W. K. Hofker, Implantation of boron in silicon. *Philips Res. Rep., Suppl.* **8**, (1975).
79. A. S. M. Salih, W. Maszara, H. J. Kim, and G. A. Rozgonyi, Extrinsic gettering with epitaxial misfit dislocations. (J. Colton, D. S. Simons, and H. W. Werner, eds.), *In* "Impurity Diffusion and Gettering in Silicon" (R. B. Fair, C. W. Pearce, and J. Washburn, eds.), pp. 61–67. Mater. Res. Soc. Pittsburgh, 1985.
80. J. C. Mikkelsen, Jr., Diffusivity of oxygen in silicon during steam oxidation. *Appl. Phys. Lett.* **40**, 336–337 (1982).
81. J. C. Mikkelsen, Jr., Oxygen solubility in silicon as a function of oxidizing temperature, ambient, and wafer orientation. *In* "Defects in Silicon" (W. M. Bullis and L. C. Kimerling, eds.), pp. 95–104. Electrochem. Soc., Princeton, New Jersey, 1983).
82. F. Shimura, W. Dyson, J. W. Moody, and R. S. Hockett, Oxygen behavior in heavily Sb-doped CZ-silicon. *In* "VLSI Science and Technology/1985" (W. M. Bullis and A. Broydo, eds.), pp. 507–516. Electrochem. Soc., Princeton, New Jersey, 1985.
83. F. Shimura, R. S. Hockett, D. A. Read, and D. H. Wayne, Direct evidence for co-aggregation of carbon and oxygen in Czochralski silicon. *Appl. Phys. Lett.* **47**, 794–796 (1985).
84. F. Shimura and R. S. Hockett, Nitrogen effect on oxygen precipitation in Czochralski silicon. *Appl. Phys. Lett.* **48**, 224–226 (1986).
85. R. S. Hockett, P. B. Fraundorf, D. A. Read, D. H. Wayne, and G. K. Fraundorf, Oxygen and carbon defect characterization in silicon by SIMS. *In* "Oxygen, Carbon, Hydrogen, and Nitrogen in Crystalline Silicon" (J. C. Mikkelsen, Jr., S. J. Pearton, J. W. Corbett, and S. J. Penny, eds.), pp. 433–438. Mater. Res. Soc., Pittsburgh, 1986.
86. F. Shimura, Carbon enhancement effect on oxygen precipitation in Czochralski silicon. *J. Appl. Phys.* **59**, 3251–5254 (1986).
87. G. J. Slusser and L. MacDowell, Source of surface contamination affecting electrical characteristics of semiconductors. *J. Vac. Sci. Technol., A* [2] **5**, 1649–1651 (1987).
88. B. F. Phillips, D. C. Burkman, W. R. Schmidt, and C. A. Peterson, The impact of surface analysis technology on the developmet of semiconductor wafer cleaning processes. *J. Vac. Sci. Technol. A* [2] **1**, 646–649 (1983).
89. S. Kawado, T. Taniguchi, and T. Maruyama, SIMS analysis of aluminum contaminants on silicon wafers. *In* "Semiconductor Silicon 1986" (H. R. Huff, T. Abe, and B. Kolbesen, eds.), pp. 989–998. Electrochem. Soc., Princeton, New Jersey, 1986.
90. D. C. Miller and G. A. Rozgonyi, Defect characterization by etching, optical microscopy and X-ray topography. *In* "Handbook of Semiconductors" (S. P. Keller, ed.), Vol. 3, pp. 217–246. North-Holland, Publ., Amsterdam, 1980.
91. E. Sirtl and A. Adler, Chromsäure-Flußsäure als spezifisches System zur Ätzgrunbenentwicklung auf Silizium. *Z. Mettalkd.* **52**, 529–531 (1961) (in German).
92. F. Secco d'Aragona, Dislocation etch for (100) planes in silicon. *J. Electrochem. Soc.* **119**, 948–951 (1972).
93. M. W. Jenkins, A new preferential etch for defects in silicon crystals. *J. Electrochem. Soc.* **124**, 757–762 (1977).

94. M. K. El Ghor, Defect delineation in oxygen denuded CZ silicon: A comparison between Secco and Write chemical etchants. M.S. Thesis, North Carolina State University, Raleigh, North Carolina (1985).

95. H. Tsuya and F. Shimura, Transient behavior of intrinsic gettering in CZ silicon wafers. *Phys. Status. Solidi A* **79**, 199–206 (1983).

96. F. Shimura, H. Tsuya, and T. Kawamura, Surface- and inner-microdefects in annealed silicon wafer containing oxygen. *J. Appl. Phys.* **51**, 269–273 (1980).

97. A. R. Lang, Studies of individual dislocations in crystals by X-ray diffraction micrography. *J. Appl. Phys.* **30**, 1748–1755 (1959).

98. B. K. Tanner and D. K. Bowen, eds. "Characterization of Crystal Growth Defects by X-ray Methods." New York, 1979.

99. N. Kato, Perfect and imperfect crystals. *In* "Characterization of Growth Defects by X-Ray Methods" (B. K. Tanner and D. K. Bowen, eds.), pp. 264–297. Plenum, New York, 1979.

100. B. K. Tanner, "X-Ray Diffraction Topography." Pergamon, Oxford, 1976.

101. A. R. Lang, A method for the examination of crystal sections using penetrating characteristic X radiation. *Acta Metall.* **5**, 358–364 (1957).

102. G. Borrmann, Über Extinktionsdiagramme von Quartz. *Phys. Z.* **42**, 157–162 (1941) (in German).

103. J. R. Patel and A. Authier, X-Ray topography of defects produced after heat treatment of dislocation-free silicon containing oxygen. *J. Appl. Phys.* **46**, 118–125 (1975).

104. E. S. Meieran, Industrial implications of crystal quality. *In* "Characterization of Crystal Growth Defects by X-Ray Methods." (B. K. Tanner and D. K. Bowen, eds.), pp. 1–27. Plenum, New York, 1979.

105. G. H. Schwuttke, New x-ray diffraction microscopy technique for the study of imperfections in semiconductor crystals. *J. Appl. Phys.* **36**, 2712–2721 (1965).

106. A. R. Lang, X-Ray detectors. *In* "Characterization of Crystal Growth Defects by X-Ray Methods." (B. K. Tanner and D. K. Bowen, eds.), pp. 320–332. Plenum, New York, 1979.

107. J. Chikawa, I. Fujimoto, and T. Abe, X-Ray topographic observation of moving dislocations in silicon crystals. *Appl. Phys. Lett.* **21**, 295–298 (1972).

108. J. Chikawa and I. Fujimoto, Video display technique for x-ray diffraction microscopy and its application. *NHK Tech. J.* **26**, 3–18 (1974) (in Japanese).

109. J. Chikawa, Laboratory techniques for transmission x-ray topography. *In* "Characterization of Crystal Growth Defects by X-Ray Methods." (B. K. Tanner and D. K. Bowen, eds.), pp. 368–400. Plenum, New York, 1979.

110. W. Hartmann, X-Ray TV imaging and real-time experiments. *In* "Characterization of Crystal Growth Defects by X-Ray Methods." (B. K. Tanner amd D. K. Bowen, eds.), pp. 497–502. Plenum, New York, 1979.

111. W. C. Dash, Copper precipitation on dislocations in silicon. *J. Appl. Phys.* **27**, 1193–1195 (1956).

112. A. J. R. de Kock, Microdefects in dislocation-free silicon crystals. *Philips Res. Rep., Suppl.* **1** (1973).

113. A. J. R. de Kock, Vacancy clusters in dislocation-free silicon. *Appl. Phys. Lett.* **16**, 100–102 (1970).

114. W. Berg, Über eine röntgenographische Methode zur Untersuchung von Gitterstörungen an Kristallen. *Naturwissenschaften* **19**, 391–396 (1931) (in German).

115. C. S. Barrett, A new microscopy and its potentialities. *Trans. Am. Inst. Min. Metall. Eng.* **161**, 15–64 (1945).

116. W. L. Bond and J. Andrus, Structural imperfections in quartz crystals. *Am. Mineral.* **37**, 622–632 (1952).

117. M. Renninger, Messungen zur Röntgenstrahl-Optik des Idealkristalls. I. Bestätigung der Darwin-Ewald-Prins-Kohler-Kurve. *Acta Crystallogr.* **8**, 597–606 (1955) (in German).

118. K. Kohra and S. Kikuta, A method of obtaining an external parallel X-ray beam by successive asymmetric diffractions and its applications. *Acta Crystallogr., Sect. A* **A24**, 200–205 (1968).

119. K. Kohra and T. Matsushita, Characterization of single crystals using multiple crystal arrangements. *J. Jpn. Assoc. Cryst. Growth* **4**, 55–72 (1977) (in Japanese).

120. Y. Takano and M. Maki, Diffusion of oxygen in silicon. *In* "Semiconductor Silicon 1973" (H. R. Huff and R. R. Burgess, eds.), pp. 469–481. Electrochem. Soc., Princeton, New Jersey, 1973.

121. A. Iida, Application of X-ray triple crystal diffractometry to studies on the diffusion-induced defects in silicon crystals. *Phys. Status Solidi A* **54**, 701–706 (1979).

122. W. L. Bond, Precision lattice constant determination. *Acta Crystallogr.* **13**, 814–818 (1960).

123. M. Ando, Lattice constant and crystal perfection of silicon. *Oyo Butsuri* **47**, 583–587 (1978) (in Japanese).

124. J. J. Hren, J. I. Goldstein, and D. C. Joy, "Introduction to Analytical Electron Microscopy." Plenum, New York, 1979.

125. W. Krakow, D. A. Smith, and L. W. Hobbs, eds. "Electron Microscopy of Materials." North-Holland Publ., Amsterdam, 1984.

126. P. B. Hirsch, A. Howie, R. B. Nicholson, and D. W. Pashley, "Electron Microscopy of Thin Crystals." Butterworth, London, 1965.

127. M. H. Loretto and R. E. Smallman, "Defect Analysis in Electron Microscopy." Chapman & Hall, London, 1975.

128. F. Shimura, H. Tsuya, and T. Kawamura, Thermally induced defect behavior and effective intrinsic gettering sinks in silicon wafers. *J. Electrochem. Soc.* **128**, 1579–1583 (1981).

129. F. Shimura and H. Tsuya, Multistep repeated annealing for CZ-silicon wafers: Oxygen and induced defect behavior. *J. Electrochem. Soc.* **129**, 2089–2095 (1982).

130. G. R. Booker and R. Stickler, Method of Preparing Si and Ge specimens for examination by transmission electron microscopy. *Br. J. Appl. Phys.* **13**, 446–448 (1962).

131. B. O. Kolbesen, K. R. Mayer, and G. E. Schuh, A new preparation method for large area electron-transparent silicon samples. *J. Phys. E* **8**, 197–199 (1975).

132. C. J. Varker and L. H. Chang, Preparation of large-area, electron-transparent silicon specimens by anisotropic etching. *Solid State Technol.* Apr., pp. 143–146 (1983).

133. H. Föll and B. O. Kolbesen, Advantages in the study of crystal defects in silicon devices by use of a high-voltage electron microscope (HVEM). *In* "Semiconductor Silicon 1977" (H. R. Huff and E. Sirtl, eds.), pp. 740–749. Electrochem. Soc., Princeton, New Jersey, 1977.

134. T. Imura, In situ high voltage electron microscope observation and its applications. *Oyo Butsuri* **12**, 1142–1158 (1979) (in Japanese).

135. H. Hashimoto and H. Endo, High resolution electron microscopic images of crystal lattices. *Oyo Butsuri* **45**, 104–124 (1974) (in Japanese).

136. A. Ourmazd, K. Ahlborn, K. Ibeh, and T. Honda, Lattice and atomic structure imaging of semiconductors by high resolution transmission electron microscopy. *Appl. Phys. Lett.* 47, 685–688 (1985).

137. K. Izui, S. Furuno, and H. Otsu, Observation of crystal structure images of silicon. *J. Electron Microsc.* **26**, 129–132 (1977).

138. A. G. Cullis and D. C. Joy, eds., "Microscopy of Semiconducting Materials/1981." Inst. Phys., Bristol, 1981.

139. J. M. Cowley, Principle of image formation. *In* "Semiconductor Silicon 1973" (H. R. Huff and R. R. Burgess, eds.), pp. 1–42. Electrochem. Soc., Princeton, New Jersey, 1973.

140. C. J. Humphreys, STEM imaging of crystals and defects. *In* "Semiconductor Silicon 1973" (H. R. Huff and R. R. Burgess, eds.), pp. 305–332. Electrochem. Soc., Princeton, New Jersey, 1973.

141. S. J. Pennycook and J. Narayan, Direct imaging of dopant distributions in silicon by scanning transmission electron microscopy. *Appl. Phys. Lett.* **45**, 385–387 (1984).

142. D. B. Williams, An overview of analytical electron microscopy. *In* "Semiconductor Silicon 1973" (H. R. Huff and R. R. Burgess, eds.), pp. 11–22. Electrochem. Soc., Princeton, New Jersey, 1973.

143. D. C. Joy. The basic principles of electron energy loss spectroscopy. *In* "Semiconductor Silicon 1973" (H. R. Huff and R. R. Burgess, eds.), pp. 223–244. Electrochem. Soc., Princeton, New Jersey, 1973.

144. K. H. Yang, R. Anderson, and H. F. Kappert, Identification of oxide precipitates in annealed silicon crystals. *Appl. Phys. Lett.* **33**, 225–227 (1978).

145. R. A. Craven, F. Shimura, R. S. Hockett, L. W. Shive, P. B. Fraundorf, and G. K. Fraundorf, Characterization techniques for VLSI silicon. *In* "VLSI Science and Technology/1984" (K. E. Bean and G. A. Rozgonyi, eds.), pp. 20–35. Electrocem. Soc., Princeton, New Jersey, 1984.

146. D. B. Holt, M. D. Muir,, P. R. Grant, and I. M. Boswara, eds., "Quantitative Scanning Electron Microscopy." Academic Press, New York, 1974.

147. D. Fathy, T. G. Sparrow, and U. Valdre, Observation of dislocations and microplasma sites in semiconductors by direct correlations of STEMBIC, STEM and ELS. *J. microsc. (Oxford)* **118**, 263–273 (1980).

Chapter 7

Grown-In and Process-Induced Defects

Silicon microelectronic circuit devices are fabricated through various processes as was shown in Fig. 5.1. After the preparation of polished wafers, the fabrication of even the simplest silicon device involves several processes in sequence or parallel.[1,2] A modern VLSI fabrication process includes hundreds of various steps. Device processing mainly involves subjecting the polished silicon wafer to a variety of chemical, physical, and thermal treatments to fabricate active and passive device elements in the wafer. Such thermal processes for silicon wafers include roughly four major steps: (1) oxidation to form silicon oxide layers, (2) chemical vapor deposition to form a silicon epitaxial layer, silicon oxide, or nitride, (3) diffusion of dopants, and (4) defect annealing. Thermal oxidation and diffusion are usually performed at temperatures around 1000°C or higher. Thus silicon crystals experience severe steps starting from the crystal growth through the complete device fabrication via wafer shaping processes. Even in recent high-quality silicon wafers, which are grown without any threading dislocations, various kinds of microdefects are induced during thermal processes. Figure 7.1 shows the steps for silicon device fabrication and the defects that can be induced into the silicon wafer. For convenience, these defects are classified into two categories: (1) grown-in defects and (2) process-induced defects, which will be further classified into surface and interior defects. The major grown-in defects, or nonuniformity, are attributable to impurity inhomogeneity due to the segregation phenomenon, which was discussed in Section 5.3.1. It should be noted that these defects may interact strongly with each other. Although dislocations or stacking faults are usually not observed in as-grown or as-polished silicon wafers used for microelectronic circuit fabrication, their origins may inherently exist in as-grown crystals and those lattice defects can be actualized by the subsequent thermal processes. In that case, the process-induced defects may have to be more suitably called *process-generated defects*.

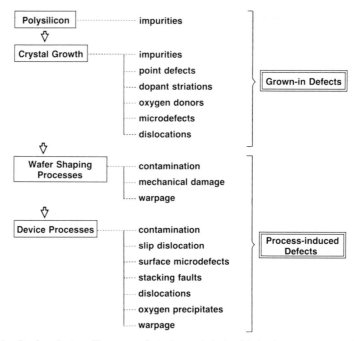

Fig. 7.1. Semiconductor silicon manufacturing and device fabrication processes, and defects induced into silicon.

There are many possible causes for device yield loss, but the interrelationship of the device processing with the properties of silicon material becomes even more important with the onset of the VLSI/ULSI era. Since crystallographic imperfections and impurities can perturb the periodicity of the lattice, the transport properties of charged carriers will be affected by their presence. The defects introduce energy levels lying within the bandgap of silicon through perturbation of the silicon lattice, which results in problems with various electrical properties.[3–6] The micrometer or submicrometer design rules in VLSI/ULSI devices increase the sensitivity to those microdefects, to say nothing of macroscopic defects such as slip dislocations and oxidation-induced stacking faults from accidental front surface mechanical damage. A charge-coupled device (CCD) is the most suitable means to evaluate and visualize the effect of various types of crystal imperfections on the electrical properties of silicon crystals.[7]

This chapter overviews first grown-in and process-induced defects. In particular, thermally induced defects are discussed in terms of their origin and behavior. Then the influences of lattice defects on electrical properties and device performance are discussed. Finally, gettering techniques, which are of eminent importance in the fabrication of semiconductor microelectronic devices, are discussed.

7.1 Grown-In Defects

7.1.1 Dislocations

The key issues in the growth of large-diameter silicon crystals is to grow the crystal without any dislocations. The generation of dislocations during crystal growth can occur through several different mechanisms; however, the dominant origin of dislocations in the case of CZ silicon crystal growth may relate to thermomechanical stresses generated in the growing crystal, particularly at the solid–melt interface.[8] Most metals are fairly closely packed in the solid state, and therefore shrink on solidification from the melt. On the other hand, silicon with the diamond structure has a rather open structure (see Section 3.4.1) and actually expands on solidification by about 10%. As was shown in Fig. 5.15, the heat flow diverges axially and radially through the growing crystal in a CZ crystal puller. The heat flow depends on several factors in the puller. When the thermal gradient is uniform throughout in the growing crystal, no thermal stress is generated. However, a nonuniform thermal gradient can result in the generation of thermal stresses in the growing crystal, since differential expansion and shrinkage can be locally generated in the growing ingot. If the thermal stress in a crystal exceeds its elastic limit, plastic deformation will take place by dislocation generation and propagation. For a given temperature gradient $\partial T/\partial r$ at a given radial position r, the thermal stress induced can be relieved by the generation of dislocations[8] of density $n(r)$:

$$n(r) = \frac{\beta}{\mathbf{b}} \left(\frac{\partial T}{\partial r} \right) \tag{7.1}$$

where β is the thermal expansion coefficient of silicon ($\sim 4.5 \times 10^{-6}/°C$ for the range 500–$850°C$[9]) and $\mathbf{b}$ is the Burgers vector.

The deposition of SiO particles at the solid–melt iterface readily generate dislocations in the growing silicon crystal. Once a dislocation is introduced, the dislocation will spread through the interface into the growing crystal, resulting in polycrystallization of silicon. Slip dislocations generated in $\langle 111 \rangle$ CZ silicon are schematically illustrated in Fig. 7.2. As noted previously, however, commercially obtained silicon crystals used for the fabrication of recent microelectronic circuits are almost exclusively so-called *dislocation-free* crystals, in which no dislocations can be observed.

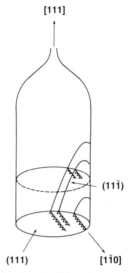

Fig. 7.2. Schematic illustration showing slip dislocations generated from the bottom of a $\langle 111 \rangle$ CZ silicon crystal ingot.

7.1.2 Microdefects

Microdefects in Dislocation-Free Silicon   The technology for the growth of dislocation-free silicon crystals enables the growth of large-diameter crystals as well; however, it has resulted in the emergence of a new type of microdefects that were not observed in *dislocated* silicon crystals. Those microdefects are apparently different from dislocations in nature, but must be related to point defects because a dislocation-free silicon crystal inherently becomes supersaturated with thermal point defects (i.e., self-interstitials and vacancies) on cooling from the growth temperature. These microdefects in as-grown silicon have been the subject of intense investigation for many years. The grown-in microdefects are commonly called *swirl defects*, and this term originates from their distribution pattern of spiraling striations revealed by chemical etching across a silicon wafer surface. The term swirl was first used by Kämper in 1970 to describe an etch pattern that was thought to be impurity striations but was actually swirl.[10] Swirl defects are observed in as-grown FZ silicon, but are usually not observed in as-grown CZ silicon.[11] In CZ silicon crystals, swirl-patterned defects such as shown in Fig. 6.30 can be observed after a certain heat treatment.

Nature of Swirl Defects   The swirl defects were first reported by Abe and co-workers in 1966, although the term "swirl" was not used to describe the

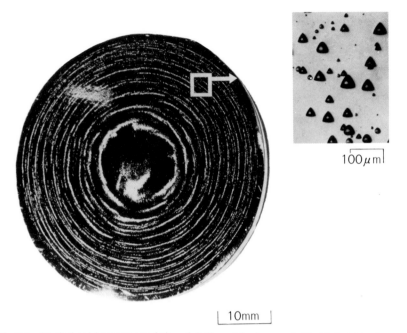

100μml

10mm

Fig. 7.3. Typical swirl pattern consisting of shallow etch pits revealed by Sirtl etching in an as-grown (111) FZ silicon crystal. (After Abe.[13] Reprinted with the permission of Academic Press, Inc. Courtesy of T. Abe, SEH.)

distribution of such grown-in defects in dislocation-free FZ silicon wafers.[12] These swirl defects were revealed as *shallow pits* by Sirtl etching. Figure 7.3 shows a typical swirl pattern consisting of shallow etch pits revealed by Sirtl etching in an as-grown (111) "dislocation-free" FZ silicon.[13] Infrared transmission microscopy with copper decoration further distinguished two different types of defects, which resulted in large elongated images and small spots. The two defect types are now commonly called *A-clusters* and *B-clusters*,[14] or more suitably *A-swirl* and *B-swirl*.[15] These defects are distributed in a swirling pattern generally related to the striations as shown in Fig. 7.4.[16] At an early stage of the investigation, these microdefects were considered to be vacancy clusters, which evolved into voids or dislocation loops.[17,18] Later, the nature of swirl defects was claimed to be ordered oxygen precipitates associated with or without vacancies.[19] Moreover, based on TEM analysis, it was reported that A-swirl defects were perfect dislocation loops both vacancy and interstitial in nature.[20] However, it has now been well established by a study with TEM that A-swirl defects are perfect dislocation loops that are interstitial-type in nature with $\mathbf{b} = \dfrac{a}{2}\langle 110 \rangle$ and that the loop planes are near

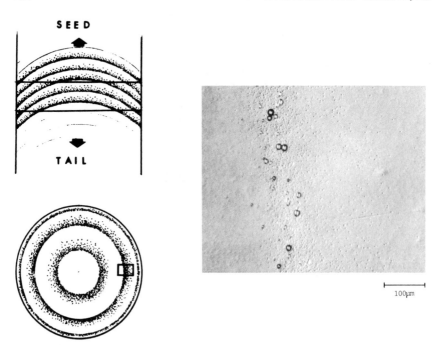

Fig. 7.4. Schematic illustration (left) showing the distribution of swirl defects in a silicon crystal: left top, longitudinal section; left bottom, transverse section. The micrograph of swirl-defect etch pits on the right corresponds to the framed rectangular area on the left bottom. (Courtesy of S. M. Hu, IBM Corporation.[16])

{111}.[15,21] On the other hand, B-swirl defects have not yet been well characterized because of their weak strain field for TEM. However, it has been reasonably assumed that B-swirl defects are not small dislocation loops but loosely packed three-dimensional agglomerates of silicon self-interstitials and some impurity atoms, most likely carbon atoms, which are less stable than A-swirl defects.[15,21]

More recently, another type of microdefect has been found in fast-grown dislocation-free FZ silicon.[22] These defects in as-grown silicon crystals can not be revealed by means of preferential etching or TEM, but can be observed using techniques such as copper or lithium decoration and XRT. The distribution of those defects, eventually called *D-type defects*, is homogeneous but does not form a striated pattern. This indicates that the formation of D-type defects does not take place on nuclei in a striated distribution, which is generally accepted to explain A- and B-swirl defects. Consequently, it has been claimed that D-type defects are agglomerates of vacancies whose dangling bonds are eliminated by the formation of double bonds.

7.2 Process-Induced Defects

7.2.1 Thermomechanically Induced Dislocations

Thermomechanical Effect Silicon and other materials with the diamond lattice structure are extremely brittle at room temperature but become ductile at temperatures above 60% of their absolute melting temperature, that is ~740°C for silicon.[3] Figure 7.5 shows a typical strain–stress curve. With increasing strain the stress increases until it reaches a value known as the *upper yield stress* (τ_{UY}), and then drops rapidly to a level referred to as the *lower yield stress* (τ_{LY}), followed by slow increase with increasing strain.[23] At the upper yield stress point, the crystal is plastically deformed, resulting in the generation of dislocations. When a silicon crystal is stressed in compression or tension at elevated temperatures higher than 600°C, the crystal can be plastically deformed accompanying a pronounced yield drop in the initial state of deformation.[24] As was discussed in Section 5.2, the upper yield value depends on the quality of silicon as well as on the deformation process conditions. That is, dissolved impurities such as oxygen and nitrogen raise the upper yield stress whereas defects such as dislocations and oxygen precipitates drop the value. Moreover, with increasing temperature the stress required to initiate plastic deformation decreases.[24] Since silicon device manufacturing involves thermal processes at high temperatures, the heating and cooling of thin silicon wafers, particularly large-diameter wafers, lead to situations where induced thermomechanical stresses are large enough to exceed the upper yield stress of silicon crystals, which results in the generation of dislocations.[25]

Thermomechanical Stresses in Silicon Wafer When a silicon wafer with a large ratio of surface area to thickness is pushed into a higher-temperature furnace, nonuniform heating will occur since the wafer must be supported with some material that can act as a heat sink or a radiation shield, as

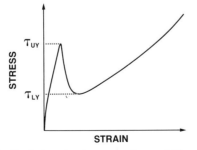

Fig. 7.5. Typical strain–stress curve for crystalline silicon.

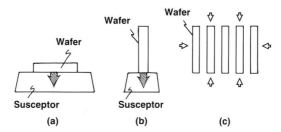

Fig. 7.6. Schematic illustration showing nonuniform heating for semiconductor wafers.

illustrated in Fig. 7.6a and b. If many wafers are positioned in a parallel row with a narrow spacing between individual wafers as shown in Fig. 7.6c, each wafer except the ones at the end of the row is shielded from the heat radiation by neighboring wafers, which also results in nonuniform heating. In the case where a wafer is placed horizontally on a susceptor (Fig. 7.6a), as in the epitaxial reactors shown in Fig. 5.64, uniform heating can be performed if the silicon wafer is ideally flat, because the heat sink touches the wafer uniformly. In order to enhance throughput and production efficiency, silicon wafers undergoing oxidation or diffusion process are usually placed vertically on a quartz or silicon carbide boat in long rows with regular narrow spacings between them. The wafers are supported at a few points around the circular wafer periphery. When the row of silicon wafers is either inserted into or withdrawn from the hot zone of a furnace, then the two factors of nonuniform heating shown in Fig 7.6b and c are superimposed, which results in a geometrical factor for radiation heat transfer.[26,27] This geometrical factor depends on the radial position on the wafer as well as on the spacing between wafers. If the spacing is small, the shielding effect will be large compared to the heat-sink effect, but as the spacing becomes larger the heat-sink effect becomes more significant. When the rates of insertion and withdrawal of the wafers are high, temperature gradients thus produced can generate high enough thermal stresses to cause plastic deformation of wafers. Particularly during cooling process, heat in the center region of a wafer tends to be trapped there, causing a substantial radial temperature gradient which leads to the generation of slip dislocations.[27] The relation between the thermal stress and induced dislocations is essentially ruled by Eq. (7.1). The thermal stress σ_T generated when silicon is cooled down from T_0 to T is generally given by

$$\sigma_T = \beta E(T_0 - T) \tag{7.2}$$

where E is Young's modulus, which is anisotropic (i.e., 1.3×10^{12} dyn/cm^2 for $\langle 100 \rangle$, 1.7×10^{12} dyn/cm^2 for $\langle 110 \rangle$, and 1.9×10^{12} dyn/cm^2 for $\langle 111 \rangle$.[28] In addition, it should be considered that the weight of a wafer that is placed vertically on a boat in a horizontal furnace (which has been used extensively for a variety of semiconductor processes) may result in an additional strain to the bottom part of the wafer. This effect obviously becomes greater for larger-diameter (thus heavier) wafers. In vertical furnaces where wafers are held horizontally either by three to four points or by a full support around their periphery, the additional strain to the bottom part of a wafer may be minimized. Moreover, another possible advantage of a vertical furnace over a conventional horizontal one is that it allows the wafer holder to be rotated during thermal processing, which can improve the thermal uniformity.[29] Vertical furnaces seem beneficial for future device technology, particularly when large-diameter wafers are used. However, critics of vertical systems claim that holding the wafers horizontally permits the wafers to sag and warp.

Nature of Slip Dislocations   As was discussed in Section 3.5.2, the slip system in the diamond lattice structure is the $\langle 110 \rangle$ directions on $\{111\}$ planes. Taking into consideration of the slip system and the anisotropy of the thermal stress, the slip dislocations generated as a consequence of thermal stresses can generally assume a threefold symmetry in (111) wafers and fourfold symmetry in (100) wafers. In all cases slip lines run toward $\langle 110 \rangle$ directions as expected. If the temperature distribution in a silicon wafer is known, it is possible to calculate the relative shear stress pattern in each slip system.[26,30] For the case of a (111) silicon wafer where three $\{111\}$ planes intersect the (111) plane, the results of these calculations are summarized in Fig. 7.7.[30] Moreover, dislocations occur in much higher densities at the wafer periphery as compared with the wafer center, reflecting the stress distribution in the wafer,[26] as well as the fact that wafer edges generally have some mechanically damaged sites that can be thermal stress-concentration

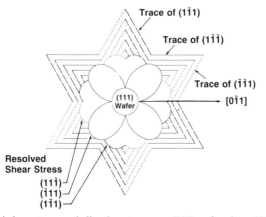

Fig. 7.7.   Resolved shear stress and slip-plane traces on (111) wafer plane. (After Bennett and Sawyer.[30] Reproduced with permission from the Bell System Technical Journal. © 1956 AT & T.)

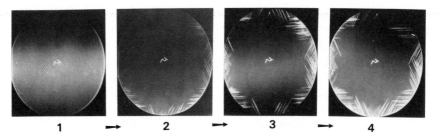

Fig. 7.8.   Lang X-ray topograph taken with $MoK\alpha_1/g_{220}$ diffraction for a (111) CZ silicon wafer subjected to heat treatment at 1000°C for three times (2, 3, and 4). (1 refers to before heat treatment.)

centers.[31] Figure 7.8 represents slip dislocations in a (111) CZ silicon wafer revealed by the Lang XRT with the $MoK\alpha_1/g_{\bar{2}20}$ diffraction. The wafer was moderately rapidly pushed in and pulled out of a furnace at 1000°C three times. The XRT was taken after each thermal step. It is observed that more slip dislocations are generated with thermal steps. The slip pattern in a (111) silicon wafer, which is commonly called the *star of David*, observed in Fig. 7.8 agrees well with the pattern of the calculated resolved shear stress and slip-plane traces shown in Fig. 7.7. It has been observed that the generator dislocation can either be a 60° dislocation or a screw dislocation.[32]

In general, the generation of slip dislocations greatly depends on several factors, such as the cooling environment, the heat treatment temperature, the wafer diameter over thickness ratio, and the amount of oxygen precipitation; slip dislocations can cause serious warpage of silicon wafers.[33]

Effect of Mechanical Damage   As described above, mechanically damaged sites can provide the nucleation centers for thermally induced dislocations.[31–34] Figure 7.9 shows the Lang XRT for a CZ silicon wafer, for which the edges were mechanically ground; the right half was then chemically etched to remove the mechanical damage, subjected to a heat treatment at 1100°C for 100 min, and pulled out of the furnace at the rate of 75 mm/min.[35] A significant difference in the generation of slip dislocations (black contrasts in the XRT) is observed between the right and left halves, which clearly indicates that mechanical damages on the wafer edge strongly affect slip generation, as expected. For the purpose of impurity gettering (see Section 7.4), mechanical damages are commonly given to the backside surface of silicon wafers. Such mechanical damages on a wafer backside can originate slip dislocations which propagate to the frontside of the wafer. Figure 7.10 shows the Lang XRT, which demonstrates how backside damages affect the generation of dislocations. The backside of the wafer was mechanically

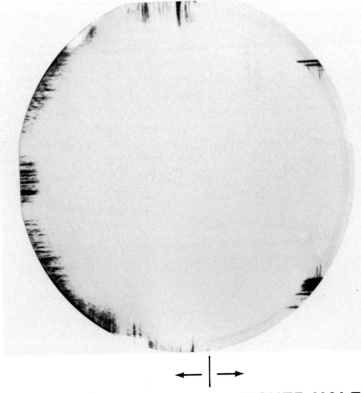

← | →

EDGE GROUND HALF   ETCHED HALF

Fig. 7.9. Lang X-ray topograph taken with $MoK\alpha_1/g_{022}$ diffraction for a (100) CZ wafer subjected to oxidation at 1100°C for 100 min. Left half edge was mechanically ground, while right half edge was chemically etched. (After Chiou.[35])

damaged with abrasion, and then the backside of the right half was chemically etched to remove the mechanical damage. The wafer was then subjected to a thermal stress by heating at 1000°C in a furnace with moderately rapid withdrawal to room temperature. It is clearly observed that dense dislocations are generated in the left half with mechanically damaged backside surface. It is thus important to reduce the thermal stress to a minimum in such an application of backside damage gettering.

Influence of Wafer Curvature   For silicon epitaxy, silicon substrates are mostly placed on the susceptor as shown in Fig. 5.64. The wafers are usually heated by the susceptor through both conduction and radiation, while they lose heat primarily by radiation. The radial temperature variation in wafers

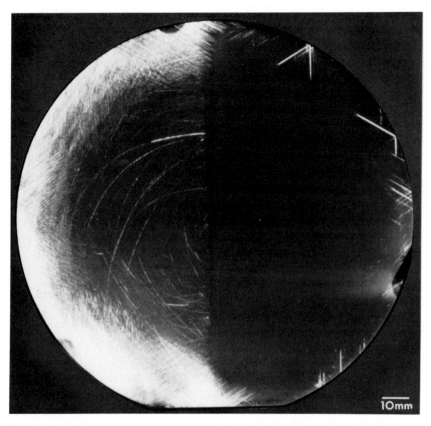

Fig. 7.10. Lang X-ray topograph taken with $MoK\alpha_1/g_{022}$ diffraction for a (100) CZ silicon wafer subjected to heat treatment at 1000°C; left half has mechanically damaged back surface, whereas right half has chemically etched back surface. (Courtesy of H-D. Chiou, Monsanto Electronic Materials Company.)

may be negligible for ideally flat ones on a uniformly heated susceptor. However, a considerable amount of radial temperature gradients can be produced in curved wafers. Radial temperature distributions for both convex and concave wafers have been calculated.[36] The theoretical radial temperature distributions in a 50-mm-diameter, 200-μm-thick wafer with 50 μm warp was shown in Fig. 7.11. The susceptor temperature was assumed to be 1230°C. It has been calculated that the radial temperature distribution is fairly insensitive to wafer thickness for the range between 100 and 400 μm. Figure 7.11 indicates that radial temperature variations on the order of 20°C are possible in even such a small-diameter wafer. It is thus reasonably assumed that the larger the diameter, the greater the temperature gradient,

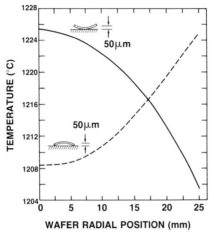

Fig. 7.11. Theoretical radial temperature distribution in a 50-mm-diameter, 200-μm-thick silicon wafer with 50 μm warp; susceptor temperature is assumed to be 1230°C. (After Huff *et al.*[36] Reproduced with the permission of The Electrochemical Society, Inc.)

resulting in more slip dislocations.[37] As previously shown, slip dislocations generally start at the wafer edge, and higher densities are found in the wafer periphery compared with the wafer center. However, plastic deformation can begin near the wafer center when the wafer has a convex shape toward a susceptor and the wafer curvature is sufficiently large.[38] For a (111) silicon wafer, Fig. 7.12 depicts a mechanism by which plastic deformation occurs near the wafer center, which contacts a hot susceptor. The slip near the center region results from compressive stresses generated by the entrapment of a hot center region of the wafer within a colder annular zone. Thus, slip generation,

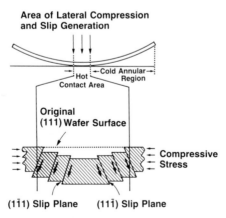

Fig. 7.12. Schematic representation showing plastic deformation caused by entrapment of hot central region within colder annular zone in a convex-shaped (111) wafer. (After Dyer *et al.*[38])

(a)

$[1\bar{1}0]$

$[\bar{1}\bar{1}2]$

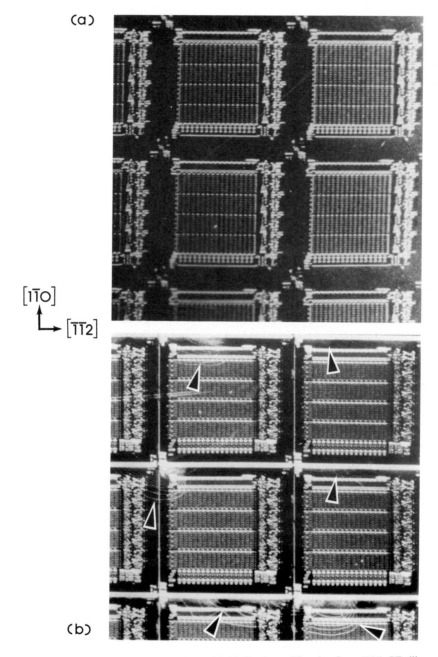

(b)

Fig. 7.13. Lang X-ray topographs taken with $MoK\alpha_1/g_{220}$ diffraction for a (111) CZ silicon wafer processed for bipolar transistor fabrication: (a) before isolation with high-concentration boron diffusion, and (b) after the isolation process. Arrows denote pattern-edge expanding dislocations.

particularly near the centered region of a wafer, may lead to considerable wafer warpage. It has also been found that warped wafers show a much larger area affected by slip on the concave side than on the convex side.[33]

Pattern-Edge Dislocations A microcircuit consists of numerous components that are fabricated through many steps of patterning. The isolation process is one of the key schemes for fabricating components that are electrically isolated from each other.[1] The isolation or patterning processes usually involve local oxidation or high-concentration impurity doping, which potentially cause the generation of pattern-edge dislocations. These pattern-edge dislocations are, in principle, generated by local thermal stresses due to the difference in volume expansion or shrinkage. Silicon oxide–nitride edge dislocations, which are usually induced during the localized oxidation of silicon (LOCOS) process, have recently been the subject of more intense investigation as the LOCOS process has become more important.[5] An example of pattern-edge dislocations generated in CCD devices was shown in Fig. 6.32b and c. Figure 7.13a and b show the Lang XRT taken with Moα_1 $g_{2\bar{2}0}$ for the same area of a (111) CZ silicon wafer processed for bipolar transistor fabrication (a) before the isolation with high-concentration boron diffusion and (b) after the isolation process. Pattern-edge dislocations marked with arrows are clearly observed after the isolation process. These dislocations have the same nature as the ones discussed above, which glide on $\{111\}$ planes toward $\langle110\rangle$ directions. Observing the pattern-edge dislocations with different diffraction conditions, one finds that the dislocations are induced dominantly from the pattern edge facing the $(1\bar{1}0)$ plane, but a few are from the $(11\bar{2})$ edge. This phenomenon can be understood as follows. Device chips are almost exclusively fabricated with a square shape using the fourfold $\langle110\rangle$ symmetry on (100) wafers and twofold symmetries of $\langle110\rangle$ and $\langle211\rangle$ on (111) wafers in order to gain process and structural controls. Figure 7.14 depicts the square shape of device chips fabricated on a (001) wafer, instead of a (100) for convenience, and a (111) wafer. Patterning either by LOCOS or by impurity doping thus results in the highest stresses in the

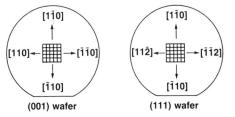

Fig. 7.14.   Schematic illustration showing square-shaped device chips fabricated on (001) and (111) wafers.

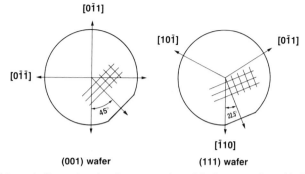

Fig. 7.15. Schematic illustration showing square-shaped device patterning with deviation from ⟨110⟩ direction for (001) and (111) wafers. (After Shimura.[39])

directions perpendicular to the chip edges or patterned lines shown in Fig. 7.14. The directions of ⟨110⟩ in which the highest stresses are induced coincide with the favored direction of slip gliding, as discussed previously. The device patterns on (001) wafers inevitably consist of four equivalent ⟨110⟩ directions that are highly stressed, whereas the patterns on (111) wafers have two. The other two pattern edges facing the ⟨211⟩ directions are less susceptible to the generation of dislocations as shown in Fig. 7.13. If the cleavage planes that are used for device chip dicing (see Fig. 3.26) can be ignored, device patterning with deviation from the ⟨110⟩ directions as shown in Fig. 7.15 may minimize the generation of pattern-edge dislocations.[39]

Prevention of Dislocations   The prevention of thermomechanically induced dislocations requires primarily a reduction of temperature gradients in the wafer. A practical and most common way to accomplish this is by programmed slow insertion and withdrawal of wafers from the furnace. Ramping is also a common practice. In the ramping procedure, the wafers are inserted at a moderate temperature around 800°C and heated up slowly to the high process temperature at a typical rate of a few to 10 degrees per minute. Cooling is performed in reverse; that is, the wafers are cooled down slowly to a moderate temperature. Wafer-boat design can also significantly affect the temperature uniformity. It has been reported that a closed-boat design can accomplish slower cooling and smaller temperature variation across a wafer.[40] These considerations to reduce the temperature gradients in a wafer are most important for larger-diameter wafers and for the thermal processes at temperatures higher than about 850°C. However, it should be noted that slow cooling may cause another problem concerning oxygen precipitation and oxygen donor formation, which will be discussed in Section 7.2.4. It is, of course, also important to reduce the density of potential dislocation sources in the substrate wafers. The mechanically damaged sites, which can act as nucleation centers for thermally induced dislocations, can be effectively reduced by introducing optimum edge rounding and polishing procedures.[41]

7.2.2 Oxidation-Induced Stacking Faults

General Remarks Stacking faults are frequently generated in the surface region of silicon wafers during thermal oxidation processes at a typical temperature range between 900 and 1200°C. These stacking faults are commonly called *oxidation-induced stacking faults* (OSFs or OISFs). Because of their detrimental effect on the electrical properties of silicon devices (see Section 7.3), great attention has been devoted to the investigation of OSFs and the prevention of their generation since the OSF was first observed in 1963.[42] Numerous publications on OSFs are cited in Refs. 3, 5, and 16.

Stacking faults are also observed in silicon epitaxial films grown on silicon substrates.[43,44] These stacking faults are commonly called *epitaxial stacking faults*, and they differ from OSFs both in structure and in the mechanisms of formation. Epitaxial stacking faults nucleate dominantly at the imperfections on the surface or in the subsurface region of the substrate, and grow into the epitaxial film. Moreover, stacking faults can be generated by oxygen precipitation in bulk silicon (see Section 7.2.5), and such stacking faults are called *bulk stacking faults*, which are also distinguished from OSFs, although the bulk stacking faults are often generated during prolonged oxidation processes.

Origin and Nature OSFs are predominantly nucleated at certain mechanical damages on the wafer surface.[43,45] Other sources of surface defects that can generate OSFs have been investigated. They include contamination with sodium[46] or metallic impurities,[47–49] and surface attack by hydrofluoric acid.[50] Moreover, swirl defects[46–51'] and oxygen precipitates[52] can also be nucleation sites for OSFs. After thermal oxidation at elevated temperatures in either dry oxygen, wet oxygen, or steam, OSFs can be delineated by preferential chemical etching as linear pits as shown in Fig. 6.22. The stacking fauts lie on $\{111\}$ planes and intersect the surface along $\langle 110 \rangle$ directions. By TEM diffraction contrast analysis, OSFs have been identified as extrinsic-type or interstitial-type stacking faults bounded by an $\frac{a}{3} \langle 111 \rangle$ Frank partial dislocation.[53,54]

Formation Mechanism Although the structure of OSFs is well established, several different models for OSF formation mechanism have been proposed. Booker and Tunstall[53] first proposed an OSF formation model following the mechanism based on the dissociation of a perfect dislocation observed in

niobium-containing autenitic stainless steels.[55] In this model, perfect dislocations with $\mathbf{b} = \dfrac{a}{2}\langle 110\rangle$ already present in the surface region due to a
mechanical damage dissociate to form a Schockley partial with $\mathbf{b} = \dfrac{a}{6}\langle 211\rangle$
and a Frank partial with $\mathbf{b} = \dfrac{a}{3}\langle 111\rangle$ during the subsequent heat treatment.
The reaction producing a (111) plane stacking fault from a dislocation with
Burgers vector $\mathbf{b} = \dfrac{a}{2}[110]$ is thus explained by

$$\frac{a}{2}[110] \longrightarrow \frac{a}{3}[111] + \frac{a}{2}[11\bar{2}] \tag{7.3}$$

The precipitation of SiO_2 accompanied by a volume change leads to gliding
of the Schockley partial and emission of vacancies from the Frank partial,
which results in the growth of the Frank partial and hence in the formation of
the stacking fault extrinsic in nature. Lawrence[56] proposed a similar model,
which is based on the Lomer–Cottrell reaction[57] among some of those
perfect dislocations induced by mechanical damages, where the dislocations
intersect with one another, resulting in the creation of partial dislocations
that can form the boundaries of a stacking fault according to the reaction

$$\frac{a}{2}[110] + \frac{a}{2}[\bar{1}01] \longrightarrow \frac{a}{3}[111] + \frac{a}{6}[\bar{1}\bar{1}2] + \frac{a}{6}[\bar{1}2\bar{1}] \tag{7.4}$$

Sanders and Dobson[58] have proposed another model where vacancies
flow continuously across the SiO_2–Si interface toward the surface during
oxidation, since oxidation takes place at the interface by the transport of
oxygen atoms by substitutional diffusion through the oxide grown. Consequently, the growing oxide continuously removes vacancies out of the silicon
surface region, and this leads to a reduced vacancy concentration, resulting in
the growth of extrinsic OSFs, which has nothing to do with oxygen
precipitates proposed by Booker and Tunstall.[53] Accordingly, it has been
shown that annealing in air causes existing stacking faults to grow, whereas
they shrink and disappear by annealing in vacuum. However, the precise
origin of the dislocations that initiate OSFs is still unclear, since the
dislocations have been induced into the wafer region by prior mechanical
damages without any heat treatment.

As an alternative to the model based on the dissociation of a preexisting
perfect dislocation, Ravi and Varker have proposed another model by which
OSFs in silicon are formed by the local collapse of excess interstitial silicon
atoms into a Frank loop lying on close-packed {111} planes.[51] Because of the

extra surface area at a mechanically damaged point, a local reduction in the vacancies would occur, leading to the precipitation of excess self-interstitials in the form of extrinsic stacking faults. Their model also suggests that the local supersaturation of silicon self-interstitials can occur at swirl defects where an inhomogeneous distribution of point defects is present.

In conjunction with the phenomenon of an enhanced diffusion of boron during the oxidation of silicon wafers, Hu[59] proposed a model that successfully explains the experimental results for both OSF growth and oxidation-enhanced diffusion of impurities. Oxidation of silicon at the SiO–Si interface is usually incomplete, which results in unoxidized silicon atoms in the vicinity of the interface. Because of their high mobility, these silicon atoms quickly enter the silicon lattice interstices, causing the lattice to be supersaturated with self-interstitials, which can form extrinsic stacking faults. Stacking-fault embryos are formed from these excess interstitials by nucleation at certain strain centers at the surface and in the bulk. Mechanically damaged sites at the surface, as well as oxygen precipitates or other kinds of sufficiently large defect clusters, can provide the nucleation centers for OSFs, although the mechanism of how such centers attach self-interstitials is not certain. In addition to the effect of incomplete oxidation on supplying self-interstitials into the silicon surface region, it is reasonable to assume that the volume increase of about 120% associated with the formation of an SiO_2 molecule from an Si atom during thermal oxidation must affect the supersaturation of silicon self-interstitials in the silicon wafer.[60]

As regards the density dependence of OSFs on the surface orientation of silicon wafers, it has been widely observed that the density in (100) wafers is generally larger than that in (111) or off-(100) wafers. As discussed previously, the density of OSFs depends primarily on the density of effective nucleation sites. It is quite understandable that (100) surfaces are more vulnerable to mechanical damages or chemical attacks than (111) surfaces, which consist of the densest packing of atoms. Moreover, the trend of OSF density in differently oriented silicon wafers can be explained with the number of surface kinks that can capture the silicon interstitials generated at the SiO_2–Si interface resulting in *surface regrowth*.[59] The number of kinks is strongly dependent on the surface orientation and decreases with the order of (111), (110), and (100).[59] That is, more self-interstitials are available for OSF formation in (100) surfaces than in (111) surfaces.

In addition, it has been observed that the OSF density is generally an order of magnitude or more higher in *n*-type silicon wafers than in *p*-type wafers.[61] This may be attributed to the lower density of active vacancies in *n*-type silicon crystals[62] or to the greater equilibrium vacancy concentration than in *p*-type silicon[58]; however, at present the mechanism that explains the fact of more OSFs in *n*-type silicon is still unresolved.

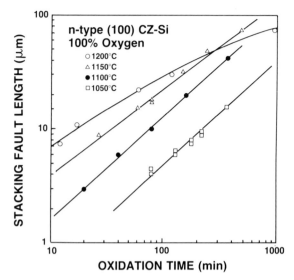

Fig. 7.16.   Stacking fault length as a function of oxidation time for oxidation in 100% oxygen at different temperatures. (After Murarka.[64])

Growth and Shrinkage   The growth of OSF depends on many factors, such as the oxidation temperature, time, ambient, crystal orientation, and impurity concentration.[63] Figure 7.16 shows a log–log plot of the stacking fault length as a function of oxidation time for oxidations carried out for $n(100)$ CZ silicon wafers in a 100% oxygen ambient at various temperatures.[64] In general, the length L (μm) of OSF is given by

$$L = At^n \exp(-Q/kT) \tag{7.5}$$

where A is a constant (which depends on many factors), t is the oxidation time (hr), n is a number exponent that is the slope of those log(OSF length)–log(time) plots and can be calculated by using the least-squares method, and Q is the activation energy of OSF growth. It is noteworthy that the values of n (~ 0.8) and Q (~ 2.3 eV) obtained for different OSF growth investigations are fairly consistent with each other, irrespective of oxidation ambients and crystal orientations of substrates.[61,63–65] The OSF growth rate increases with oxidation rate[63] as well as with oxidation temperature. The growth of OSFs as a function of temperature during oxidation for 3 hr in a dry oxygen ambient is shown in Fig. 7.17 for n-type CZ silicon wafers with different orientations.[65] Two different growth kinetic regions are clearly observed; the *growth region* and the *retrogrowth region*. The growth kinetics of OSF in a (100) wafer and 5°-off-(100) wafer appear to be the same, although the

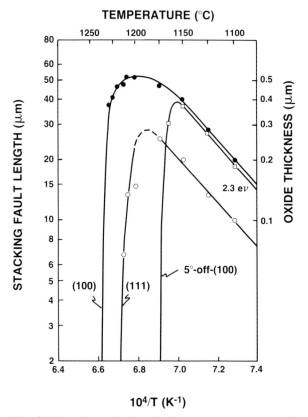

Fig. 7.17.   Stacking fault length as a function of oxidation temperature for silicon wafers with different surface orientations. (After Hu.[65])

retrogrowth of OSF occurs at a considerably lower temperature for a 5°-off-(100) wafer. The temperature at which retrogrowth occurs decreases with decreasing partial pressure of oxygen in the oxidizing ambient.[66] It will also be observed in Fig. 7.18 that the OSF growth rate is about a factor of two larger on (100) than on (111). The dependence of OSF growth rate on the wafer orientation is well established.[67–68] The values of A in Eq. (7.5) for (100) and (111) silicon wafer surfaces in the cases of oxidation both in dry and wet oxygen ambients are listed in Table 7.1.[16]

It has been widely observed that a heat treatment at a high temperature in nonoxidizing ambients results in the shrinkage of OSF.[58,69,70] Figure 7.18 shows the OSF shrinkage as functions of annealing time and temperature for $n(100)$ and $n(111)$ CZ silicon wafers subjected to heat treatments in a nitrogen ambient.[69] The figure points out that (1) OSFs shrink linearly with the

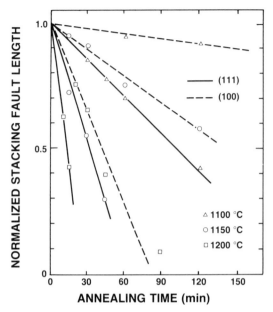

Fig. 7.18. Stacking fault shrinkage as functions of annealing time and temperature for $n(100)$ and $n(111)$ CZ silicon wafers annealed in nitrogen. (After Sugita *et al.*[69])

annealing time, (2) the higher temperature results in the greater shrinkage rate, and (3) OSFs in (111) wafers shrink faster than those in (100) wafers. The results shown in Fig. 7.18 obey the Arrhenius equation, and the activation energies for the shrinkage process have been obtained as 4.1 ± 0.3 eV and 4.9 ± 0.3 eV for (111) and (100) silicon wafer surfaces, respectively. Similar results have been obtained for silicon wafers subjected to heat treatments in hydrogen and argon ambients.[70]

Two different mechanisms have been proposed for the shrinkage of OSF in silicon: the climb process, involving either the absorption of vacancies or emission of interstitials,[58,69,70] and the unfaulting reaction involving the nucleation and motion of Schockley partials.[71] Since the activation energy

Table 7.1 Values of OSF Growth Constant A in $L = At^n \exp(-Q/kT)$ for Silicon[a]

Orientation	Dry O_2 oxidation	Wet O_2 oxidation
100	2.28×10^9	4.16×10^9
111	1.09×10^9	1.53×10^9

[a] After Hu.[16]

for the shrinkage obtained lies between 4.1 and 4.9 eV and is consistent with the activation energy for the self-diffusion of silicon (i.e., 4.78 eV[72] to 5.13 eV[73]), it is likely that the shrinkage is governed by the climb process of the bounding Frank partial loop rather than by the unfaulting reaction. An interstitial-type OSF grows by the emission of vacancies or absorption of self-interstitials, and shrinks by the absorption of vacancies or emission of silicon atoms forming the fault. These processes will occur at jogs on the bounding Frank loop. The flow of point defects will be determined by their concentration gradients between the fault and its surroundings. The fact that the activation energy for the shrinkage is in good agreement with that for the self-diffusion indicates that the OSF shrinkage is controlled by the motion of self-interstitials. It is thus assumed that an undersaturation of interstitials arises at the SiO_2–Si interface during heat treatment in nonoxidizing ambients. The difference in the activation energies for OSF growth (2.3 eV) and shrinkage (4.1–4.9 eV) suggests the different kinetics for agglomeration and dissociation of silicon interstitial atoms.

Oxidation in Chlorine-Containing Ambients The addition of a few mole percent of HCl or Cl_2 to the oxidizing ambient has been established to improve significantly the electrical stability of SiO_2 films due to the reduction of the surface state density at the SiO_2–Si interface.[74] It has been also widely recognized that the addition of chlorine-containing compound such as HCl,[75] trichloroethylene (TCE),[76] and trichloroethane[77] drastically increases the shrinkage rate of OSF and can completely suppress the generation of OSF.[78] The effect of HCl on OSF growth and shrinkage has been investigated for the wide range of HCl content in the oxidizing ambient and oxidation temperature.[78] For example, Fig. 7.19 plots OSF lengths in $n(100)$ CZ silicon wafers as a function of annealing time for the oxidation at 1150°C in ambients with various concentrations of HCl.[78] It is generally observed in HCl oxidation that:

1. The growth rate of OSF is significantly decreased with the addition of a small amount of HCl into the oxidizing ambient.

2. With the addition of a certain amount of HCl into the ambient, the OSF size increases as the first stage of oxidation reaches a maximum, and then shrinks.

3. The generation of OSF is completely suppressed by adding HCl at a concentration higher than a critical value (e.g., 1% for oxidation at 1150°C) as shown in Fig. 7.19.

4. The critical concentration decreases with the increase in oxidation temperature.

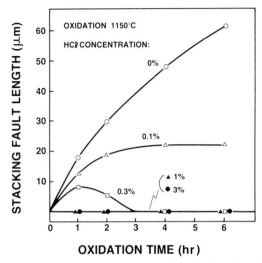

Fig. 7.19. Stacking fault length in $n(100)$ CZ silicon wafers as a function of annealing time for oxidation at 1150°C in ambients with various concentrations of HCl. (After Shiraki.[78])

The effect of HCl or chlorine-containing compound on OSF shrinkage and suppression has been first attributed to the reduction in the concentration of excess silicon interstitials,[75] which results from injection of vacancies into silicon due to the interaction between silicon and chlorine atoms forming an SiCl compound.[79] Second, the chlorine compounds form volatile complexes with impurities, particularly metallic impurities, which can be nucleation centers for OSF, and thereby eliminate the OSF nucleation sites in the surface region of the wafer.[80]

7.2.3 Surface Microdefects

General Remarks   Microdefects that manifest themselves as small saucer pits (S-pits) with a typical density of about $10^6/cm^2$ as shown in Fig. 6.24 are commonly generated in the surface of silicon wafers or epitaxial silicon films subjected to heat treatment at temperatures higher than 1100°C in a "not clean" furnace.[81] These surface microdefects have been attributed to contamination with transition metals during thermal processes.[47–49] Figure 7.20 shows etched figures of surface defects revealed by the Sirtl etchant for CZ silicon wafers subjected to different heat treatments in a "not clean" furnace. These figures indicate that (1) microdefects generate by wet O_2 oxidation are larger in size than those generated by dry O_2 oxidation, and (2) these microdefects grow into large stacking faults by prolonged and repeated heat treatment at high temperatures. Moreover, it has been found that the density of surface microdefects strongly depends on the cleanliness of the furnace

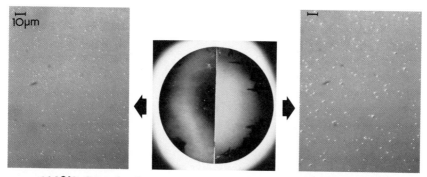

1140°C, 2 hr , dry O₂   **1140°C, 2 hr , wet O₂**

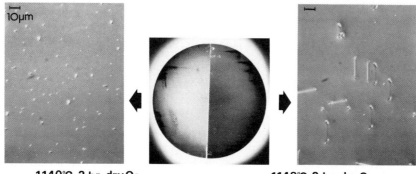

1140°C, 2 hr , dry O₂   **1140°C, 2 hr , dry O₂**
+1230°C, 2 hr , wet O₂   **+1230°C, 2 hr , wet O₂**
                                    +1140°C, 2 hr , wet O₂

Fig. 7.20.   Etched figures of surface defects delineated by Sirtl etching for CZ silicon wafers subjected to different heat treatments. The scale bar denotes 10 μm.

used, and the generation of these surface microdefects can be drastically suppressed by gettering techniques that will be discussed in Section 7.4.

Nature   The characterization by means of TEM and AEM suggests three different stages of surface microdefects that manifest themselves as S-pits, but not as linear etch pits which corresponds to stacking faults. The first stage is a tiny cluster of predominantly transition metals, which does not show any visible contrast in the TEM image.[81] In order to show the existence of these clusters, TEM observation has been carried out for the thin specimen prepared from the surface region of a (111) silicon wafer slightly etched using the Sirtl etchant for 5 sec. In this sample, S-pits are not observed by an optical microscope, although they can be observed after 30-sec etching, but TEM reveals triangular-shaped shallow pits as shown in Fig. 7.21.[81] The shallow

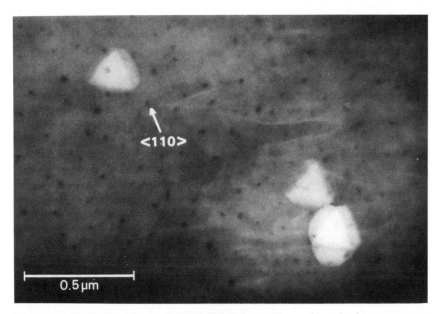

Fig. 7.21. TEM micrograph of surface microdefects in CZ silicon subjected to heat treatment at 1140°C for 2 hr in wet O_2. TEM sample was first etched by Sirtl etching for 5 sec. (After Tsuya and Shimura.[81])

pits, which can be correlated with microprecipitates of metallic impurities, are recognized to be surrounded by $\langle 100 \rangle$-oriented edges of about 0.15 μm length. The second stage is a small stacking fault with impurity clusters at the central region of the fault plane as shown in Fig. 7.22.[48] The stacking fault is extrinsic in nature and is bounded by an $\frac{a}{3} \langle 111 \rangle$ Frank partial, the same as the OSF previously discussed. Stacking faults associated with larger precipitates of impurities, such as shown in Fig. 7.23, are occasionally observed in heavily contaminated silicon wafers oxidized at a high temperature. The third stage is a small stacking fault whose surrounding Frank partial is heavily decorated with impurity clusters, occasionally whiskers, as shown in Fig. 7.24.[48] These stacking faults also have the same nature as those mentioned above. The impurities, which are located in the central region of the stacking fault and decorate the Frank partial, have been identified as copper or copper-containing compound by STEM–EDX analysis.[48] Other transition metals such as Ni, Fe, Co, and Cr have also been observed to cause different types of surface microdefects.[82,83] Figure 7.25a also shows a TEM micrograph of two different types of surface microdefects, marked with arrows A and C.[84] The EDX spectrum of the defect A and the EELS spectrum of the

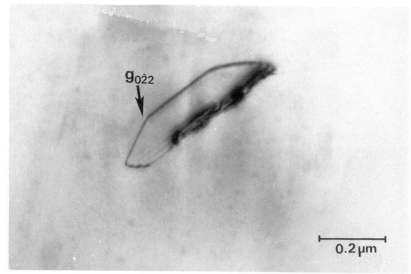

Fig. 7.22.   TEM micrograph of surface stacking fault in CZ silicon subjected to heat treatment at 1100°C for 2 hr in wet O_2. (After Shimura *et al.*[48])

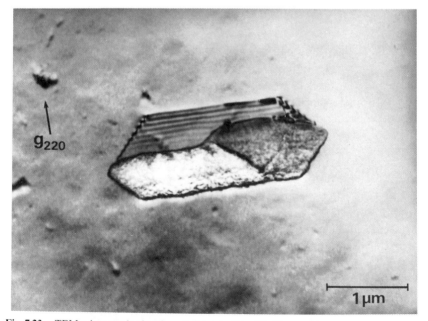

Fig. 7.23.   TEM micrograph of surface stacking fault in CZ silicon subjected to heat treatment at 1100°C for 2 hr in wet O_2.

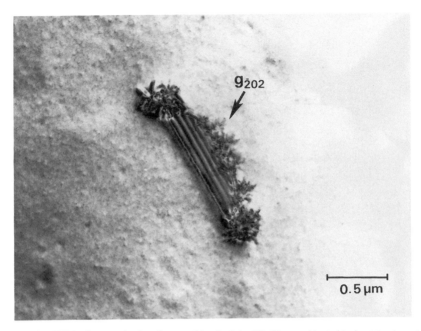

Fig. 7.24.   TEM micrograph of surface stacking fault in CZ silicon subjected to heat treatment at 1100°C for 2 hr in wet O_2. Frank partial is decorated with whisker precipitates. (After Shimura *et al.*[48])

defect C are shown with those spectra for the background region B in Fig. 7.25b and c, respectively. The EDX spectrum indicates the association of copper with the defect A, while the EELS spectrum indicates that the defect C includes both carbon and oxygen. The carbon–oxygen association will be discussed in more detail in the next subsection.

Formation Mechanism   The morphology of surface microdefects depends on the degree of contamination and on the nature of contaminants, as well as on the heat treatment conditions. A schematic model for the formation and growth of surface microdefects that results in the three different stages is shown in Fig. 7.26.[49] Transition metals are supplied from the heat treatment environment, and agglomerate in the surface region of a wafer during a thermal process at a higher temperature. At this stage, these agglomerates may cause elastic strain around them, but lattice defects such as stacking faults are not formed yet. After further agglomeration of impurities, with the resultant formation of larger clusters, and proceeding oxidation, extrinsic stacking faults are generated at these sites as in the way of OSF formation discussed previously. If the contamination continues after the formation of stacking faults, the contaminants will be trapped preferentially at the Frank

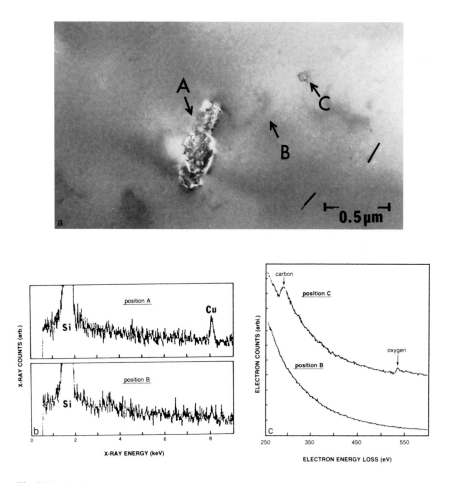

Fig. 7.25. Surface microdefects in CZ silicon subjected to heat treatment at 1100°C for 2 hr in wet O_2: (a) TEM micrograph, (b) EDX spectra for positions A and B in the TEM micrograph, and (c) EELS spectra for positions B and C in the TEM micrograph. (After Craven *et al.*[84] Reproduced with the permission of The Electrochemical Society, Inc.)

partial loop, or decorate the stacking fault, resulting in stabilization of both the stacking fault and the contaminants themselves. By absorbing self-interstitials, these small stacking faults can grow into large ones, which manifest themselves as linear etch pits by chemical etching.

7.2.4 Oxygen Precipitation

General Remarks As repeatedly described, microelectronic circuits are fabricated mostly on CZ silicon wafers containing oxygen on the order of

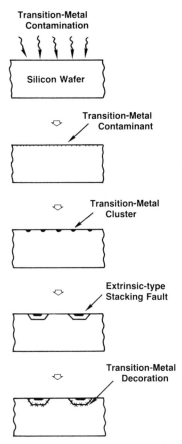

Fig. 7.26. Schematic illustration showing formation and growth of surface microdefects. (After Shimura and Craven.[49])

10^{18} atoms/cm^3. No matter how much oxygen is incorporated in the CZ silicon wafers used for the fabrication, resulting impurities critically affect the properties and yield of electronic devices because of the following three factors: (1) bulk defects generated by oxygen precipitation benefit the intrinsic gettering (IG) effect (see Section 7.4.3); (2) mechanical strength of silicon wafers greatly depends on the oxygen concentration and state; and (3) oxygen donors are formed by annealing at about 450°C and in the range between 550 and 850°C.[85] Under the circumstances, a great deal of attention has been devoted to the investigation on the behavior of oxygen in silicon for the last three decades, particularly the decade when the beneficial effect of oxygen on device performance was found. Consequently, it is very important to understand the behavior of oxygen and to control not only the oxygen concentration but also the oxygen precipitation.

Precipitation Kinetics   Although the growth process and the related phenomena of oxygen precipitates have been extensivey investigated, the nucleation process has not been established yet. *Homogeneous nucleation* is nucleation from a homogeneous phase, as it is called, in which nucleation occurs randomly; catalyzing nucleation where discontinuities such as lattice defects and second-phase particles in the matrix supply the nucleation sites is called *heterogeneous nucleation*. Heterogeneous nucleation requires far less energy than homogeneous nucleation, and is by far the more commonly observed in any system. Mainly based on the analyses of experimental results, it has been proposed by different investigators that the nucleation for oxygen precipitation in silicon would be a homogeneous process,[86,87] a heterogeneous process,[88,89] or a combination of homogeneous and heterogeneous processes.[90] That is, all the nucleation processes seem to be possible. The major difficulties in this argument may lie in the definition of *homogeneous* and *heterogeneous*,[91] although the physical difference between them is very clear,[92] and in uncertainty on whether the process experimentally observed is the *nucleation* or *growth* process of oxygen precipitates. Putting the definition aside, it has been widely accepted that oxygen precipitation does not occur by a homogeneous nucleation process based on simple oxygen supersaturation and the simple agglomeration of oxygen atoms in silicon,[86] but does require a *nucleation active center*[93] and depends on various factors including heterogeneous factors and the oxygen supersaturation ratio as well.[94] It would be a realistic explanation that the oxygen supersaturation plays the determining role in oxygen precipitation in higher supersaturated circumstances at a low temperature, while at low supersaturations (namely, at a high temperature) heterogeneous factors play the dominant role.[90] As regards the growth of oxygen precipitates, it is well established that the growth is governed by a diffusion-limited process.[91,95]

Infrared Spectra of Oxygen   The interpretation of infrared (IR) absorption spectra of oxygen in silicon proposed by Hrostowski and Kaiser,[96] as shown in Fig. 6.11, has been accepted for many years. At room temperature, an absorption observed at 1106 cm^{-1} (v_{03}) and another weaker one at 515 cm^{-1} (v_{02}) due to interstitial oxygen in a CZ silicon crystal have been shown in Fig. 6.12. The linear relationship of the relative absorption intensity ratio of v_{03} to v_{02} (i.e., $v_{03}/v_{02} = 3.9 \pm 0.6$) has been obtained for silicon crystals that have different oxygen concentration but have no oxygen precipitation.[97] However, it has been reported that the v_{02} absorption at 515 cm^{-1} loses its intensity more quickly than does the v_{03} aborption at 1106 cm^{-1} following oxygen precipitation heat treatments.[98,99] This has led to a new configuration model for oxygen in the silicon lattice, which alternates the configuration of oxygen proposed by Hrostowski and Kaiser. The IR spectra ranging from 300 to 1700 cm^{-1} for as-grown and heat-treated CZ silicon crystals are shown in

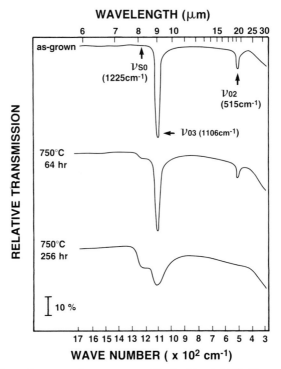

Fig. 7.27. IR absorption spectra for as-grown and for heat-treated CZ silicon crystals at 750°C for 64 or 256 hr. (After Shimura *et al.*[99])

Fig. 7.27.[99] It is evident that as the precipitation of interstitial oxygen increases, the v_{02} absorption at 515 cm^{-1} decreases anomalously when compared with the v_{03} absorption at 1106 cm^{-1}, and v_{02} disappears entirely after a prolonged heat treatment for 256 hr. Figure 7.28 summarizes the changes in the ratio v_{03}/v_{02} as a function of precipitated oxygen concentration for CZ silicon crystals subjected to various heat treatments at temperatures ranging from 600 to 1230°C for different times.[99] These figures just discussed indicate that interstitial oxygen atoms in the silicon lattice do not distribute in the configuration of only an Si–O–Si molecular unit as previously assigned by Hrostowski and Kaiser, but that some distribute differently. The analysis of the perturbation by silicon or oxygen atoms linked in the Si–O–Si unit has led to a hypothesis that interstitial oxygen atoms may be forming chain-like structures as the forestage of SiO$_2$ precipitates in silicon crystals where oxygen precipitation occurred.[99] The formation of Si–O chain structures will not greatly modify the v_{03} absorption due to antisymmetric stretching but will greatly suppress the vibration due to symmetric bending, which is responsible for the v_{02} aborption. On the other hand, another

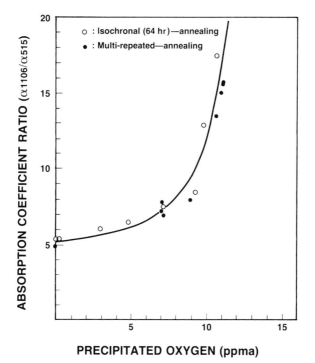

Fig. 7.28.   Absorption coefficient ratio (v_{03}/v_{02}) change as a function of precipitated oxygen for CZ silicon subjected to various heat treatments. (After Shimura *et al.*[99])

hypothesis has been suggested: that the v_{02} absorption at 515 cm^{-1} is not due to interstitial oxygen at all but rather to distinct substitutional oxygen species.[100] Moreover, a stress-induced dichroism study of v_{02} and v_{03} modes of interstitial oxygen in silicon has proposed that the v_{02} mode is due to the symmetric stretching, but not bending, motion of the Si–O–Si unit.[101] However, the latter two models have not explained the mechanism for the phenomenon shown in Fig. 7.28.

Precipitation and Redissolution   The behavior of oxygen precipitation and dissolution in CZ silicon crystals has been extensively investigated. Figure 7.29a shows the IR spectra ranging from 1000 to 1300 cm^{-1} for as-grown and heat-treated CZ silicon crystals at different temperatures for 16 h in a dry O_2 ambient.[89] It clearly shows that oxygen precipitation occurs, and SiO_2 precipitates resulting in the v_{SO} absorption at 1225 cm^{-1} are formed during the heat treatment. The amount of SiO_2 precipitates, or precipitated oxygen, strongly depends on the heat-treatment temperature. If those silicon samples with oxygen precipitates are subjected to a subsequent heat treatment at a

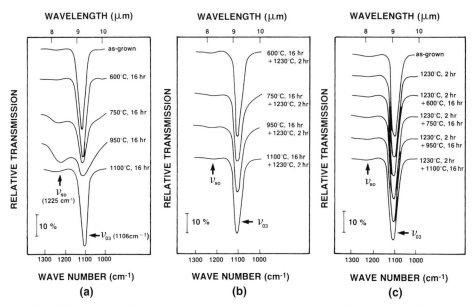

Fig. 7.29. IR absorption spectra for as-grown and heat-treated CZ silicon at different temperatures in dry O_2. (After Shimura *et al.*[89])

higher temperature such as 1230°C, SiO_2 precipitates are dissolved and oxygen atoms redistribute at interstitial sites[102] as shown in Fig. 7.29b. The IR spectra for the samples subjected to heat treatments of an inverse order to those shown in Fig. 7.28b are shown in Fig. 7.29c. No oxygen precipitation is observed in the samples that were subjected to a high-temperature heat treatment first, even after the heat treatment that resulted in oxygen precipitation as shown in Fig. 7.29a. That is, it is indicated that oxygen precipitation is not governed only by the supersaturation ratio of oxygen, since the first-step high-temperature heat treatment does not affect the oxygen concentration of the samples shown in Fig. 7.29c. Accordingly, it has been suggested that oxygen precipitation occurs by the heterogeneous nucleation process and that the active precipitation seeding sites are grown-in SiO_2 precipitates,[89,103,104] which result in weak absorption at 1225 cm^{-1} in as-grown silicon.[89] If the radius of grown-in precipitate r_0 (r_0 must be smaller than ~ 25 Å according to TEM weak-beam observation) becomes smaller than the critical radius r_c for growing as the result of dissolving or shrinking by high-temperature heat treatment, oxygen precipitation hardly occurs during subsequent heat treatment. Even when such shrunken grown-in precipitates are larger than r_c, prolonged annealing must be required for oxygen precipitation. In the proposed grown-in precipitate model depicted in Fig. 7.30, the original nucleus for an SiO_2 precipitate formed during the

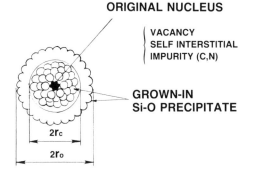

Fig. 7.30.   Schematic illustration showing grown-in precipitate. (After Shimura *et al.*[89])

silicon crystal growth process has been considered to be the condensation of vacancies,[62,88,93] self-interstitials,[105] and/or impurities such as carbon[104,106–110] and nitrogen.[111,112]

The change in interstitial oxygen concentration $[O_i]$ for CZ silicon crystals subjected to multistep heat treatments is shown in Fig. 7.31. Group I samples were heat treated at low (750°C) or medium (1000°C) temperature first, then at a high temperature (1230°C), while Group II samples were heat treated inversely. In Group I samples, oxygen precipitation (odd-numbered steps) and oxygen redissolution (even-numbered steps) repeat regularly from step 1 to step 8. The $[O_i]$ in samples after even-numbered steps, namely, oxygen redissolution steps, is consistent around 10 ppma, which approximates the effective oxygen solubility in silicon at 1230°C (see Fig. 5.36). Oxygen precipitation behavior in Group II samples distinctly differs from that in Group I samples. It is noteworthy that even during a high-temperature heat treatment, considerable oxygen precipitation occurs in later steps, such as step 3, when the size of precipitates grown by previous heat treatment is larger than the critical size at 1230°C and the existing oxygen concentration exceeds the solubility at 1230°C. Consequently, the necessary condition for oxygen precipitation in a silicon crystal is the existence of both nucleation active centers and supersaturated oxygen atoms. If either is missing, precipitated oxygen redissolves, and then oxygen precipitation does not occur. Yet the oxygen precipitation rate depends on various factors once this necessary condition has been satisfied.[94]

Oxygen Precipitation Factors   Oxygen precipitation predominantly depends on the initial oxygen concentration $[O_i]_0$, annealing temperature, and time. The change in interstitial oxygen concentration $[O_i]$ for CZ silicon crystals with different $[O_i]_0$ as functions of annealing temperature and time is

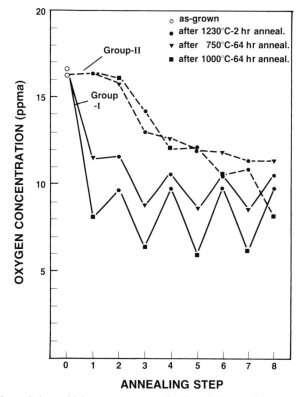

Fig. 7.31.   Change in interstitial oxygen concentration for CZ silicon subjected to multistep heat treatment in dry O_2. (After Shimura and Tsuya.[13] Reprinted with the permission of The Electrochemical Society, Inc.)

shown in Fig. 7.32. The figure indicates first that oxygen precipitation does not occur or rarely occurs when the $[O_i]_0$ is less than a certain critical concentration $[O_i]_c$ (~ 14 ppma), and second that the oxygen precipitation rate does not simply depend on the annealing temperature; that is, two peaks are observed at 750°C and 1000°C in the curve with an asterisk. The oxygen precipitation rate primarily depends on the diffusion coefficient and solubility, and in turn the supersaturated ratio, of oxygen in silicon (see Figs. 5.35 and 5.36). The $[O_i]$ change in a plate sample, cut parallel to the growth direction, that was subjected to a heat treatment at 1000°C for 64 hr is shown in Fig. 7.33. It is quite evident that the higher $[O_i]_0$ results in the greater oxygen precipitation when heat-treatment conditions are constant. The $[O_i]$ is plotted against the square root of annealing time for silicon crystals annealed at 750, 1000, and 1150°C in Fig. 7.34. An incubation period of about 16 hr is evident preceding oxygen precipitation at 750°C, while the period for

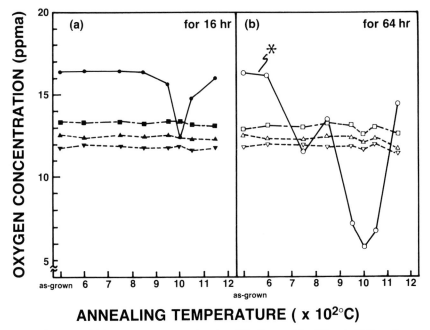

ANNEALING TEMPERATURE (x 10^{2}°C)

Fig. 7.32. Interstitial oxygen concentration for CZ silicon with different $[O_i]_0$ values as functions of annealing temperature and time in dry O_2. The curve with an asterisk corresponds to the sample shown in Fig. 6.25. (After Shimura and Tsuya.[94] Reprinted with the permission of The Electrochemical Society, Inc.)

the annealing at 1000°C seems to be quite short. Oxygen precipitation thus proceeds after an incubation period that may depend on both the diffusion coefficient and the supersaturated ratio of oxygen. In addition, TEM observation has shown that thermally induced lattice defects such as dislocations and stacking faults, which will be discussed in the next subsection, can be very effective heterogeneous seeding sites for oxygen precipitation during a subsequent heat treatment,[113] as shown in Figs. 7.35 and 7.36, indicating the effect of dislocations and stacking faults, respectively.

As depicted in Fig. 5.19, every portion of a CZ silicon crystal is grown at a different time, under different growth conditions. Moreover, every portion has been exposed to a different thermal history as a result of its different position along the crystal length. This difference in thermal history greatly affects the oxygen precipitation behavior in silicon wafers prepared from different portions of a crystal ingot.[114,115] The effect of prior thermal history of a silicon crystal in the crystal puller on oxygen precipitation is strikingly demonstrated in Fig. 7.37 for sample wafers prepared from different portions of an n(100) CZ silicon ingot whose $[O_i]_0$ is very uniform from the seed to tail

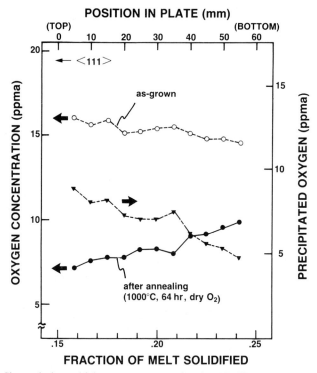

Fig. 7.33. Change in interstitial oxygen concentration for CZ silicon plate subjected to heat treatment at 1000°C for 64 hr in dry O_2. (After Shimura and Tsuya.[94] Reprinted with the permission of The Electrochemical Society, Inc.)

ends. In the cases of a single-step heat treatment at 750 or 1000°C, the difference in precipitated oxygen concentration between the seed-end and the tail-end wafers is dramatic. Applying a two-step heat treatment to these wafers, the amount of precipitated oxygen becomes highly uniform; however, in general, oxygen precipitation in wafers prepared from closer to the seed end—namely, those that have longer and gradual thermal history in the puller—is larger than that in wafers prepared from closer to the tail end (those that were cooled more rapidly and have a shorter thermal history in the puller). The difference in the oxygen precipitation shown in Fig. 7.37 can be simply attributed to the difference in the sample position in the crystal ingot, that is, the prior thermal history, since the $[O_i]_0$ of each wafer is identical and the heat-treatment conditions are common in the wafers. The difference in thermal history will result in the variation in the size and density distribution of grown-in SiO_2 precipitates, impurity clusters, and intrinsic

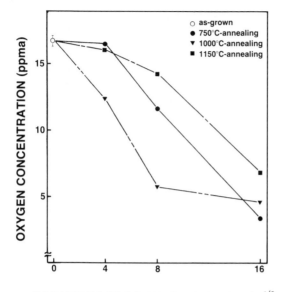

SQUARE ROOT OF ANNEALING TIME (hr$^{1/2}$)

Fig. 7.34.   Interstitial oxygen concentration for CZ silicon subjected to heat treatment at 750, 1000, or 1150°C in dry O_2 as a function of annealing time. (After Shimura and Tsuya.[94] Reprinted with the permission of The Electrochemical Society, Inc.)

point defects, which all can be effective heterogeneous nucleation sites for oxygen precipitation.

Consequently, the amount of oxygen precipitation $\Delta[O_i]_T$ during finite annealing conditions can be qualitatively described by the simplified function[94,116]

$$\Delta[O_i]_T = f(D_T, S_T, N_T, F, t) \tag{7.6}$$

where T is the annealing temperature, D_T the diffusion coefficient of oxygen in silicon at T, S_T the supersaturated oxygen ratio at T, N_T the number of nuclei precipitates larger than the effective critical radius at T, F the factor characterized by the number and type of existing lattice defects and impurity clusters, and t the annealing time. By increasing the annealing temperature T, D_T increases, while S_T and N_T decrease. If S_T or N_T were the dominant factor for oxygen precipitation, then annealing at lower temperatures should result in higher oxygen precipitation rates ($\Delta[O_i]_T/t$). If, on the other hand, D_T were dominant, then annealing at higher temperatures should result in higher precipitation rates. Neither of these can be valid, since Fig. 7.32 shows the highest and second highest precipitation rates occur at 1000 and 750°C,

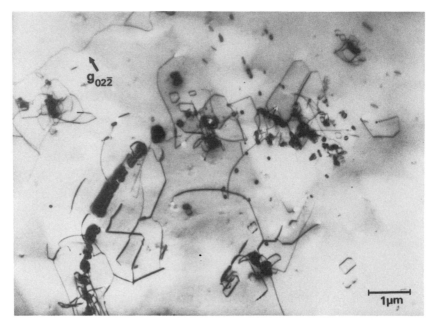

Fig. 7.35. TEM micrograph of oxygen precipitates in CZ silicon subjected to three-step heat treatment ($1000°C/64\,h + 1230°C/2\,hr + 1000°C/64\,hr$) in dry O_2 (refer to Group I samples shown in Fig. 7.31). (After Shimura and Tsuya.[113] Reprinted with the permission of The Electrochemical Society, Inc.)

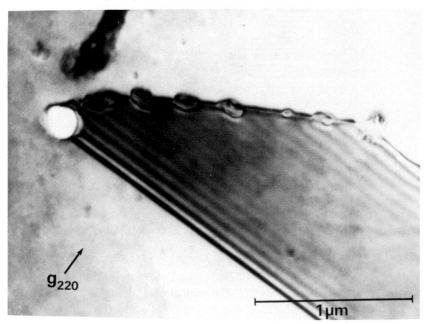

Fig. 7.36. TEM micrograph of oxygen precipitates at Frank partial of stacking faults in CZ silicon subjected to two-step heat treatment ($1150°C/2\,hr + 1000°C/16\,hr$) in dry O_2.

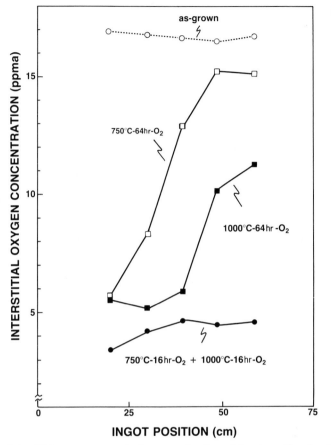

Fig. 7.37. Interstitial oxygen concentration as a function of ingot position for CZ silicon subjected to different heat treatments in dry O_2.

respectively. This therefore suggests that all the factors related straightforwardly to annealing temperature can be involved in the oxygen precipitation to different extents. Moreover, in particular, from the experimental results it can be reasonably assumed that heterogeneous factors (i.e., N_T and F) have a great influence on oxygen precipitation in silicon.

Figure 7.38 shows the precipitated oxygen $\Delta[O_i]$ as a function of $[O_i]_0$ for CZ silicon subjected to a two-step heat treatment (800°C/2 hr + 1050°C/16 hr).[117] The data in the figure were collected from two furnace runs that included 161 sample wafers from various portions of 51 crystal ingots grown with different processes using different crystal pullers. Thus Fig. 7.38 represents the practical correlation between $[O_i]_0$ and $\Delta[O_i]$ in commercially available silicon wafers. The generic curves for this situation are given in

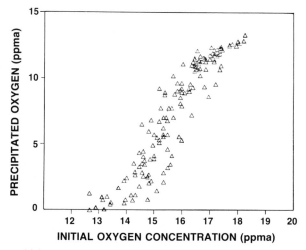

Fig. 7.38. Interstitial oxygen concentration as a function of $[O_i]_0$ for CZ silicon subjected to two-step heat treatment (800°C/2 hr + 1050°C/16 hr). (After Chiou and Shive.[117] Reproduced with the permission of The Electrochemical Society, Inc.)

Fig. 7.39; which indicates three characteristic regions as a function of $[O_i]_0$: (1) zero precipitation, (2) partial precipitation, and (3) 100% precipitation.[118] Hopefully, $\Delta[O_i]$ during a finite thermal cycle could be controlled simply by $[O_i]_0$ as a homogeneous nucleation process model claims; however, unfortunately there is considerable variation in $\Delta[O_i]$ for wafers with identical $[O_i]_0$ in the partial precipitation region, which mostly occurs in practical device fabrication processes. The variation is, as discussed

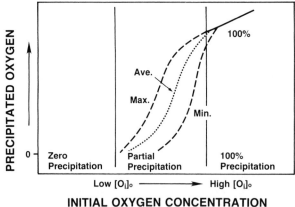

Fig. 7.39. Generic curve showing change in interstitial oxygen concentration as a function of $[O_i]_0$ for heat-treated CZ silicon. (After Chiou.[118] Reproduced with the permission of Solid State Technology.)

above, indisputably due to the heterogeneous factors. In order to minimize the variation in $\Delta[O_i]$ due to heterogeneous factors, conducting a homogenizing heat treatment at a high temperature ($> 1200°C$) on as-grown silicon crystals has been proposed.[85] Recently it has been shown that the effect of prior thermal history in the crystal puller can be erased to a large extent by a short-time heat treatment at 1320°C for silicon wafers.[115]

Oxygen Redissolution Rate   In practical device fabrication processes, which usually include thermal processes at a high temperature ($> 1200°C$), redissolution of precipitated oxygen may also occur in processing silicon wafers. It is important to understand the redissolution phenomenon as well as oxygen precipitation, since both can critically affect the properties of silicon wafers. Figure 7.40 shows the change in $[O_i]$ as a function of length of high-temperature redissolution heat treatment at 1230°C for CZ silicon crystals

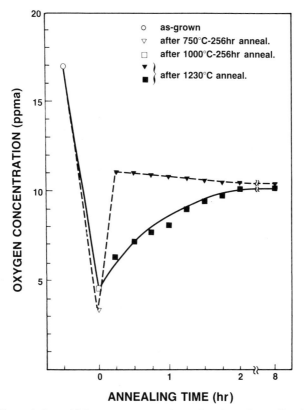

Fig. 7.40.   Change in interstitial oxygen concentration as functions of annealing time at 1230°C in dry O_2 for CZ silicon preannealed at 750 or 1000°C for 256 hr in dry O_2. (After Shimura.[119])

preannealed at 750 or 1000°C for 256 hr in order to precipitate the same amount of oxygen.[119] As a result of high-temperature annealing, precipitated oxygen redissolves and the oxygen atoms redistribute into interstitial site, and thus $[O_i]$ increases up to the effective solubility limit at 1230°C, as discussed previously. However, the oxygen redissolution behavior in the 750°C preannealed silicon distinctly differs from that in the 1000°C preannealed silicon. That is, in the former silicon, the dissolution completes in only 15 min of annealing with a dissolution rate estimated at $\sim 3 \times 10^{18}$ atoms/cm^3 hr. In the latter silicon, however, the dissolution proceeds gradually for 2 hr with the average dissolution rate estimated at $\sim 3 \times 10^{17}$ atoms/cm^3 hr. In both cases, the oxygen dissolution rate is much larger than the oxygen precipitation rate (e.g., $\sim 5 \times 10^{15}$ atoms/cm^3 hr at 750°C). The TEM observations have revealed that dense {100} platelet microprecipitates of approximately $800 \times 800 \times 200$ Å in size were generated in the silicon sample annealed at 750°C for 256 hr, whereas in the sample annealed at 1000°C for 256 hr, precipitation–dislocation complexes (PDCs) consisting of {100} platelike precipitates of about $4800 \times 4800 \times 2000$ Å in size were generated. Based on the TEM results, the difference in the oxygen redissolution rate in the CZ silicon first annealed at 750 and 1000°C has been described mainly by the difference in the total surface area of oxygen precipitates, which dominate the flux of oxygen diffusion, and partly by the dislocation pinning effect on dissolving oxygen atoms. It should be emphasized again that the dissolution rate of precipitated oxygen strongly depends on the precipitation conditions but not on $[O_i]$ itself.

Carbon Effect on Oxygen Precipitation   Subsidiary impurities such as carbon[104,106–110] and nitrogen[111,112] greatly affect oxygen precipitation in CZ silicon crystals when they are subjected to heat treatment. In particular, a significant enhancement effect of oxygen precipitation due to the presence of carbon atoms has been widely observed. Figure 7.41 shows the change in $[O_i]$ for low-carbon-content (L[C]) and high-carbon-content (H[C]) $n(100)$ CZ silicon subjected to heat treatment for 64 hr at temperatures shown in the horizontal axis. The H[C] silicon crystal was grown with intentional carbon doping, but otherwise with the same growth conditions as for the L[C] crystal. The substitutional carbon concentrations $[C_s]$ of as-grown L[C] silicon and H[C] silicon are < 0.1 ppma and 5.5 ppma, respectively. The change in $[C_s]$ for H[C] silicon is also shown in Fig. 7.41. Although the initial oxygen concentrations $[O_i]_0$ of L[C] and H[C] silicon are identical (~ 15 ppma), compared with L[C] silicon, enormous oxygen precipitation is observed in H[C] silicon at temperatures ranging from 600 to 1000°C. A dramatic change in $[C_s]$ for the H[C] silicon is also observed. That is, a reduction of substitutional carbon is accompanied by oxygen precipitation at

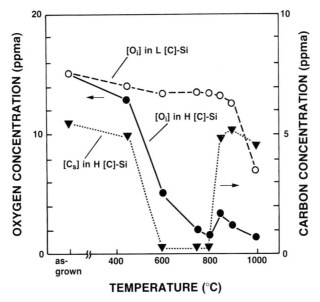

Fig. 7.41.   Change in interstitial oxygen concentration and substitutional carbon concentration for L[C] and H[C] CZ silicon subjected to heat treatment at various temperatures for 64 hr in dry O_2. (After Shimura.[120])

temperatures lower than a critical temperature T_c, which is estimated between 800 and 850°C,[120] but no carbon reduction is observed in oxygen precipitation at temperatures higher than T_c.[109,120–122] The significant difference in the behavior of oxygen and carbon is qualitatively indicated with the IR spectra. The IR spectra for as-grown L[C] and H[C] silicon have been shown in Figs. 6.12a and 6.13, respectively. Figures 7.42a and c show the IR spectra for H[C] silicon subjected to heat treatment at 750°C ($< T_c$) and 900°C ($> T_c$), respectively, for 64 hr. It is quite evident that v_c absorption due to substitutional carbon (C_s) entirely disappears in Fig. 7.42a, but absorption v_{P1} (~ 850 cm^{-1}) and v_{P2} (~ 1100 cm^{-1}) appear instead. These two bands have been attributed to a perturbed C(3) center,[123] which consists of multiple oxygen atoms around a C(3) center, namely, a complex $[O_i\text{–}C_i]$ of interstitial oxygen O_i and interstitial carbon C_i.[124] The correlation between the reduction in $[O_i]$ and $[C_s]$ for H[C] silicon subjected to heat treatment at 750°C (i.e., $< T_c$) is shown in Fig. 7.43. The number ratio of oxygen atoms that do not occupy isolated interstitial sites to carbon atoms that do not occupy substitutional sites is nearly constant, estimated between 1.8 and 2.7 in the annealing time range of 8–64 hr. That is, a perturbed C(3) center may consist of a complex with two to three oxygen atoms per carbon atom. This

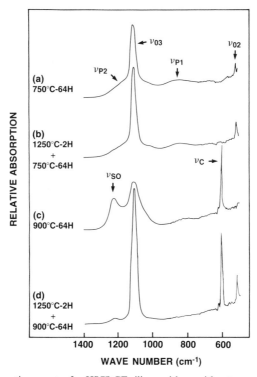

Fig. 7.42.   IR absorption spectra for H[C] CZ silicon with or without preannealing at 1250°C for 2 hr: (a) after heat treatment (750°C/64 hr), (b) after heat treatment (1250°C/2 hr + 750°C/64 hr), (c) after heat treatment (950°C/64 hr), and (d) after heat treatment (1250°C/2 hr + 950°C/64 hr). (After Shimura.[120])

number ratio of oxygen to carbon roughly coincides with recent estimation; that is, most of the oxygen and carbon in H[C] CZ silicon subjected to heat treatment at 750°C for 64 hr was found in subcritical clusters with as few as two oxygen atoms per carbon.[125]

The effect of preannealing at 1250°C for 2 hr on the precipitation behavior of oxygen and carbon in H[C] silicon has further clarified the enhancement mechanism that explains the role of carbon. Figure 7.44 shows the change in $[O_i]$ and $[C_s]$ for the H[C] silicon, with or without preannealing at 1250°C, when subjected to heat treatment for 64 hr at temperatures given on the horizontal axis. The IR spectra for the samples, with or without the preannealing, subjected to heat treatment at 750°C ($< T_c$) or 900°C ($> T_c$) for 64 hr are shown in Fig. 7.42. Two significant phenomena should be recognized. First, the formation of perturbed $[O_i–C_i]$ C(3) centers is not affected by the preannealing at all. This implies that the formation of $[O_i–C_i]$

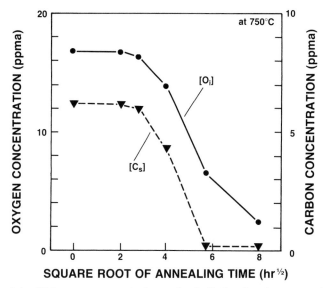

Fig. 7.43.   Interstitial oxygen concentration and substitutional carbon concentration as a function of square root of annealing time at 750°C for H[C] CZ silicon subjected to heat treatment at 750°C in dry O_2. (After Shimura.[120])

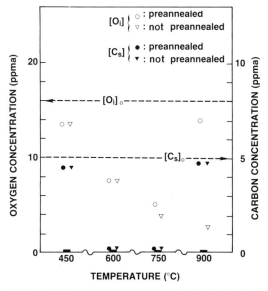

Fig. 7.44.   Change in interstitial oxygen concentration and substitutional carbon concentration for preannealed and not preannealed H[C] CZ silicon subjected to heat treatment at various temperature for 64 hr in dry O_2; preannealing was carried out at 1250°C for 2 hr in dry O_2. (After Shimura.[120])

complexes is independent of previous thermal history or point defects, and is quite different from the fact that oxygen precipitation resulting in SiO_2 formation is significantly affected not only by preannealing at a high temperature but also by the prior thermal history of the crystal. Second, the effect of preannealing at 1250°C on oxygen precipitation in H[C] silicon is very characteristic. Contrary to previous observation for low-carbon-content, or commercial, CZ silicon such as shown in Fig. 7.29c, Figs. 7.42 and 7.44 indicate for H[C] silicon that the preannealing does not affect oxygen precipitation during subsequent heat treatment at temperatures lower than T_c. However, the preannealing significantly suppressed oxygen precipitation in H[C] silicon during subsequent heat treatment at temperatures higher than T_c, as previously observed for low-carbon-content CZ silicon. The suppression effect of high-temperature preannealing on subsequent oxygen precipitation in low-carbon-content CZ silicon has been explained in terms of the dissolution and shrinkage of grown-in precipitate nuclei. Again, the necessary condition for oxygen precipitation in silicon is the coexistence of both heterogeneous seeding sites for oxygen precipitation and supersaturated oxygen atoms. Dissolution of growth-in SiO_2 precipitates probably occurs even in H[C] CZ silicon subjected to preannealing at a high temperature. However, the results shown in Figs. 7.42 and 7.44 imply that carbon atoms directly supply heterogeneous seeding sites, as $[O_i-C_i]$ complexes, for oxygen precipitation at temperatures lower than T_c, since formation of the complexes is not affected by annealing at a high temperature. Carbon atoms, however, do not provide seeding sites for oxygen precipitation at temperatures higher than T_c, although they do enhance oxygen precipitation at those temperatures as shown in Fig. 7.41 if the necessary condition for oxygen precipitation is satisfied. It is noteworthy to point out that those carbon atoms that enhance oxygen precipitation at temperatures higher than T_c coaggregate with oxygen atoms, as shown in Figs. 6.17 and 6.18, keeping their substitutional sites. Consequently, the two different explanations of carbon enhancement of oxygen precipitation have been confirmed. Carbon atoms directly provide heterogeneous seeding sites for oxygen precipitation at temperatures lower than T_c, while carbon plays a catalytic role by modifying the interfacial energy or the point defect atmosphere at the oxygen precipitation surface at temperatures higher than T_c. Oxygen precipitate growth involves emission of silicon self-interstitials (I_{Si}) into the surrounding silicon matrix.[93] The high diffusivity of C_s, compared with that of I_{Si}, in silicon and the I_s-enhanced C_s-diffusion[126] may reasonably lead carbon atoms to an oxygen precipitate growing area. As a next step, these I_{Si} may interact with the diffused C_s, resulting in the formation of $C_s I_{Si}$.[127] The formation of $C_s I_{Si}$ may reduce the interfacial energy of oxygen precipitates. This catalytic effect of carbon on oxygen precipitation will be negligibly small at temperatures lower than T_c, since the carbon diffusivity in silicon at such a low temperature (e.g., $D_c \approx 10^{-15}$ cm^2/sec at 750°C) is much smaller than that at temperatures higher than T_c (e.g., $D_c \approx 10^{-12}$ cm^2/sec at 1000°C).[128]

Finally, it should be noted that this carbon enhancement effect on oxygen precipitation is limited to CZ crystals with intentionally doped carbon of a high concentration. The carbon concentration of commercially available silicon crystals is usually less than 0.1 ppma (i.e., the detection limit by an IR absorption measurement). Although carbon of concentration less than 0.1 ppma may affect oxygen precipitation to some extent, the effect is unknown because of the inability to detect carbon at such a low concentration.

Oxygen Precipitation in Heavily Doped Silicon   The demand for epitaxial silicon wafers using heavily doped CZ silicon substrates, that is, p/p^+ or n/n^+, for advanced CMOS devices has recently increased with decreasing device dimension. It is known that oxygen precipitation is affected by electrically active dopants. In particular, it has been widely observed that oxygen precipitation is significantly suppressed in heavily doped n-type, or n^+, CZ silicon crystals.[129–132] This has been a major disadvantage of n^+ silicon substrates when intrinsic gettering (IG) is considered.[130,132] The density of bulk (interior) defects induced by oxygen precipitation for p-type (B-doped) and n-type (Sb-doped) (100) CZ silicon wafers as a function of specific resistivity, and in turn of dopant concentration, are shown in Fig. 7.45.[131] The defect density gradually increases with the doping concentration of boron, while the density drastically decreases with increasing doping concentration of antimony. For P-doped n-type CZ silicon, a similar tendency has been observed.[131] The suppression of bulk defects in wafers heavily doped with n-type dopant has been primarily attributed to the lack of heterogeneous seeding sites for oxygen precipitation.[62,129] De Kock and van de Wijgert[129] have interpreted the mechanism for this precipitation suppression as one that involves the Coulomb attraction of positively charged dopant and negatively charged silicon self-interstitials. Since they assume that self-interstitials play a crucial role in creating heterogeneous nuclei for oxygen precipitation in silicon,[105] they attribute the oxygen precipitation suppression in n^+ silicon to decreasing the number of self-interstitials through the formation of complex $Sb^+I_{Si}^-$. Although agglomerates of I_{Si} resulting in dislocations or extrinsic-type stacking faults supply effective heterogeneous seeding sites for oxygen precipitation in silicon, as was shown in Figs. 7.35 and 7.36, oxygen precipitates growth might be retarded by the solid solution of silicon self-interstitials, since oxygen precipitate growth involves emission of silicon self-interstitials (I_{Si}) into the surrounding silicon matrix,[93] as previously described. Moreover, vacancies, not self-interstitials, are the most likely effective

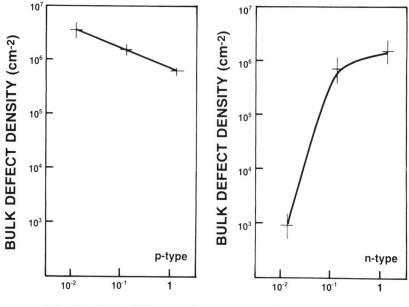

Fig. 7.45. Density of bulk defects as a function of specific resistivity for p-type (B-doped) and n-type (Sb-doped) Cz silicon subjected to three-step heat treatment. (After Tsuya et al.[13])

heterogeneous seeding sites for interstitial oxygen precipitation.[93,133,134] Although the charge state of intrinsic point defects has not yet been established, taking into account a positively charged self-interstitial (I_{Si}^{+}) and negatively charged vacancy (V^{-}),[135-137] the following point defect interactions have been proposed to consistently explain the oxygen precipitation phenomena in silicon[62]:

$$\text{Recombination:}\quad I_{Si}^{+} + V^{-} \longrightarrow Si_{sub} \tag{7.7}$$

$$\text{Complex formation:}\quad V^{-} + Sb^{+} \longrightarrow Sb^{+}V^{-} \tag{7.8}$$

$$I_{Si}^{+} + B_{-} \longrightarrow I_{Si}^{+}B^{-} \tag{7.9}$$

Thus, the number of active vacancies decreases via interaction (7.8), resulting in increasing self-interstitials in n-type silicon crystals, while active vacancies increase in p-type silicon. In addition, the greater equilibrium vacancy concentration would be expected in n-type silicon compared with p-type silicon.[58] Consequently, the precipitation of oxygen in CZ silicon crystals is suppressed by a high concentration of an n-type dopant via the following mechanisms: (1) the concentration of negatively vacancies, which act as effective heterogeneous seeding sites for interstitial oxygen, decreases via interaction involving the Coulomb attraction; (2) the growth of oxygen precipitate is retarded by decreased vacancies, namely increased self-interstitials; and possibly (3) the diffusion of interstitial oxygen in silicon is retarded by decreased vacancies. These mechanisms can be inversely applicable to oxygen precipitation in p-type silicon. In addition, although the suppression of oxygen precipitation in heavily Sb-doped silicon is independent of the initial oxygen concentration, the incorporation of oxygen into heavily Sb-doped CZ silicon tends to be decreased as a result of Sb_2O_3 evaporation from the melt during crystal growth.[62,138,139]

7.2.5 Interior Defects

Origin Thermally induced interior defects or bulk defects in CZ silicon crystals are primarily caused by oxygen precipitation. As was discussed in the preceding subsection, since oxygen is usually supersaturated in CZ silicon at modern processing temperatures, heat treatment leads to oxygen precipitation, which results in the formation of SiO_x ($x \approx 2$) precipitates. Oxygen precipitates consist of amorphous or crystalline SiO_x with a volume V_{ox} per SiO_2 unit of roughly two times the atomic volume V_{Si} in the silicon lattice. Accordingly, the precipitate growth can proceed either by relieving the excessive stresses by inducing plastic deformation of the silicon matrix, or by emitting one silicon self-interstitial for every two oxygen atoms incorporated into the precipitate in the surrounding silicon matrix.[59,134,140] The process of self-interstitial emission is, in principle, similar to the process that occurs during surface oxidation at the SiO_2–Si interface. The main difference is that the overwhelming part of the volume expansion during SiO_2 formation due to surface oxidation is accommodated by viscoelastic flow toward the surface of the oxide film,[60] whereas such a process is not possible within a silicon crystal. As in the case of surface oxidation, self-interstitials generated by oxygen precipitates may condense into dislocations or extrinsic-type stacking faults.

The straightforward correlation between oxygen precipitation and interior defect generation is made by referring to Figs. 6.25 and 7.32b. The $[O_i]$ change curve marked with an asterisk shown in Fig. 7.32b was obtained for the same CZ silicon samples that indicated interior defects revealed by chemical etching shown in Fig. 6.25. It is quite obvious that a higher oxygen precipitation results in a higher density or larger volume of interior defects.

Nature The nature of interior microdefects depends primarily on the annealing temperature and heat-treatment sequence, and has been extensively characterized by TEM.[48,113,140–144] The first stage of thermally induced interior microdefects in CZ silicon crystals is oxygen precipitates, which are

categorized into roughly three groups in terms of the precipitation temperature: (1) low temperature range ($< 750°C$), (2) medium temperature range
(850–$1000°C$), and (3) high temperature range (1100–$1200°C$).

After a long period of heat treatment at 650–$750°C$, microprecipitates such
as shown in Fig. 6.38 are generated in CZ silicon. The TEM contrast analysis
characterizes the shape of these microprecipitates as tiny platelets. The phase
of these microprecipitates has been considered to be crystalline cristobalite[48]
or amorphous SiO_2,[143,145] but has not been established. Among dense
microprecipitates, dislocation dipoles such as shown in Fig. 7.46 are frequently observed. The dipoles consist of two parallel dislocations in the
$\langle 110 \rangle$ directions. Both 60° and 90° dislocation dipoles have been observed in
silicon annealed in the temperature range between 550 and 750°C, but the
latter type forms only in the minority.[144] In addition to such microprecipitates as shown in Figs. 6.38 and 7.46, elongated precipitates frequently
associated with rod-like defects along $\langle 110 \rangle$ have been observed.[143,146] The
microdiffraction pattern analysis characterized the elongated precipitates as
coesite, which is a dense high-pressure form of silica.[143] Ribbon-like microprecipitates observed after a prolonged heat treatment at a temperature as

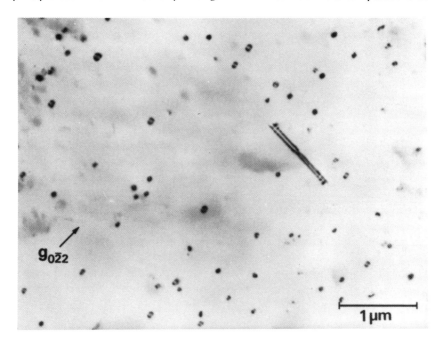

Fig. 7.46. TEM micrograph of oxygen precipitates and dislocation dipoles in CZ silicon
subjected to heat treatment at 750°C for 64 hr in dry O_2. (After Shimura and Tsuya.[113]
Reprinted with the permission of The Electrochemical Society, Inc.)

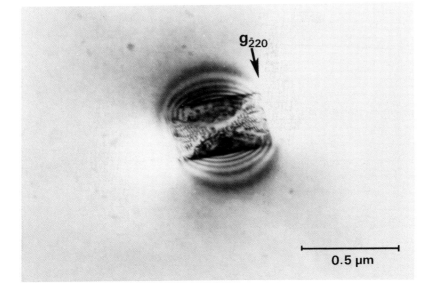

Fig. 7.47. TEM micrograph of plate-like oxygen precipitate in CZ silicon subjected to heat treatment at 950°C for 16 hr in dry O_2. (After Shimura *et al.*[48])

low as 485°C have also been proposed to be coesite.[147] However, the precipitate once characterized as *coesite* has recently been reinterpreted as *hexagonal silicon*, instead of a Si-O complex, by the same investigator.[147a]

Heat treatment at a temperature in the medium range (850–1000°C) for CZ silicon generates large square-shaped plate-like precipitates on {100} planes with ⟨110⟩ edges such as shown in Fig. 7.47.[48] According to IR spectrum and TEM image analyses, these precipitates have been identified as cristobalite.[89,145,148] Selected area microdiffraction experiments have been attempted, but no diffraction spots from such precipitates have been observed. Large precipitates often give rise to prismatic punching of dislocation loops and precipitate–dislocation complex (PDC). Figure 7.48 shows a TEM micrograph of a plate-like precipitate observed in CZ silicon annealed at 950°C for 16 hr: the precipitate is about to generate prismatic dislocation loops. A typical TEM micrograph of PDC observed in CZ silicon annealed at 1000°C for 64 hr is shown in Fig. 6.39. In this temperature range, stacking faults associated with oxygen precipitates at the central region are also frequently observed.

Heat treatment for CZ silicon at a temperature in the range between 1100 and 1200°C forms large octahedral precipitates, which occasionally generate dislocations.[48,145,149] Figure 7.49 shows TEM micrographs of five different views for an octahedral precipitate observed in CZ silicon annealed at 1150°C for 64 hr.[150] It is very characteristic that, at the precipitate/matrix interface,

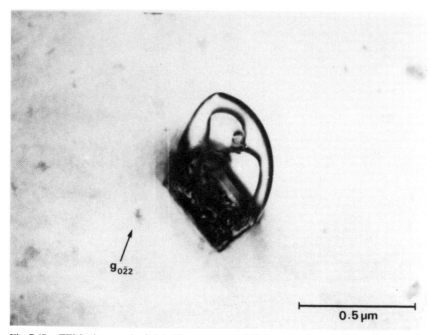

Fig. 7.48.   TEM micrograph of plate-like oxygen precipitate with prismatic dislocation loops in CZ silicon subjected to heat treatment at 950°C for 16 hr in dry O_2.

the precipitate does not cause strain due to misfit, which is commonly observed as black/white lobes for a crystalline precipitate such as shown in Fig. 7.47. The octrahedral precipitate is primarily bounded by eight {111} planes but is slightly truncated by {100} planes. The IR spectrum resulting from these octahedral precipitates indicated that they are amorphous SiO_2 and give rise to broad absorption bands around 470 and 810 cm^{-1}.[151] Later, high-resolution TEM (HRTEM) directly gave evidence that the structure of an identical octahedral precipitate is amorphous.[142] The HRTEM micrograph of an octahedral precipitate heavily truncated by two {100} planes observed in CZ silicon annealed at 1175°C for 64 hr is shown in Fig. 7.50. Clear thickness fringes observed in the TEM micrographs shown in Fig. 7.49 are caused by an electron diffraction effect in the silicon matrix, but not in the precipitate since the octahedral precipitates has been identified to be amorphous. The octahedral shape of amorphous SiO_2 is due to the negative crystal form of the silicon matrix, because the favorite growth habit of silicon is octahedral bounded by {111} planes.[152]

As described above, the nature of oxygen precipitates formed in CZ silicon strongly depends on the annealing temperature; however, for the formation of different types of oxygen precipitates, the variation of Gibbs free energy of

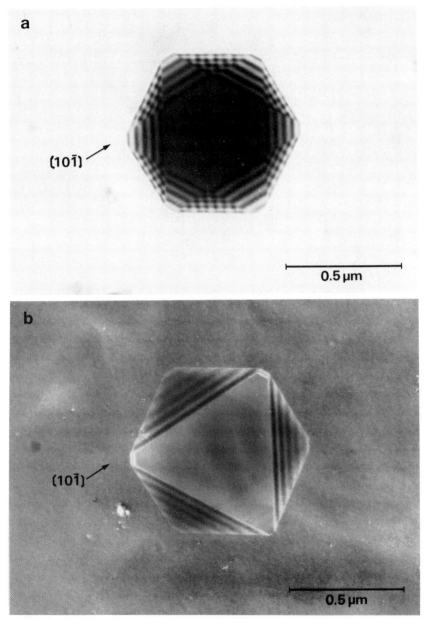

Fig. 7.49. TEM micrograph of octahedral oxygen precipitate in CZ silicon subjected to heat treatment at 1150°C for 64 hr in dry O_2: (a) bright-field image viewed from [111], (b) dark-field weak-beam image viewed from [111], (c) bright-field image viewed from [110], (d) bright-field image viewed from [101], and (e) bright-field image viewed from [011]. (After Shimura.[150] Reprinted with the permission of North-Holland Publishing Company.) (*Figure continues.*)

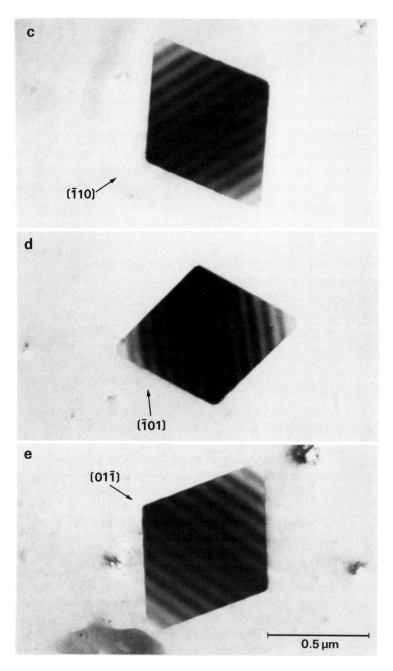

Fig. 7.49 (*Continued*)

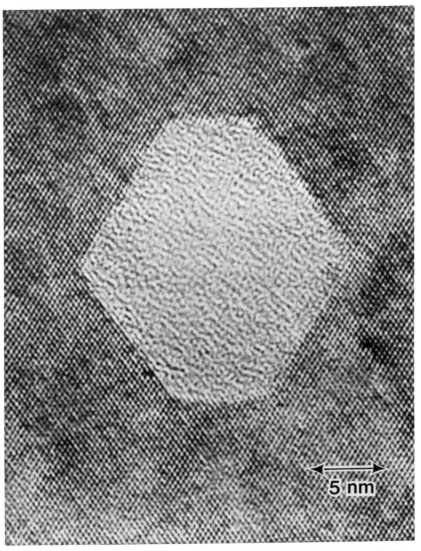

Fig. 7.50.   High-resolution TEM lattice image of amorphous oxygen precipitate in CZ silicon subjected to heat treatment at 1175°C for 64 hr in N_2. (Courtesy of F. A. Ponce, Hewlett-Packard.[142])

the crystal during the heat treatment should be considered by taking into account the elastic strains due to precipitates and the relaxation from emitting silicon self-interstitials.[153] Thus, it may be reasonable to expect that point defects or subsidiary impurities such as carbon and nitrogen can modify the morphology of oxygen precipitates formed at different temperatures. Moreover, subsequent heat treatment at a high temperature such as 1230°C leads to phase transformation from preexisting crystalline to amorphous,[113] as well as leading to conversion of defect structure, such as from a stacking fault to a perfect dislocation loop.[116] For example, plate-like crystalline SiO_2 precipitates generated by annealing at 1000°C for 64 hr (see Fig. 6.39) are transformed to globular amorphous-like precipitates such as shown in Fig. 7.51 during repeated high-temperature heat treatment, which corresponds to step 8 in the sequence shown in Fig. 7.31 for a Group I sample subjected to heat treatment at 1000°C first. The IR spectrum of those precipitates shown in Fig. 7.51 shows that the absorption at $1225\,cm^{-1}$ becomes obscure and, instead, absorption bands appear around 470 and $810\,cm^{-1}$.

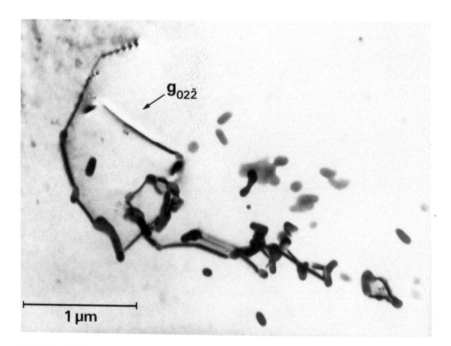

Fig. 7.51. TEM micrograph of oxygen precipitates and dislocations in CZ silicon subjected to multistep heat treatment (1000°C/64 hr + 1230°C/2 hr × 4, see Fig. 7.31) in dry O_2. (After Shimura and Tsuya.[113] Reprinted with the permission of The Electrochemical Society, Inc.)

Consequently, the type of secondary defects—namely, dislocations or stacking faults—induced in CZ silicon crystals containing oxygen by heat treatment strongly depends on the density and size of initial oxygen precipitates, which in turn decides the distribution of silicon self-interstitials.

Formation of Dislocations   The TEM micrograph of prismatic punched-out dislocation rows, denoted with D_1, D_2, D_3, and D_4, observed in CZ silicon annealed at 1100°C for 16 hr are shown in Fig. 7.52a.[48] The corresponding Burgers vectors are $\frac{a}{2}[0\bar{1}1]$, $\frac{a}{2}[01\bar{1}]$, $\frac{a}{2}[10\bar{1}]$, and $\frac{a}{2}[1\bar{1}0]$. Generator dislocations G_1 and G_2 with $\mathbf{b} = \frac{a}{2}[0\bar{1}1]$ and $\frac{a}{2}[1\bar{1}0]$ are also shown in the figure; an enlarged one with a view from [111] is shown in Fig. 7.52b. Oxygen precipitates in silicon can exist either alone, as shown in Figs. 7.47 and 7.49, or with the generation of dislocations as shown in Figs. 6.38 and 7.52. Dislocations are primarily generated by the mechanism of prismatic punching by a compressive misfit stress introduced into the silicon matrix.[140] The stresses arise from oxygen precipitation and from differential contraction between precipitates and the silicon matrix during a cooling process after heat treatment. When upper critical stress is introduced at the interface, interstitial dislocation loops are punched out along the glide directions ⟨110⟩, thereby reducing misfit strain. For the case of a spherical precipitate geometry, the relationship of dislocation generation to the precipitate size and to the extent of precipitate shear stress has been theoretically discussed and the critical size of the precipitate that may introduce punched-out dislocations has been proposed.[154,155] However, a notable correlation of punched-out dislocation generation and the size of oxygen precipitates has not been verified in CZ silicon crystals. Since punched-out dislocations are generated to release the stress introduced at the precipitate–matrix interface, the generation must depend on the consequence of various combined conditions such as the degree of noncoherency, the size of a precipitate, the cooling rate, and the existence of other defects near the precipitate. The shape of loops and the spacing between them along the ⟨110⟩ directions shown in Fig. 7.52 are very consistent. However, punched-out dislocation loops can climb when a local abundance of either interstitials or vacancies exists in the environment. Entangling of dislocations that climb during prolonged annealing leads to irregular-shaped dislocation networks as shown in Fig. 6.39.

Formation of Stacking Faults   Stacking faults are another type of common interior defects observed in CZ silicon crystals subjected to heat treatment at

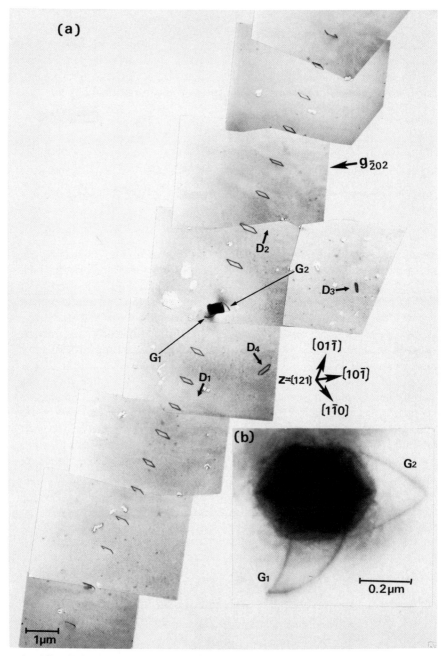

Fig. 7.52. TEM micrograph of octahedral oxygen precipitate and punched-out prismatic dislocation loops in CZ silicon subjected to heat treatment at 1150°C for 16 hr in dry O_2: (a) view from [121], and (b) view from [111]. The terms D_1, D_2, D_3, and D_4 denote punched-out dislocation rows, while G_1 and G_2 are generator dislocations of D_1 and D_2, respectively. (After Shimura *et al.*[48])

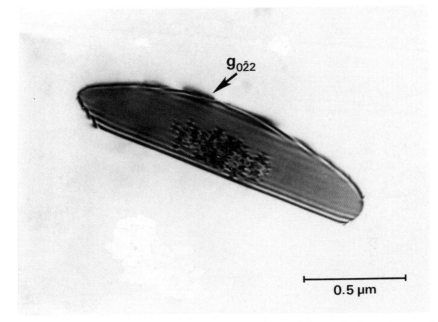

0.5 µm

Fig. 7.53. TEM micrograph of stacking fault with microprecipitate colonies at the central region in CZ silicon subjected to heat treatment at 950°C for 16 hr in dry O_2. (After Shimura *et al.*[48])

a temperature higher than $\sim 900°C$. Those stacking faults thermally generated in CZ silicon crystals are extrinsic in nature and bounded by $\frac{a}{3}\langle 111 \rangle$ Frank partials, and are generally accompanied by oxygen precipitates at the central region as shown in Fig. 6.40. Figure 7.53 shows the TEM micrograph of a stacking fault, observed in CZ silicon annealed at 950°C for 16 hr, which has microprecipitate colonies at the central region.[48] These microprecipitates are essentially circular disks with a diameter of about 250 Å. The TEM observation by tilting the specimen has proved that these precipitates lie on or very close to the plane of the stacking fault. Among those models for formation of stacking faults associated with oxygen precipitates in silicon that have proposed to date, the model proposed by Mahajan *et al.*[156] best explains the TEM observations. That is, the first stage in the origin of stacking faults is assumed to involve the formation of silicon–oxygen clusters on the {111} planes. As these clusters grow, a compression of the adjoining silicon matrix along the $\langle 111 \rangle$ direction would result in emission of silicon interstitials into the adjoining lattice. At a certain critical stage in their growth sequence, the local concentration of silicon interstitials reaches a level

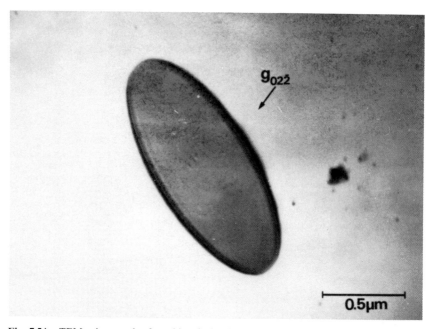

Fig. 7.54. TEM micrograph of stacking fault without precipitate association in CZ silicon subjected to two-step heat treatment (750°C/64 hr + 1230°C/2 hr) in dry O_2.

where they must form a sequence of plates and thus an extrinsic stacking fault is formed. Further growth of the fault is presumed to occur by the repeated precipitation of SiO_2 clusters. Octahedral precipitates of amorphous SiO_2 bounded by {111} planes can also generate stacking faults in the same sequence just described.[157]

Stacking faults that are not directly associated with oxygen precipitates, such as shown in Fig. 7.54, have been observed in CZ silicon subjected to a low–high two-step heat treatment, such as 750°C/64 hr + 1230°C/2 hr.[113] When CZ silicon is subjected to heat treatment at a low temperature (e.g., 750°C) for a long time, dense microprecipitates are generated, resulting in dense emitted silicon interstitials, which distribute randomly in the silicon matrix. However, these interstitials are not condensed enough to form dislocations or stacking faults at this stage. During subsequent heat treatment at a high temperature, the silicon interstitials agglomerate around certain nuclei into extrinsic stacking faults in order to reduce the Gibbs free energy of the silicon matrix. Observation by TEM reveals that stacking faults are formed during heat treatment at 1230°C for only 15 min following heat treatment at 750°C for 64 hr. Consequently, a model has been suggested such that these stacking faults are formed via two stages: (1) generation of dense silicon self-interstitials that distribute randomly in the silicon matrix, and (2) agglomeration of the interstitials into extrinsic stacking faults.[113]

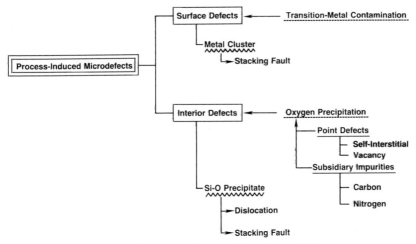

Fig. 7.55. Summary of process-induced microdefects.

7.2.6 Summary of Process-Induced Microdefects

Process-induced microdefects, excluding OSFs and slip dislocations, have been classified as either surface or interior (or blulk) defects, as summarized in Fig. 7.55. Contamination, particularly with transition metals, during thermal processing initiates surface microdefects. On the other hand, interior microdefects are exclusively caused by oxygen precipitation, which depends on various factors. Oxygen precipitates either can be interior microdefects by themselves, or can originate secondary lattice defects such as dislocations and stacking faults by emitting excess silicon self-interstitials. As discussed in Section 7.4.3, these interior defects play a key role as intrinsic gettering sinks for surface impurities that would otherwise limit device performance or initiate surface microdefects. On the other side of this double-edged knife, these interior defects may degrade the mechanical strength of silicon wafers when too much oxygen precipitation occurs; this results in serious warpage of silicon wafers.[33,158,159] Consequently, it is essential to control oxygen precipitation to an optimum level during the thermal processes in order to maximize the device performance and device fabrication yield.

7.3 Effects of Defects on Electrical Properties

7.3.1 Impurities

Metallic Impurities As was discussed in Chapter 5, the contamination with metallic impurities is pervasive at many different silicon process steps. In particular, silicon device fabrication processing with furnace operation at high temperatures increases the contamination. The primary electical effect of metallic impurities is the introduction of energy levels close to the center of the bandgap of silicon. Since these levels act as recombination centers, metallic impurities cause a decrease in minority carrier lifetime and an increase in the leakage currents of *p-n* junctions. When the impurities exceed their solubility limits in silicon, metallic clusters generate and represent a dielectric constant discontinuity, and reduce the breakdown strength of the oxide.[160-162] Second, as was discussed previously, metallic impurities generate some other types of defect, such as stacking faults[47-49] or other precipitates.[163] Third, metallic impurities make dislocations and stacking faults electrically active.[164,165]

Oxygen and Carbon   Under most circumstances, oxygen and carbon have no direct effect on the electrical properties of silicon crystals, but rather induce secondary defects such as dislocations and stacking faults, which can strongly degrade the electrical performance of silicon devices. They may also act as preferential condensation sites for metallic impurities.[163] Consequently, oxygen and carbon potentially lead to an enhancement of the junction leakage currents and a degradation of minority carrier lifetime.

From the other viewpoint, oxygen in silicon can uniquely form donors at up to $2 \times 10^{16}/cm^3$ when the silicon crystal is heated at a temperature between 300 and 500°C, with the highest formation rate at 450°C.[166-168] This oxygen donor formation may cause variation in the resistivity of silicon crystals, and can even convert *p*-type silicon into *n*-type silicon. Fortunately, the oxygen donors can be annihilated by heat treatment in the temperature range between 650 and 800°C. Thus the resistivity stabilization of silicon crystals is usually achieved by a thermal process called *donor annihilation*, wherein silicon crystals are typically maintained at a temperature between 650 and 700°C for about 60 min in an inert ambient, followed by quenching in air. This process effectively removes oxygen donors from the silicon crystal, which results in the resistivity being governed only by the dopant concentration. Recent investigation has shown, however, more complicated behavior of oxygen-related carrier concentration, which strongly depends on the heat-treatment temperature.[169] Eventually, another type of oxygen donor called "new donors"[170] can be formed in silicon subjected to heat treatment in the temperature range between 500 and 900°C. For convenience, the oxygen donors formed around 450°C are occasionally referred to as "old donors." The new donors can be annihilated by heat treatment at a high temperature, $>1000°C$.[171] Thus the standard heat treatment for the annihilation of old donors can cause the generation of new donors. In order to avoid new-donor generation, therefore, rapid thermal processing (RTP) at 650°C for a short time, on the order of seconds, has been suggested as an effective alternative donor-annihilation step.[172] As regards the effect of carbon on oxygen donor formation, it has been reported that carbon strongly inhibits the formation of old donors,[173,174] whereas it enhances the generation of new donors,[161,172] which are related to oxygen precipitates.

7.3.2 Dislocations and Stacking Faults

Dislocations The primary effect of dislocations on the electrical properties of silicon electronic devices can be described as an enhancement of dopant diffusion, which causes *diffusion pipe* or *diffusion spikes* in the *p-n* junction region as depicted in Fig. 7.56. It has also been observed that dislocations directly or indirectly enhance junction leakage current when they cross the *p-n* junction, and degrade the minority carrier lifetime.[176–179] There is, however, conflicting argument on whether dislocations without the association of metallic precipitates can be harmful. That is, it has been suggested that dislocations can be electrically active and can result in junction leakage current only when they are decorated by metallic impurities, while undecorated "clean" dislocations have negligible electrical effect on the leakage.[4,179] The effect of metallic impurities on electrical properties is straightforward; however, there are some experimental results that suggest a primary effect of dislocations, without any involvement of metallic impurities, on the leakage current.[177,178] Apart from the effect of associated metallic impurities, the effect of dislocations on the electrical properties of silicon would be explained by the bandgap energy variation in the dislocatioin region. This change in the bandgap can be implied by a part of the lattice in the vicinity of a dislocation being compressed while another part is expanded, and the change would have two effects: (1) the electron energy required for impact ionization may be reduced, and hence the ionization rate would be increased, and (2) channeling of carriers into the dislocation core would occur.[180] Thus, the bipolar transistors developed for high-speed operation are particularly susceptible to dislocations because of their thin active regions. Moreover, dislocations act effectively as generation–recombination centers and introduce surface states; thus dislocations decrease the minority carrier lifetime.[181] It has been found

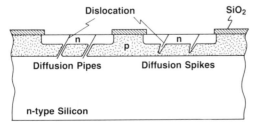

Fig. 7.56. Schematic illustration showing diffusion pipes and diffusion spikes caused by dislocations in *p-n* junction region.

that a dislocation density of 10^7–10^8/cm^2 leads to interface charges Q_{it} of $\sim 1 \times 10^{12}$/cm^2, and dislocation densities of $\sim 10^9$/cm^2 lead to $Q_{it} > 10^{13}$/cm^2; the effect is insignificant when the dislocation density is below 10^6/cm^2.[182] However, no clear explanation for the phenomenon is available at present.

Stacking Faults   Every stacking fault is associated with a dislocation loop, which bounds the fault plane. Hence, the electrical effects of stacking faults are very similar to those of a dislocation. However, compared with dislocations, stacking faults have received much more attention in the literature because OSFs are the defects most commonly observed in silicon device regions. Stacking faults that penetrate *p-n* junctions greatly enhance the recombination current, and consequently enhance the junction leakage currents.[183–186] Low leakage current is one of the most important conditions for reliable operation of all dynamic MOS devices such as DRAM and CCD. Stacking faults in MOS capacitors result in a deterioration of the refresh behavior of dynamic memories. The density of stacking faults in MOS capacitors has been inversely correlated with the refresh time of MOS devices.[187] The excess reverse current is harmful in charge storage type imaging devices, such as silicon vidicon and CCD, because the current results in bright spot image defects in a video display[188,189] and leads to dark-current nonuniformity.[189,190] It has been shown that high-density OSFs generated in MOS structures drastically increase the surface generation velocity and decrease the bulk lifetime.[191] This oxide MOS capacitors that exhibit high leakage and low breakdown strength have been also correlated with the stacking faults located in the silicon substrate near the SiO$_2$/Si interface.[192] It should be noted that, similar to the case of dislocations, most of the experimental results suggest the contribution of metallic impurities, which decorate stacking faults, to the effect on electrical properties.[183,184,186,187,192]

7.4 Gettering

7.4.1 Gettering Phenomena

General Remarks   As has been discussed in a preceding section, a variety of lattice defects and impurities can be introduced during crystal growth, the wafer shaping process, and subsequent device fabrication process. The electronic device properties are greatly degraded by these defects and impurities, particularly transition metals, when they are located in the device regions. It is thus indispensable to eliminate the detrimental effect of these defects and impurities in order to ensure the high performance of electronic devices.

In the fabrication of VLSI/ULSI devices, dry etching processes have been replacing wet etching processes in which silicon wafers are immersed in liquid etching reagents. Wet etching offers a low-cost, reliable, high-throughput process with excellent selectivity for the most wet etch processes; however, it is not capable of reproducible and controllable transfer of patterns in the micrometer or submicrometer range, which is required for VLSI/ULSI fabrication. Dry etching processes, which are primarily based on physical sputtering, ion-beam etching, or plasma etching, offer several advantages over the counterpart wet processes. However, dry processing has a tremendous contaminating capability and provides local heating and kinetic energy for contamination of the silicon wafer surface. Since reactive-ion etching (RIE) can selectively etch one chemical species in favor of another, it is possible to concentrate the residual species on the surface of the substrate, which is then driven into the circuit. In particular, metallic impurities (e.g., Fe, Ni, Cu, Cr) can be sputtered from the chamber surfaces or components that consist of these materials, and can then be deposited on the surfaces being etched.[2] The intentional components of thin films (e.g., metal films or silicides) can also be the source of metallic contamination. Furthermore, reduced dimensions, particularly when a trench structure is used, are proving to be very difficult to clean, both from a particulate point of view and from a chemical solubility point of view, since the surface tension of the cleaning fluids is quite high in their pure states. Consequently, the contamination problem due to impurities has become more serious in the VLSI/ULSI era.

The elimination of the effects of defects and impurities can be achieved through three steps: (1) suppression of the sources that may generate defects, (2) annihilation of existing defects, and (3) removal of impurities from the device regions in a silicon wafer. The process that accomplishes (2) and (3), particularly (3) in a narrow sense, is generally referred to as *gettering*. The term "gettering" was originally used by Goetzberger and Shockley for the process of removing metallic impurities from the device region by a predeposited surface layer of either boron oxide or phosphorus pentoxide on a silicon wafer.[193] Since metallic impurities are highly mobile, as listed in Table 7.2, for a 1-hr diffusion period at various temperatures of interest from the device fabrication process point of view,[194] they diffuse from the surface of a wafer through the silicon lattice into the device regions very easily at processing temperatures. Gettering thus concerns mainly the removal of transition metals that diffuse quickly, cause surface microdefects, and make lattice defects electrically active. The purpose of gettering is primary to create a defect-free surface region in a silicon wafer used for electronic device fabrication. The gettering process involves three steps: (1) impurities are removed from the surface of a wafer, (2) they then diffuse through the silicon lattice into certain *gettering sinks* at a position away from the device region, and (3) they are gettered or captured by the gettering sinks.

Table 7.2   One-Hour Diffusion Length of Impurities in Silicon at Several Temperatures

Element	Diffusion length (μm)[a]					
	700°C	800°C	900°C	1000°C	1100°C	1200°C
H	3.3×10^3	4.3×10^3	5.4×10^3	6.5×10^3	7.6×10^3	8.7×10^3
Li	5.8×10^2	8.4×10^2	1.1×10^3	1.5×10^3	1.8×10^3	2.2×10^3
Cu	3.2×10^3	4.0×10^3	4.9×10^3	5.8×10^3	6.7×10^3	7.5×10^3
Ag	1.9	4.6	9.7	1.8×10	3.1×10	4.9×10
B	3.5×10^{-4}	2.2×10^{-3}	1.1×10^{-2}	3.9×10^{-2}	1.2×10^{-1}	3.1×10^{-1}
Al	1.1×10^{-3}	7.4×10^{-3}	3.6×10^{-2}	1.3×10^{-1}	4.1×10^{-1}	1.1
C	7.5×10^{-3}	4.2×10^{-2}	1.8×10^{-1}	5.9×10^{-1}	1.6	4.0
Ge	2.0×10^{-5}	2.7×10^{-4}	2.3×10^{-3}	1.4×10^{-2}	6.6×10^{-1}	2.5×10^{-1}
Ti	3.4×10^{-1}	7.9×10^{-1}	1.6	2.8	4.7	7.2
N	1.6×10^{-3}	1.0×10^{-2}	4.6×10^{-2}	1.7×10^{-1}	5.0×10^{-1}	1.3
P	3.7×10^{-4}	2.9×10^{-3}	1.6×10^{-2}	6.5×10^{-2}	2.2×10^{-1}	6.3×10^{-1}
As	1.9×10^{-4}	1.4×10^{-3}	7.3×10^{-3}	2.9×10^{-2}	9.6×10^{-2}	2.7×10^{-1}
Sb	9.4×10^{-5}	7.1×10^{-4}	3.9×10^{-3}	1.6×10^{-2}	5.4×10^{-2}	1.5×10^{-1}
O	6.5×10^{-2}	2.7×10^{-1}	8.9×10^{-1}	2.4	5.6	1.2×10
Cr	1.6×10^2	2.7×10^2	4.2×10^2	6.2×10^2	8.7×10^2	1.2×10^3
Mn	9.6×10^1	2.0×10^2	3.6×10^2	6.0×10^2	9.2×10^2	1.3×10^3
Fe	3.7×10^2	5.4×10^2	7.4×10^2	9.7×10^2	1.2×10^3	1.5×10^3
Co	9.9	4.7×10	1.7×10^2	5.1×10^2	1.3×10^3	2.9×10^3
Ni	1.6×10^3	2.1×10^3	2.6×10^3	3.1×10^3	3.7×10^3	4.2×10^3

[a] Diffusion length $L = (Dt)^{1/2}$ was calculated following Table 5.2.

Gettering Techniques   A gettering technique may therefore eventually be defined as a method to supply effective sinks for harmful impurities. Until now, various types of gettering techniques have been investigated for application to silicon wafers. These techniques are classified into three categories for convenience: (1) *extrinsic* or *external gettering* (EG), (2) *intrinsic* or *internal gettering* (IG), and (3) *chemical gettering* (CG). Extrinsic gettering involves the use of external means to introduce gettering sinks into a silicon lattice, predominantly at the back surface region of a silicon wafer. Hence, EG is occasionally called backside or back surface gettering. The gettering sinks are formed in the back surface by mechanical damage, diffusion of impurities such as boron and phosphorus, or deposition of films such as silicon nitride and polysilicon. The stresses induced in the backside damaged layer can be a source of driving force in the gettering mechanism, and they result in the creation of lattice defects or chemically active sites at which mobile impurities are captured. These EG sinks should be introduced

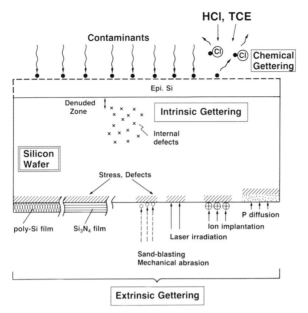

Fig. 7.57. Schematic illustration showing various gettering techniques for a silicon wafer.

in a silicon wafer prior to the first oxidation step, or concurrently with processing steps. In contrast, intrinsic gettering uses thermally induced interior defects as gettering sinks in a silicon wafer. The interior defects due to oxygen precipitation are generated by IG heat treatment prior to the first oxidation, or during concurrent oxidation or diffusion processes. On the other hand, chemical gettering does not provide gettering sinks as EG and IG do, and instead is performed during oxidation or heat treatment in chlorine-containing ambients. In this case, the elimination of metallic impurities is by their evaporation as a result of chemical reaction with chlorine resulting in volatile metal chlorides.[80] Moreover, as was discussed already, CG contributes to shrinkage of OSF by supplying vacancies near the surface region of a silicon wafer. Figure 7.57 depicts gettering techniques that have been categorized into three groups. Since the effect of oxidation in chlorine-containing ambients on shrinkage or elimination of OSF has already been discussed, the discussion in this section will focus on EG and IG.

Phenomena of Gettering Interaction   It should be noted that most gettering phenomena may involve more than one category, since the phenomena depicted in Fig. 7.57 can affect each other more or less. For example, oxygen precipitation, which results in IG sinks, can be affected by externally induced damage and deposited films on the wafer back surface. The change in $[O_i]$ as

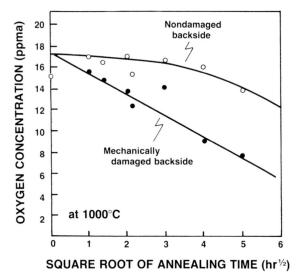

Fig. 7.58. Interstitial oxygen concentration as a function of oxidation time at 1000°C for (100) CZ silicon wafers with or without backside mechanical damage. (After Takano *et al.*[195] Reproduced with the permission of The Electrochemical Society, Inc.)

a function of the square root of oxidation time at 1000°C for (100) CZ silicon wafers with or without backside mechanical damage is shown in Fig. 7.58.[195] Interstitial oxygen atoms precipitate proportionally to the square root of oxidation time in a damaged wafer, whereas they precipitate much more slowly in an undamaged wafer. Figure 7.59 shows the change in oxygen precipitate density as functions of distance from the back surface with a deposited polysilicon film of 1.6 μm thickness and of oxidation time at 1000°C for (100) CZ silicon wafers whose $[O_i]_0$ is around 15 ppma.[196] Both the figures clearly indicate that oxygen precipitation is considerably enhanced by externally introduced damage and a film deposited on the back surface of a silicon wafer. From the practical point of view, those gettering techniques categorized for convenience should be used complementarily or in duplicate. Among the gettering techniques depicted in Fig. 7.57, the most commonly used techniques in the silicon industry are backside mechanical damage, polysilicon deposition, and IG. The silicon wafers with these most common gettering treatments are commercially available, and have been widely used for microelectronic circuit fabrication.

Elimination of Contamination Source   Although various gettering techniques have been investigated and some of them have been extensively utilized in the silicon semiconductor industry, it should be noted at this point that the primary effort undertaken to eliminate the detrimental effect of

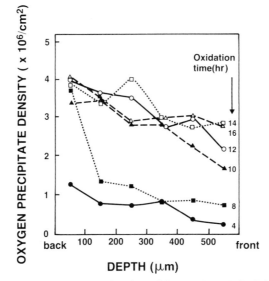

Fig. 7.59. Oxygen precipitate density as functions of distance from the backside polysilicon film and oxidation time at 1000°C. (After Shirai *et al.*[196])

metallic impurities is to remove the sources of the harmful impurities that contaminate silicon wafers during the entire fabrication processes. Such contaminant sources that have been identified in silicon processing include (1) chemicals used in etching and cleaning processes, (2) stainless steel parts of processing equipment, (3) heating coils, (4) graphite susceptors, and (5) metallic tools such as tweezers. In order to minimize the level of metallic contamination from these sources, the following procedure has been established as effective: (1) reducing process temperatures,[197] since the diffusivity of impurities and permeability of quartz tubes to the impurities decrease with decreasing temperature; (2) subjecting furnace tubes to frequent chemical cleaning and high temperature HCl gas cleaning; and (3) use of double tube configuration in furnaces to protect impurity diffusion into processing silicon wafers. In addition, cassette-to-cassette operations in handling of wafers has reduced the contamination originated from human body and human handling.[198] The ideal situation would be the establishment of perfectly clean processes, which do not need any gettering techniques; however, the practical silicon device fabrication processes at present seem to require some gettering treatment for the processing wafer, since gettering studies in increasing number have consistently shown that gettering operations are capable of overcoming defect and impurity problems arising during device processing. In fact, the gettering techniques used in IC processing have been widely found to be beneficial to device performance and device manufacturing yield.[85,199–206] In addition, it has been recognized that no single gettering may be adequate for all processes, and a tailored gettering program is required for the particular technology that utilizes a sequence of many processes. In any case, however, the limit of gettering capability should be recognized.

7.4.2 Extrinsic Gettering

Mechanical Damage Mechanical damage in a near-surface region of a silicon wafer back surface can be induced by rotary abrasive lapping, sand blasting, or scribing[207,208] and impacting with small tungsten balls (300 μm in diameter) under acoustic stressing.[209] Usually, the mechanical damage treatment is performed prior to the wafer polishing process. During subsequent thermal processes, dislocations or stacking faults are generated as a result of relieving the stresses caused by mechanical damage consisting mainly of microcracks. These defects then serve as gettering sinks. Figure 7.60 demonstrates the effect of backside mechanical damage gettering on reduction in the formation of surface defects. The CZ silicon wafers with or without backside damage were subjected to heat treatment at 1100°C for 2 hr in wet O_2. Sirtl etching revealed S-pit densities of $\sim 10^5/\text{cm}^2$ and $< 2 \times 10^2/\text{cm}^2$ for nongettered and gettered wafers, respectively. That is, contaminants diffused from the surface of a wafer with backside damage through the silicon lattice toward the back surface region where the impurities were captured. It has been observed that backside sinks can getter not only surface impurity atoms but also surface microdefects, most likely transition-metal clusters that manifest themselves as S-pits by chemical etching, which were induced

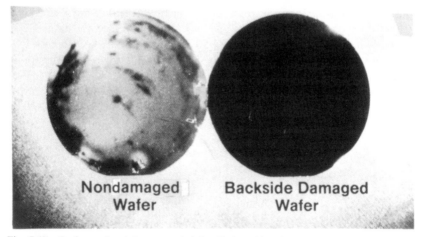

Fig. 7.60. Surface microdefects revealed by Sirtl etching for nondamaged and backside mechanically damaged wafers subjected to oxidation at 1100°C for 2 hr in wet O_2.

previously in the wafer surface region.[81] The mechanical damaging process for silicon wafers is simple and less costly; however, the major disadvantages of this technique are that the silicon dust created during the damaging process is difficult to remove completely and may cause additional defects during subsequent thermal processes, and that the degree of damage, and in turn the density of induced defects for gettering sinks, is difficult to control with high reproducibility. In addition, the damage-inducing process itself is dirty and may contaminate the silicon wafers with the abrasive used.

Laser-Irradiation Damage In order to overcome the shortcomings of the traditional gettering technique using mechanical backside damage, laser irradiation has been investigated to introduce backside damage for gettering sinks.[210] The principal technological advantages of the laser irradiation method are (1) that the damage can be precisely controlled with high reproducibility, and (2) that the laser damaging process can be performed at room temperature in a clean environment. The damage is induced on the backside (rarely on the frontside) of a silicon wafer by scanning a focused laser beam of usually a Q-switched Nd:YAG laser[210–212] or an argon laser.[213] Recently, the KrF excimer laser has also been successfully used for backside damaging.[214] Using Q-switched Nd:YAG laser pulses with a pulse spot of 40 μm diameter for (111) CZ silicon wafers, it has been found that several different-shaped dislocations with Burgers vectors of $\frac{a}{2}\langle 110 \rangle$ on {111} planes are generated, depending on the applied laser energy density higher than 8 J/cm^2.[211] After subsequent oxidation, these dislocations move far from laser-induced grooves and form dislocation networks as shown by the XRT of Fig. 7.61. These dislocation networks have been shown to be effective gettering sinks that significantly suppress the generation of surface microdefects. Chemical etching for the cross section of the silicon wafer shown in Fig. 7.61 has shown that the dislocation propagation from the back surface stops two-thirds of the way to the front surface. A mechanism for the formation of dislocation networks induced by high-power laser pulses above 20 J/cm^2 is illustrated in Fig. 7.62. When a silicon wafer is irradiated by laser pulses, the near-surface region melts and the melt front moves rapidly into the bulk. In the recrystallization of a melt, dislocations are generated to relieve thermal stresses at the interface. These dislocations propagate into the bulk, resulting in dislocation networks during subsequent oxidation to relieve the stress induced by laser irradiation. These primary dislocations and dislocation networks can act as effective gettering sinks for contaminant impurities and point defects as well. The amount of laser damage, which is critical to induce optimum gettering sinks without any unwanted wafer distortion, can be controlled primarily by changing the energy density of the laser pulses,

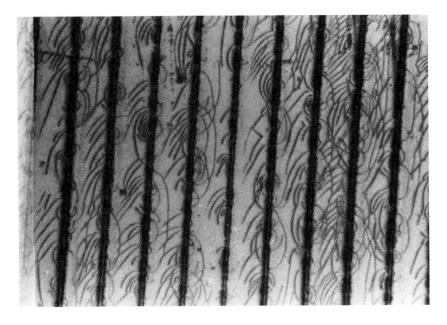

⊢————————⊣
500μm

Fig. 7.61. X-ray topograph of dislocation networks induced by laser-irradiation and following oxidation. (Courtesy of Y. Hayafuji, SONY Corporation. After Hayafuji *et al.*[211] Reprinted with the permission of The Electrochemical Society, Inc.)

and secondarily by varying the pitch of the scanning laser beam and the amount of overlap between spots. However, the major obstacles of the laser irradiation gettering method may be the throughput and wafer distortion due to lattice damage generated by high-power pulsed laser irradiation.

Ion-Implantation Damage   As an alternative to mechanical damaging techniques, the use of ion implantation to introduce wafer backside damage has also been investigated.[215] The process of ion-implantation gettering basically consists of introducing the atomic displacement resulting in an amorphous lattice by ion-implantation damage, and of subsequent annealing which recrystallizes the amorphous region leaving numerous crystallographic defects. These resultant defects then trap the migrating impurities. Gettering treatment by ion-implantation damage is usually performed on the wafer back surface, although gettering implantation in noncritical areas on the front surface has been effectively used.[216] Various ion species (i.e., Ne, Ar, Kr, Xe, O, P, Si, As, and B) have been used and compared in terms of the induced damage and gettering capability.[215,217–219] Depending on ion species, dose,

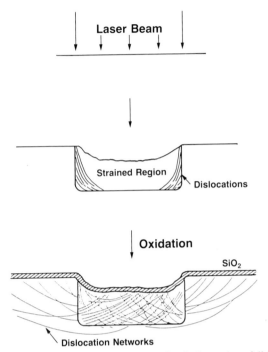

Fig. 7.62. Schematic illustration showing mechanism for the formation of dislocation network by laser irradiation and following oxidation. (After Hayafuji *et al.*[211] Reproduced with the permission of The Electrochemical Society, Inc.)

and energies, various types of crystallographic defects such as grain boundaries, microtwins, stacking faults, dislocation loops, and dislocation networks are produced by subsequent annealing conditions. It has been found that the state of initial damage affects the final lattice disorder after subsequent annealing,[219] and the gettering sinks are the direct result of ion-implantation damage regardless of whether n^+, p^+, or electrically neutral regions are created.[215] In general, gettering studies find that argon ion implantation results in more effective gettering than other species. This may be partly attributed to the large size of an argon ion. Ion-implantation gettering treatments can also be performed through oxide layers,[220,221] which results in the formation of stable dislocations produced by recoil of oxygen atoms from SiO_2 as an important by-product of ion implantations.[222] It also appears that the formation of microtwins, polycrystals, and other defects is promoted by the presence of implanted inert gas and the formation of gas bubbles[219] such as discussed in Section 3.4.4. In addition, it has been found that perfect dislocations with $\mathbf{b} = \dfrac{a}{2} \langle 110 \rangle$ are critical in providing effective gettering sinks, but $\frac{a}{3}\langle 111\rangle$ Frank partials surrounding stacking faults contribute less compared with perfect dislocations.[218] Finally, it should be noted that the conditions of thermal treatment following ion implantation play an important role in determining the defect structure, and in turn the gettering capability, since the damage induced by ion implantation may anneal out at high temperatures leaving no gettering sinks.

Phosphorus Diffusion The gettering of metallic impurities by phosphorus or boron diffused oxide layers on wafer surfaces has been studied extensively since the first "gettering" work was made by Goetzberger and Shockley.[193] The introduction of a backside heavily phosphorus-diffused layer for gettering sinks has been proposed as one of the so-called *preoxidation gettering at other side* (POGO) techniques by Rozgonyi *et al.*[223] Since then, the effect of backside phosphorus diffusion gettering on the reduction of surface microdefects and OSF has been widely recognized. Phosphorus using $POCl_3$ as the phosphorus source is usually diffused into the back surface of a silicon wafer at temperatures around 1100°C for several hours. The concentration of phosphorus in the diffused area reaches the order of 10^{19}–10^{21} atoms/cm³.[224] During this diffusion process, a dense array of misfit dislocations is introduced to a depth of 2 μm or more below the phosphorus-diffused surface. It has been assumed that misfit dislocations induced by heavily doped impurities in the backside region of the wafer getter impurities that cause surface microdefects or OSFs.[223] In addition, it has been verified recently that the gettering of gold also occurs in the phosphorus-diffused region where no dislocations are introduced because of diffusion at a low temperature.[225] The gettering effect becomes significant at concentrations of phosphorus exceeding 10^{20} atoms/cm³. This ability of phosphorus diffusion to getter gold in the absence of dislocations is attributable to a strong interaction between gold atoms and *E-centers* (phosphorus-vacancy pairs) to create a P–Au pair,[224,225] or to an enhanced metal diffusivity in heavily doped n^+ silicon due to pairing of the substitutional metal acceptor with donors at high temperatures.[226] Thus, the effects of phosphorus diffusion are due primarily to a drastic reduction of the concentration of metallic impurities in the wafer bulk and front surface, and secondarily to introduction of misfit dislocations. A major disadvantage of the phosphorus diffusion gettering is, however, that the high concentration of phosphorus in the wafer backside can cause the problems of contamination or autodoping during susequent thermal cycles.

Film Deposition A thin film of polysilicon[227–230] and/or silicon nitride (Si_3N_4)[227,231–233] deposited on the back surface of silicon wafers has been demonstrated to supply effective gettering sinks for impurities resulting in the reduction of surface microdefects and OSFs. A polysilicon film of a thickness around 1 μm is usually deposited on the back surface of silicon wafers by a CVD process at a temperature around 650°C. TEM observation indicates[228] that an unoxidized polysilicon film is composed of grains of less than 0.1 μm in size. After a single oxidation at 1100°C for 2 hr, the grains increase in size and small stacking faults ($<1\,\mu$m in size) are induced. During subsequent oxidation processes, the grain size continues to increase until the polysilicon layer is gone. At this point the stacking faults grow rapidly, approaching 8 μm in size, and dislocations also become apparent. The gettering capabilities of mechanical damaging and polysilicon deposition techniques have been compared with each other for CZ silicon wafers subjected to multiple oxidation at 1100°C for 2 hr in wet O_2[228] Figure 7.63 shows the density of surface microdefects, which manifest themselves as S-pits by chemical etching, as a function of oxidation sequential time for the CZ silicon wafers whose back surfaces were mechanically damaged by abrasion or deposited with a polysilicon film. Both gettering techniques are able to reduce a maximum density of surface microdefects to less than $10^2/cm^2$ after one oxidation. However, the surface microdefect density increases up to about $10^5/cm^2$ in

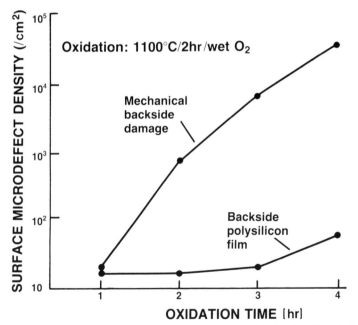

Fig. 7.63.   Density of surface microdefects as a function of oxidation (1100°C/2 hr/wet O_2) sequential time for CZ silicon wafers with mechanical backside damage or backside polysilicon film. (After Keefe-Fraundorf *et al.*[228])

mechanically damaged silicon wafers with repeated oxidation. Polysilicon-deposited silicon wafers in contrast, show a maximum density of less than $3 \times 10^2/cm^2$ after four sequential oxidation. This result indicates a general limitation of extrinsic gettering using mechanical damage. Mechanically induced damage in the backside of wafers will be annealed out during repeated thermal cycles, and the lattice defects that provide gettering sinks can no longer trap newly introduced impurities nor recapture the impurities released back into the silicon lattice from gettered sites. In the case of polysilicon gettering, on the other hand, the polysilicon grains grow larger during repeated oxidation steps, resulting in the generation of fresh gettering sinks for impurities. In addition it has been found that oxygen precipitation to induce IG sinks is enhanced in a silicon wafer with a backside polysilicon film.[196] Hence, polysilicon gettering has been commonly called *enhanced gettering.*[228–230,234] This enhancement effect is attributed primarily to the formation of SiO_2 microprecipitate embryos during the polysilicon CVD process, and secondarily to the creation of vacancy-rich circumstances in the silicon lattice close to the backside film.[196] The latter effect is evidenced by Fig. 7.64, the depth profile of intentionally induced OSF on the cleaved plane that was sand-blasted and followed by oxidation at 1100°C for 2 hr. The density of OSF becomes less in the region closer to the backside polysilicon film, which is considered to absorb silicon self-interstitials.[198] In addition, a backside polysilicon film has generally been shown to make the silicon wafer resistant to warpage during oxidation.[234]

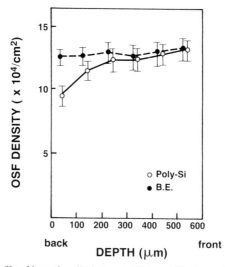

Fig. 7.64. Depth profile of intentionally induced OSF on CZ silicon cleaved surface that was treated with sand-blasting followed by oxidation at 1100°C for 2 hr. Symbols: ○, wafers with backside polysilicon; ●, wafers with chemically etched back surface. (After Shirai *et al.*[196])

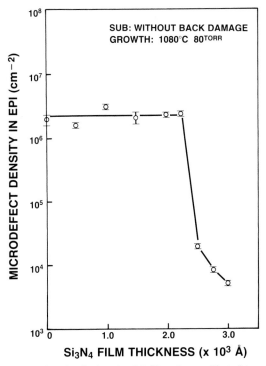

Fig. 7.65. Surface microdefect density in epitaxial silicon layer subjected to oxidation at 1140°C for 1 hr in wet O_2 as a function of backside Si_3N_4 film thickness. (After Tanno *et al.*[233] Reproduced with the permission of The Electrochemical Society, Inc.)

An Si_3N_4 film of about 0.3–0.4 μm in thickness is usually deposited on the wafer back surface by CVD at a temperature in the range between 700 and 800°C.[227,231–233] Figure 7.65 shows the correlation between the backside Si_3N_4 film thickness and the surface microdefect density in silicon epitaxial layers of 2 μm thickness deposited at 1080°C under a reduced pressure of 80 torr.[233] The surface microdefects were evaluated by Sirtl etching after annealing at 1140°C for 1 hr in wet O_2. The microdefect density in an epitaxial layer drastically decreases from the range of $10^7/cm^2$ to 10^3–$10^4/cm^2$ when the Si_3N_4 film thickness exceeds 2500 Å. When an Si_3N_4 film is deposited on a mechanically damaged back surface, it has been found that the microdefect density in an epitaxial layer decreases remarkably from $10^6/cm^2$ to $10^3/cm^2$ with the 200-Å-thickness backside Si_3N_4 film, and an Si_3N_4 film of 1000 Å thickness reduces the surface microdefect density to $5 \times 10^2/cm^2$.[233] The defects induced by a backside Si_3N_4 film and following thermal cycles have been identified as dislocations or stacking faults, which provide gettering sinks for impurities. In addition, the chemical analysis for impurities of gettered wafers has indicated that the Si_3N_4 film actively traps metallic impurities such as Au and Cu. The gettering mechanism that effectively eliminates surface microdefects and OSFs can be explained in two ways[231]: (1) the impurities that generate surface microdefects and supply nucleation centers for OSFs are gettered by the internal stresses of an Si_3N_4 film and lattice defects introduced by the Si_3N_4 film during subsequent heat treatment, and (2) Si_3N_4 deposited is considered to not be stoichiometric in general, being deficient in Si, and thus diffusion of silicon interstitials across the Si–Si_3N_4 interface may take place to form the stoichiometric Si_3N_4.

Future Trend of Extrinsic Gettering   All the EG techniques discussed in this subsection utilize gettering sinks formed in the backside region of a silicon wafer. For the formation of defect-free surface regions, impurity atoms must migrate from the surface through the silicon lattice toward the gettering sinks. As was discussed earlier, the diameter of silicon wafers used for electronic device fabrication has been steadily increased with the development of silicon technology. The increasing diameter is generally accompanied by the increasing thickness; that is, the contaminant impurities to be gettered have to migrate a longer path from the wafer front surface toward the backside gettering sinks in a larger-diameter wafer. In addition, the potential advantage of reduced-temperature device processing has been widely recognized for the VLSI/ULSI technology. However, the diffusivity of impurities to be gettered drastically decreases with a decreasing temperature. Under the circumstances, it is desired that the gettering sinks be located in close vicinity to the wafer surface, but far enough from the critical device regions, as is performed by intrinsic gettering, which is discussed in the next subsection. Consequently, the disadvantage of traditional EG techniques, in addition to various limitations of individual EG techniques, will become more apparent as the VLSI/ULSI technology is realized.

As an alternative, or complement, a new EG technique using a uniform network of interfacial misfit dislocations that are deliberately introduced at an epitaxial-layer/substrate interface has been recently developed.[235] The lattice dilation required for misfit dislocation formation is obtained by incorporation of electrically inactive germanium during the silicon epitaxial growth. This is readily accomplished by adding GeH_4 to a flowing gas mixture of SiH_4/H_2 in a CVD reactor. The controllability of the misfit dislocation density is achieved by adjusting the Ge content in the silicon matrix after a critical thickness of the lattice mismatched layer is surpassed. The structure consisting of the silicon substrate, Ge-doped epitaxial silicon, and an epitaxial silicon layer where electronic devices are fabricated is depicted in Fig. 7.66. The misfit dislocation networks spreading on the plane parallel to the wafer surface can effectively getter the impurities that originate

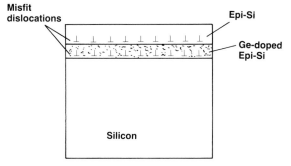

Fig. 7.66. Schematic illustration showing the structure of Si/Ge-doped Si/Si substrate and misfit dislocations induced at the interfaces.

surface defects. The effectiveness has been demonstrated by SIMS depth profiles of the gettered impurity and Ge in the epitaxial layer in Fig. 6.15.

7.4.3   Intrinsic Gettering

Correlation between Surface and Interior Defects   An example of the correlation between surface and interior microdefects is shown in Fig. 7.67.[48] A (111) CZ silicon wafer was divided into two halves, and each half was subjected to one-step or two-step heat treatment as shown in the figure. After the heat treatment, surface and interior defects were delineated by chemical etching. For the left half, dense surface microdefects ($2 \times 10^6/cm^2$) and no interior defects are observed, as shown in Fig. 7.67a and b, respectively. The phenomenon of surface microdefects but no interior defect is schematically illustrated by the vertical sectional view of the half in Fig. 7.67c. For the other half subjected to two-step heat treatment, there are no surface microdefects except in the region marked by a circle where the density is $2 \times 10^5/cm^2$. The micrograph of the defects on this section is shown in Fig. 7.67d. However, interior defects with a swirl pattern are generated with a density of $5 \times 10^5/cm^2$ as shown in Fig. 7.67e. The surface microdefects observed in the region marked by a circle in Fig. 7.67d have no corresponding interior defects at the place marked by a cross. The area with no surface microdefects corresponds to the area where dense interior defects are generated. This situation showing the correlation between surface and interior microdefects is illustrated by schematic vertical section in Fig. 7.67f. That is, interior defects getter the surface microdefects and/or their origin. This phenomenon was first reported as "*in situ* gettering" by Rozgonyi *et al.*[236] in 1976. Later, Tan *et al.* further clarified this phenomenon and termed it "intrinsic gettering" so as to distinguish it from "extrinsic gettering," which had been commonly realized.[237] Since then, the term *intrinsic gettering* (IG) has been commonly

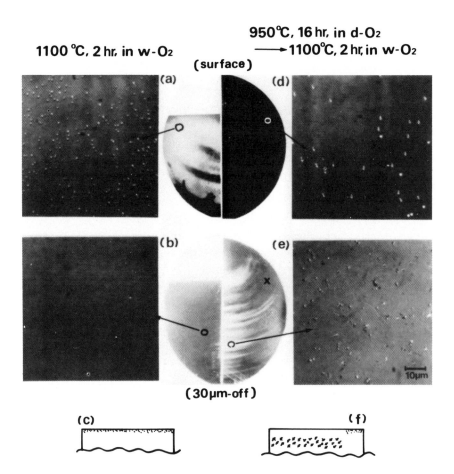

Fig. 7.67. Correlation between surface and interior microdefects in (111) CZ silicon wafer: (a and b) (1100°C/2 hr/wet O$_2$) treatment, (c) schematic cross section of (a) and (b), (d and e) (950°C/16 hr/dry O$_2$ + 1100°C/2 hr/wet O$_2$) treatment, (f) schematic cross section of (d) and (e). (After Shimura *et al.*[48])

used and the IG technique has received increasing attention in the silicon industry.[238] The IG effectiveness depends on the type and density of interior defects, which in turn depend on the initial oxygen concentration $[O_i]_0$ of the silicon wafer, annealing temperature, and time, as was discussed earlier. Figure 7.68 shows surface and interior microdefects observed by chemical etching both in the surface and cleaved plane of (111) CZ silicon wafers subjected to the two-step heat treatment noted in the caption. These three silicon wafers have different $[O_i]_0$ values shown in the figure; that is, they are called low $[O_i]_0$, medium $[O_i]_0$, and high $[O_i]_0$ wafers for convenience. In

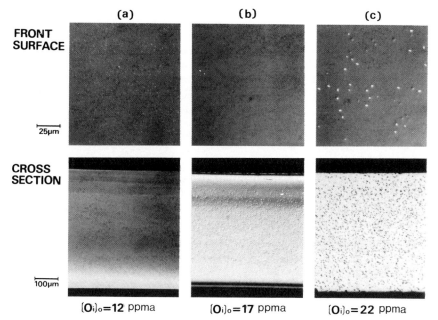

Fig. 7.68. Surface and interior microdefects in (111) CZ silicon wafer subjected to two-step heat treatment (950°C/16 hr/dry O_2 + 1100°C/2 hr/wet O_2); (a) $[O_i]_0$ = 12 ppma, (b) $[O_i]_0$ = 17 ppma, and (c) $[O_i]_0$ = 22 ppma.

the low $[O_i]_0$ wafer shown in Fig. 7.68a, no interior defect is generated but dense microdefects appear in the surface because of no IG effect. In the high $[O_i]_0$ wafer, dense interior defects are generated through the front surface to back surface. Because of the strong IG effect, no surface microdefect appears; however, the interior defects themselves do appear in the wafer surface. On the other hand, in the medium $[O_i]_0$ wafer, a considerable amount of interior defects and a denuded zone (DZ) of about 30 μm depth where no defect is observed are generated as shown in Fig. 7.68b. As a result, no defect is observed in the surface of the medium $[O_i]_0$ wafer. This indicates that the initial oxygen concentration of wafers must be carefully considered when IG is applied to device fabrication processes.

Oxygen Outdiffusion and Denuded-Zone Formation   The key scheme in IG is thus to form sufficient, but not too many, interior defects under the optimum depth of denuded zone where electronic devices are fabricated. Since interior defects are generated by oxygen precipitation, a denuded zone can be formed by oxygen outdiffusion from the wafer surface, resulting in an oxygen-lean region whose oxygen concentration is not high enough to generate oxygen precipitates that introduce interior lattice defects. Such oxygen outdiffusion occurs in the wafer surface region during heat treatment in any ambient. The driving force for this outdiffusion is assumed to be the lower oxygen solubility in silicon at processing temperatures.[239] Assuming that the thickness of the wafer is much larger than the corresponding thickness of oxygen outdiffusion and that there exists neither oxygen precipitates nor any point defects, the oxygen concentration at depth χ from the surface for annealing at temperature T for time t can be given by the error function[239]:

$$[O](\chi, t) = [O]_s + ([O]_0 - [O]_s)\, \mathrm{erf}(\chi/2\sqrt{D_T t}) \tag{7.10}$$

where $[O]_s$ is the solid solubility of oxygen in silicon at T, $[O]_0$ the initial oxygen concentration in the wafer, and D_T the diffusion coefficient of oxygen at T. The oxygen concentration at the SiO_2/Si interface reaches $[O]_s$ in the case of annealing in O_2 ambient, and oxygen outdiffuses to the interface where the oxygen may help SiO_2 growth, while the oxygen concentration at the wafer surface reaches zero in inert ambients. The outdiffusion of oxygen for both cases is depicted in Fig. 7.69, where $[O_e]_{Si}$, $[O_e]_{SiO_2/Si}$, $[O_e]_{SiO_2/A}$, and $[O_e]_{A/SiO_2}$ are the equilibrium concentration of oxygen in silicon, in SiO_2 at the SiO_2/Si interface, in SiO_2 at the ambient/SiO_2 interface, and in

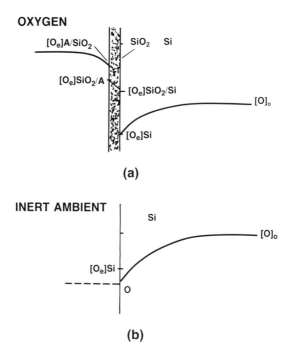

Fig. 7.69. Schematic illustration showing outdiffusion of oxygen: (a) oxygen ambient, and (b) inert gas ambient. (After Tice and Tan.[238])

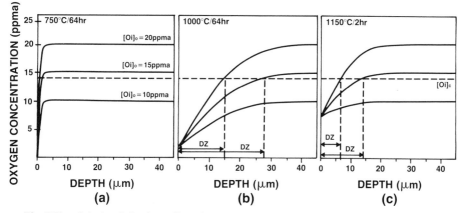

Fig. 7.70. Calculated depth profiles of oxygen concentration for silicon wafers with initial oxygen concentration 10, 15, and 20 ppma: (a) (750°C/64 hr/O_2) heat treatment, (b) (1000°C/ 64 hr/O_2) heat treatment, and (c) (1150°C/2 hr/O_2) heat treatment. (Courtesy of T. Higuchi, Toshiba Ceramics.)

ambient at the ambient/SiO_2 interface, respectively.[238] Thus, strictly speaking, $[O]_s$ in Eq. (7.10) should be replaced with $[O_e]_{Si}$ and zero for heat treatment in oxygen and inert ambients, respectively.

Figure 7.70 shows the depth profiles of oxygen concentration calculated using Eqs. (5.40), (5.41), and (7.10) in three different heat-treatment cases for silicon wafers with three different $[O_i]_0$ values. Assuming a certain critical oxygen concentration $[O_i]_c$ (e.g., 14 ppma) for the occurrence of oxygen precipitation, the denuded zone depths depending both on $[O_i]_0$ and on annealing conditions are obtained as depicted in Fig. 7.70b and c. Oxygen precipitation occurs in the wafer region deeper than the denuded zone—in other words, in the region whose $[O_i]$ is higher than $[O_i]_c$.

It is obvious that the DZ depth is infinite (i.e., no oxygen precipitation occurs) in silicon wafers whose $[O_i]_0$ is less than $[O_i]_c$. Electronic devices are fabricated in the denuded zone or in an epitaxial layer formed on the denuded zone, which is located right above the IG sinks as shown in Fig. 7.71.[85] However, it should be noted that, as discussed earlier, a critical concentration $[O_i]_c$ may vary with various heterogeneous oxygen precipitation factors and with the dopant concentration on which oxygen solubility depends. Impurities such as carbon may significantly decrease the $[O_i]_c$, resulting in shallower DZ for the identical $[O_i]_0$ and annealing conditions. In addition, the diffusivity of interstitial oxygen can be influenced by the presence of point defects[240] and by the dopant species and concentration.[241]

Thermal Cycles For the purpose of forming optimum denuded zone and interior defects, several thermal cycles for IG have been investigated.[242–244]

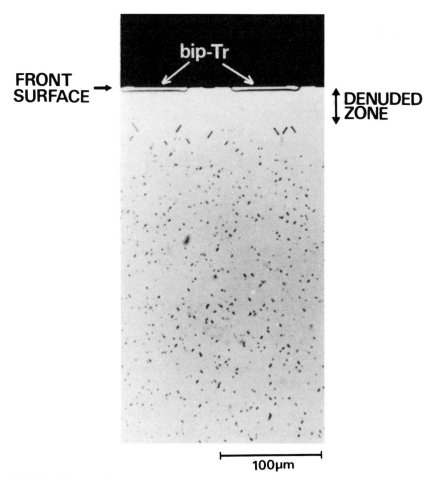

Fig. 7.71. Cross section of multistep IG-treated wafer used for fabrication of shallow-junction bipolar transistors. (After Shimura.[85] Reprinted with the permission of The Electrochemical Society, Inc.)

The IG thermal cycle commonly used is called a *high–low–high* (or-*medium*) sequence, which in principle consists of three steps:

1. Oxygen outdiffusion heat treatment at a high temperature ($>1100°C$) for DZ formation. In order to prevent OSF generation, this heat treatment is usually carried out in an inert ambient.

2. Heterogeneous SiO_2 seeding site formation at a low temperature (600–750°C). Since preannealing at a high temperature suppresses oxygen precipitation during subsequent heat treatment, annealing at a low temperature is required to grow SiO_2 embryos.

3. Gettering-sink introduction at a medium or high temperature (1000–1150°C). During this heat treatment, SiO_2 precipitates grow larger and lattice defects as IG sinks are induced in the region under the denuded zone.

Furthermore, a multistep heat treatment consisting of several steps at temperatures from low (~ 500°C) to high (~ 1150°C) has proved to be effective on IG in low $[O_i]_0$ or heavily doped n^+ silicon wafers where oxygen precipitation rarely occurs.[85,242] However, in general, IG heat treatment requires a long period of furnace operation in addition to the time for device fabrication processes. This time-consuming heat treatment is a major disadvantage of an IG technique. In order to utilize IG in practical device processing, silicon wafers must be treated so that intrinsic gettering occurs simultaneously during device processes without any additional IG heat treatment.

Gettering Mechanism   It has been observed that a low density of dislocations or stacking faults results in effective IG, but a dense microprecipitate of SiO_2 does not.[116] Accordingly, it has been recognized that dislocations or stacking faults are required for effective IG sinks for impurities to be gettered.[116,237] The effect of these lattice defects on gettering impurities is explained primarily by the Cottrell effect, in which the solubility of a foreign atom will be greater in the vicinity of a dislocation.[245] An atom that tends to expand the crystal will dissolve preferentially in the expanded region near a dislocation, while a small atom will tend to dissolve preferentially in the contracted region near the dislocation. Moreover, dangling bonds introduced by edge dislocations or stacking faults have been considered to be effective gettering sites for impurities.[116] The dislocations can act as rows of closely spaced acceptors[246] since a dangling bond has an unpaired electron in contrast to a vacancy or vacancy agglomerates, which have incomplete covalent bonds by linking with neighboring atoms.[247] Therefore, dislocations may directly attract negatively charged species. Contrary to the previous observation, however, it has recently been reported that oxygen precipitates themselves can act as effective gettering sinks for iron atoms by forming a new phase directly connected to a redissolution of oxygen precipitates.[248] In addition, more strikingly, the denuded-zone formation and intrinsic gettering in "oxygen-free" FZ silicon wafers have been recently reported.[249] Eventually the process has been explained with the outdiffusion and precipitation of silicon self-interstitials, instead of oxygen, which can act as gettering sites for metallic impurities.[250] It has been identified by analytical TEM observation that intrinsically gettered centers are three-dimensional butterfly-shaped complexes consisting of multiply extended dislocation loops and a high density of microprecipitates, which are metal silicides—mostly copper, occasionally nickel, and very rarely iron silicides.[251] Moreover, an intrinsic

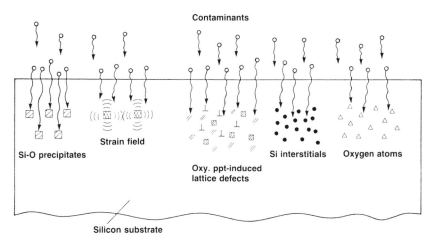

Fig. 7.72. Schematic illustration showing intrinsic gettering mechanism.

gettering phenomenon has been found in oxygen-lean MCZ silicon wafers
where oxygen precipitates are rarely found, and there is a mechanism by
which metallic impurities can be gettered by combining with interstitially
dissolved oxygen.[252] Figure 7.72 schematically summarizes possible mechanisms by which contaminant impurities can be intrinsically gettered.

Guideline for Intrinsic Gettering   The intrinsic gettering process is basically
clean and can provide effective gettering sinks in the region close to the
surface where electronic devices are fabricated, as was shown in Fig. 7.71.
These schemes of IG will be favorable to the VLSI/ULSI technology.
However, IG requires the strict control of various oxygen-related processes in
order to perform uniform and consistent gettering. For this goal, the silicon
wafers processed must meet the following major requirements: (1) they must
have specific $[O_i]_0$ with uniform radial distribution across the wafer diameter, and (2) they must result in a uniform and reproducible denuded zone,
oxygen precipitation, and interior defects. As was discussed earlier, the
control of oxygen incorporation into growing silicon crystal is possible;
however, oxygen precipitation and resulting interior defects are greatly
influenced by various factors, primarily $[O_i]_0$ and heat-treatment conditions.
Under the circumstances, a cooperative effort between the silicon wafer
supplier and the user (i.e., device manufacturer) might be indispensible to
establish successful IG operation for the VLSI/ULSI technology. In order to
eliminate the cumbersome oxygen-related phenomena, an approach that uses
silicon wafers with $[O_i]_0$ far below $[O_i]_c$ needed for oxygen precipitation
might be viable if an alternative gettering technique will remain effective
throughout entire device fabrication processes, or if the entire process is clean

enough to require no gettering technique. It should be emphasized again that the primary effort undertaken to eliminate the detrimental effect of contamination is to remove the sources.

References

1. S. K. Ghandhi, "VLSI Fabrication Principles." Wiley, New York, 1983.
2. S. Wolf and R. N. Tauber, "Silicon Processing for the VLSI Era." Lattice Press, Sunset Beach, California, 1986.
3. K. V. Ravi, "Imperfections and Impurities in Semiconductor Silicon." Wiley, New York, 1981.
4. J. R. Monkowski, Gettering processes for defect control. *Solid State Technol.* July, pp. 44–51 (1981).
5. B. O. Kolbesen and H. P. Strunk, Analysis, electrical effects, and prevention of process induced defects in silicon integrated circuits. *In* "VLSI Electronics Microstructure Science (N. G. Einspruch and H. Huff, eds.), Vol. 12, pp. 143–222. Academic Press, New York, 1985.
6. L. C. Kimerling and J. R. Patel, Silicon defects: Structure, chemistry, and electrical properties. *In* "VLSI Electronics Microstructure Science" (N. G. Einspruch and H. Huff, eds.), pp. 223–267. Academic Press, New York, 1985.
7. L. Jastrzebski, R. Soydan, G. W. Cullen, W. N. Henry, and S. Vecrumba, Silicon wafers for CCD imagers *J. Electrochem. Soc.* **134**, 212–221 (1987).
8. E. Billig, Some defects in crystals grown from the melt. I. Defect caused by thermal stesses. *Proc. R. Soc. London, Ser. A* **235** 37–65 (1956).
9. L. Maissel, Thermal expansion of silicon. *J. Appl. Phys.* **31**, 211 (1960).
10. M. Kämper, A new striation etch for silicon. *J. Electrochem. Soc.* **117**, 261–262 (1970).
11. T. Kawamura, Research history of microdefects in silicon crystals. *J. Jpn. Assoc. Cryst. Growth* **7**, 161–196 (1980) (in Japanese).
12. T. Abe, T. Samizo, and S. Maruyama, Etch pits observed in dislocation-free silicon crystals. *Jpn. J. Appl. Phys.* **5**, 458–459 (1966).
13. T. Abe, Crystal fabrication. *In* "VLSI Electronics Microstructure Science (N. G. Einspruch and H. Huff, eds.), Vol. 12, pp. 3–61. Academic Press, New York, 1985.
14. A. J. R. de Kock, The elimination of vacancy-cluster formation in dislocation-free silicon crystals. *J. Electrochem. Soc.* **118**, 1851–1856 (1971).
15. H. Föll and B. O. Kolbesen, Formation and nature of swirl defects in silicon. *Appl. Phys.* **8**, 319–331 (1975).
16. S. M. Hu, Defects in silicon substrates. *J. Vac. Sci. Technol.* **14**, 17–31 (1977).
17. T. S. Plaskett, Evidence of vacancy custers in dislocation-free float-zone silicon. *Trans. Metall. Soc. AIME* **233**, 809–812 (1965).
18. A. J. R. de Kock, Vacancy clusters in dislocation-free silicon. *Appl. Phys. Lett.* **16**, 100–102 (1970).
19. K. V. Ravi and C. J. Varker, Growth 'striations' and 'swirls' in float-zone single crystals. *In* "Semiconductor Silicon 1973" (H. R. Huff and R. R. Burgess, eds.), pp. 136–149. Electrochem. Soc., Princeton, New Jersey, 1973.
20. P. M. Petroff and A. J. R. de Kock, Characterization of swirl defects in floating-zone silicon crystals. *J. Cryst. Growth* **30**, 117–124 (1975).
21. H. Föll, U. Gösele, and B. O. Kolbesen, The formation of swirl defects in silicon by agglomeration of self-interstitials. *J. Cryst., Growth* **40**, 90–108 (1977).
22. P. J. Roksnoer and M. M. B. van den Boom, Microdefects in a non-striated distribution in floating-zone silicon crystals. *J. Cryst. Growth* **53**, 563–573 (1981).
23. R. E. Smallman, "Modern Physical Metallurgy," 2nd ed. Butterworth, London, 1963.

24. W. D. Sylwestrowicz, Mechanical properties of single crystals of silicon. *Philos. Mag.* [8] **7**, 1825–1845 (1962).
25. P. Penning, Generation of imperfections in germanium crystals by thermal strain. *Philips Res. Rep.* **13**, 79–97 (1958).
26. K. Morizane and P. S. Gleim, Thermal stress and plastic deformation of thin silicon slices. *J. Appl. Phys.* **40**, 4104–4107 (1969).
27. S. M. Hu, Temperature distribution and stresses in circular wafers in a row during radiative cooling. *J. Appl. Phys.* **40**, 4413–4423 (1969).
28. J. J. Wortman and R. A. Evans, Young's modulus, shear stress and Poisson's ratio in silicon and germanium. *J. Appl. Phys.* **36**, 153–156 (1965).
29. P. H. Singer, Trends in vertical diffusion furnaces. *Semicond. Int.* Apr., pp. 56–60 (1986).
30. D. C. Bennett and B. Sawyer, Single crystals of exceptional perfection and uniformity by zone leveling. *Bell Syst. Tech. J.* **35**, 637–660 (1956).
31. P. Rai-Choudhury and W. J. Takei, Thermally induced dislocations in silicon. *J. Appl. Phys.* **40**, 4980–4982 (1969).
32. S. M. Hu, S. P. Klepner, R. O. Schwenker, and D. K. Seto, Dislocation propagation and emitter edge defects in silicon wafers. *J. Appl. Phys.*, **47**, 4098–4106 (1976).
33. B. Leroy and C. Plougonven, Warpage of silicon. *J. Electrochem. Soc.* **127**, 961–970 (1980).
34. Y. Tsunekawa and S. Weissmann, Importance of microplasticity in the fracture of silicon. *Metall. Trans.* **5**, 1585–1593 (1974).
35. H.-D. Chiou, Effects of mechanical damage on the slip of CZ wafers. *Ext. Abstr., Electrochem. Soc. Fall. Meet.* Vol. 83-2, pp. 538–539 (1983).
36. H. R. Huff, R. C. Bracken, and S. N. Rae, Influence of silicon slice curvature on thermally induced stresses. *J. Electrochem. Soc.* **118**, 143–145 (1971).
37. J. Bloem and A. H. Goemans, Slip in silicon epitaxy. *J. Appl. Phys.* **43**, 1281–1283 (1972).
38. L. D. Dyer, H. R. Huff, and W. W. Boyd, Plastic deformation in central regions of epitaxial silicon slices. *J. Appl. Phys.* **42**, 5680–5688 (1971).
39. F. Shimura, Manufacturing method of semiconductor devices. Japanese Patent 56-123522 (1981) (in Japanese).
40. E. W. Hearn, E. H. teKaat, and G. H. Schwuttke, The closed boat: A new approach for semiconductor batch processing. *Microelectron. Reliab.* **15**, 61–66 (1976).
41. A. C. Borona, Silicon wafer process technology: Slicing, etching, polishing. *In* "Semiconductor Silicon 1977" (H. R. Huff and E. Sirtl, eds.), pp. 154–169. Electrochem. Soc., Princeton, New Jersey, 1977.
42. D. J. D. Thomas, Surface damage and copper precipitation in silicon. *Phys. Status Solidi* **3**, 2261–2273 (1963).
43. R. H. Finch and H. J. Queisser, Structure and origin of stacking faults in epitaxial silicon. *J. Appl. Phys.* **34**, 406–415 (1963).
44. S. Mendelson, Stacking fault nucleation in epitaxial silicon on variously oriented silicon substrates. *J. Appl. Phys.* **35**, 1570–1581 (1964).
45. H. J. Queisser and P. G. G. van Loon, Growth of lattice defects in silicon during oxidation. *J. Appl. Phys.* **35**, 3066–3069 (1964).
46. D. I. Pomerantz, Effect of grown-in and process-induced defects in single crystal silicon. *J. Electrochem. Soc.* **119**, 255–165 (1972).
47. C. W. Pearce and R. G. McMahon, Role of metallic contamination in the formation of 'saucer' pit defects in epitaxial silicon. *J. Vac. Sci. Technol.* **14**, 40–43 (1977).
48. F. Shimura, H. Tsuya, and T. Kawamura, Surface- and inner-microdefects in annealed silicon wafer containing oxygen. *J. Appl. Phys.* **51**, 269–273 (1980).
49. F. Shimura and R. A. Craven, Process-induced microdefects in VLSI silicon wafers. *In* "The Physics of VLSI" (J. C. Knight, ed.), pp. 205–219. Am. Inst. Phys., New York, 1984.

50. C. M. Drum and W. van Gelder, Stacking faults in (100) epitaxial silicon caused by HF and thermal oxidation and effects on p-n junctions. *J. Appl. Phys.* **43**, 4465-4468 (1972).
51. K. V. Ravi and C. J. Varker, Oxidation-induced stacking faults in silicon. I. Nucleation phenomenon. *J. Appl. Phys.* **45**, 263-271 (1974).
52. M. L. Joshi, Stacking faults in steam-oxidized silicon. *Acta Metall.* **14**, 1157-1172 (1966).
53. G. R. Booker and W. J. Tunstall, Diffraction contrast analysis of two-dimensional defects present in silicon after annealing. *Philos. Mag.* [8] **13**, 71-83 (1966).
54. R. J. Jaccodine and C. M. Drum, Extrinsic stacking faults in silicon after heating in wet oxygen. *Appl. Phys. Lett.* **8**, 29-30 (1966).
55. J. M. Silcock and W. J. Tunstall, Partial dislocations associated with NbC precipitation in austenitic stainless steels. *Philos. Mag.* [8] **10**, 361-389 (1964).
56. J. E. Lawrence, Stacking faults in annealed silicon surfaces. *J. Appl. Phys.* **40**, 360-365 (1969).
57. A. H. Cottrell, "Dislocations and Plastic Flow of Crystals." Oxford Univ. Press, London and New York, 1953.
58. I. R. Sanders and P. S. Dobson, Oxidation, defects and vacancy diffusion in silicon. *Philos. Mag.* [8] **20**, 881-893 (1969).
59. S. M. Hu, Formation of stacking faults and enhanced diffusion in the oxidation of silicon. *J. Appl. Phys.* **45**, 1567-1573 (1974).
60. T. Y. Tan and U. Gösele, Growth kinetics of oxidation-induced stacking faults in silicon: A new concept. *Appl. Phys. Lett.* **39**, 86-88 (1981).
61. S. P. Muraka and G. Quintana, Oxidation induced stacking faults in n- and p-type (100) silicon. *J. Appl. Phys.* **48**, 46-51 (1977).
62. F. Shimura, W. Dyson, J. W. Moody, and R. S. Hockett, Oxygen behavior in heavily Sb-doped CZ-silicon. *In* "VLSI Science and Technology/1985" (W. M. Bullis and S. Broydo, eds.), pp. 507-516. Electrochem. Soc., Princeton, New Jersey, 1985.
63. Y. Sugita, Oxidation and oxidation-induced lattice defects in silicon. *Oyo Butsuri* **46**, 1056-1068 (1977) (in Japanese).
64. S. P. Muraka, Oxygen partial-pressure dependence of the oxidation-induced surface stacking faults in (100) n silicon. *J. Appl. Phys.* **48**, 5020-5026 (1978).
65. S. M. Hu, Anomalous temperature effect of oxidation stacking faults in silicon. *Appl. Phys. Lett.* **27**, 165-167 (1975).
66. S. P. Murarka, Oxygen pressure dependence of the retrogrowth of oxidation-induced stacking faults in (100) silicon. *J. Appl. Phys.* **49**, 2513-2516 (1978).
67. A. W. Fisher and J. A. Amick, Defect structure on silicon surfaces after thermal oxidation. *J. Electrochem. Soc.* **113**, 1054-1060 (1966).
68. Y. Sugita, T. Kato, and M. Tamura, Effect of crystal orientation on stacking fault formation in thermally oxidized silicon. *J. Appl. Phys.* **42**, 5847-5849 (1971).
69. Y. Sugita, H. Shimizu, A. Yoshinaka, and T. Aoshima, Shrinkage and annihilation of stacking faults in silicon. *J. Vac. Sci. Technol.* **14**, 44-46 (1977).
70. C. L. Claeys, G. J. Declerck, and R. J. Van Overstraeten, The influence of annealing ambient on the shrinkage kinetics of oxidation-induced stacking faults in silicon. *Appl. Phys. Lett.* **35**, 797-799 (1979).
71. K. V. Ravi, On the annihilation of oxidation induced stacking faults in silicon. *Philos. Mag.* [8] **30**, 1081-1090 (1974).
72. R. F. Peart, Self-diffusion of intrinsic silicon. *Phys. Status Solidi* **15**, K119-K122 (1966).
73. J. M. Fairfield and B. J. Masters, Self-diffusion in intrinsic and extrinsic silicon. *J. Appl. Phys.* **38**, 3148-3154 (1967).
74. R. J. Kriegler, Y. C. Cheng, and D. R. Colton, The effect of HCl and Cl_2 on the thermal oxidation of silicon. *J. Electrochem. Soc.* **119**, 388-392 (1972).

75. H. Shiraki, Elimination of stacking faults in silicon by HCl added dry O_2 oxidation. *Jpn. J. Appl. Phys.* **14**, 747–752 (1975).

76. T. Hattori, Elimination of stacking faults in silicon by trichloroethylene oxidation. *J. Electrochem. Soc.* **123**, 945–946 (1976).

77. C. L. Claeys, E. E. Laes, G. L. Declerck, and R. J. Van Overstraeten, Elimination of stacking faults for charge-coupled device processing. *In* "Semiconductor Silicon 1977" (H. R. Huff and E. Sirtl, eds), pp. 773–784. Electrochem. Soc., Princeton, New Jersey, 1977.

78. H. Shiraki, Stacking fault generation supression and grown-in defect elimination in dislocation-free silicon wafers by HCl oxidation. *Jpn. J. Appl. Phys.* **15**, 1–10 (1976).

79. T. Y. Tan and U. Gösele, Kinetics of silicon stacking fault growth/shrinkage in an oxidizing ambient containing a chlorine compound. *J. Appl. Phys.* **53**, 4767–4778 (1982).

80. P. H. Robinson and F. P. Heiman, Use of HCl gettering in silicon device processing. *J. Electrochem. Soc.* **118**, 141–143 (1971).

81. H. Tsuya and F. Shimura, Transient behavior of instrinsic gettering in CZ silicon wafers. *Phys. Status Solidi* A **79**, 199–206 (1983).

82. W. T. Stacy, D. F. Allison, and T.-C., Wu, The role of metallic impurities in the formation of haze defects. *In* "Semiconductor Silicon 1981" (H. R. Huff and R. J. Kriegler, eds.), pp. 344–353. Electrochem. Soc., Princeton, New Jersey, 1981.

83. G. Keefe-Fraundorf and R. A. Craven, Microdefects near the surface of oxidized silicon wafers. *In* "Defects in Silicon" (W. M. Bullis and L. C. Kimerling, eds.), pp. 406–413. Electrochem. Soc., Princeton, New Jersey, 1982.

84. R. A. Craven, F. Shimura, R. S. Hockett, L. W. Shive, P. B. Fraundorf, and G. Keefe-Fraundorf, Characterization techniques for VLSI silicon. *In* "VLSI Science and Technology/1984" (K. E. Bean ad G. A. Rozgonyi, eds.), pp. 20–35. Electrochem. Soc., Princeton, New Jersey, 1984.

85. F. Shimura, Behavior and role of oxygen in silicon wafers for VLSI. *In* "VLSI Science and Technology/1982" (C. J. Dell'Oca and W. M. Bullis, eds.), pp. 17–32. Electrochem. Soc., Princeton, New Jersey, 1982.

86. P. E. Freedland, K. A. Jackson, C. W. Lowe, and J. R. Patel, Precipitation of oxygen in silicon. *Appl. Phys. Lett.* **30**, 31–33 (1977).

87. J. Osaka, N. Inoue, and K. Wada, Homogeneous nucleation of oxide precipitation in Czochralski-grown silicon. *Appl. Phys. Lett.* **36**, 288–290 (1980).

88. K. V. Ravi, The heterogeneous precipitation of silicon oxides in silicon. *J. Electrochem. Soc.* **121**, 1090–1098 (1974).

89. F. Shimura, H. Tsuya, and T. Kawamura, Precipitation and redissolution of oxygen in Czochralski-grown silicon. *Appl. Phys. Lett.* **37**, 483–486 (1980).

90. V. V. Batavin, Solid solutions of oxygen in decomposition of supersaturated dislocation-free silicon. *Sov. Phys.—Crystalogr. (Engl. Transl.)* **25**, 100–107 (1970).

91. S. M. Hu, Oxygen precipitation in silicon. *In* "Oxygen, Carbon, Hydrogen, and Nitrogen in Crystalline Silicon" (J. C. Mikkelsen, Jr., S. J. Pearton, J. W. Corbett, and S. J. Pennycook, eds.), pp. 249–267. Mater. Res. Soc., Pittsburgh, 1986.

92. W. D. Kingery, H. K. Bowen, and D. R. Uhlmann, "Introduction to Ceramics," 2nd ed. Wiley, New York, 1976.

93. S. M. Hu, Precipitation of oxygen in silicon: Some phenomena and a nucleation model. *J. Appl. Phys.* **52**, 3974–3984 (1981).

94. F. Shimura and H. Tsuya, Oxygen precipitation factors in silicon. *J. Electrochem. Soc.* **129**, 1062–1066 (1982).

95. F. S. Ham, Theory of diffusion-limited precipitation. *Phys. Chem. Solids* **6**, 335–351 (1958).

96. H. J. Hrostowski and R. H. Kaiser, Infrared absorption of oxygen in silicon. *Phys. Rev.* **107**, 966–972 (1957).

97. B. Pajot, H. J. Stein, B. Cales, and C. Nand, Quantitative spectroscopy of interstitial oxygen in silicon. *J. Electrochem. Soc.* **132**, 3034–3037 (1985).
98. P. E. Freeland, Oxygen precipitation in silicon at 650°C. *J. Electrochem. Soc.* **127**, 754–756 (1980).
99. F. Shimura, Y. Ohnishi, and H. Tsuya, Heterogeneous distribution of interstitial oxygen in annealed Czochralski-grown silicon crystals. *Appl. Phys. Lett.* **38**, 867–870 (1981).
100. W. C. O'Mara, Oxygen in silicon. *In* "Defects in Silicon" (W. M. Bullis and L. C. Kimerling, eds.), pp. 120–129. Electrochem. Soc., Princeton, New Jersey, 1984.
101. M. Stavola, Infrared spectrum of interstitial oxygen in silicon. *Appl. Phys. Lett.* **44**, 514–516 (1984).
102. T. S. Glowinke and J. B. Wagner, Jr., The effect of oxygen on the electrical properties of silicon. *J. Phys. Chem. Solids* **38**, 963–970 (1977).
103. N. Inoue, J. Osaka, and K. Wada, Oxide microprecipitates in as-grown CZ silicon. *J. Electrochem. Soc.* **129**, 2780–2788 (1982).
104. T. Usami, Y. Matsushita, and M. Ogino, Embryo formation during crystal growth. *J. Cryst. Growth* **70**, 319–323 (1984).
105. A. J. R. de Kock and W. M. van de Wijgert, The influence of thermal point defects on the precipitation of oxygen in dislocation-free silicon crystals. *Appl. Phys. Lett.* **38**, 888–890 (1981).
106. S. Kishino, Y. Matsushita, and M. Kanamori, Carbon and oxygen role for thermally induced microdefect formation in silicon crystals. *Appl. Phys. Lett.* **35**, 213–215 (1979).
107. G. S. Oehrlein, D. J. Challou, A. E. Jaworowski, and J. W. Corbett, The role of carbon in the precipitation of oxygen in silicon. *Phys. Lett.* **86**, 117–119 (1981).
108. M. Ogino, Suppression effect upon oxygen precipitation in silicon by carbon for a two-step thermal anneal. *Appl. Phys. Lett.* **41**, 847–849 (1982).
109. C. Y. Kung, L. Forbes, and J. D. Peng, The effect of carbon on oxygen precipitation in high carbon CZ silicon crystals. *Mater. res. Bull.* **18**, 1437–1441 (1983).
110. F. Shimura, R. S. Hockett, D. A. Read, and D. H. Wayne, Direct evidence for co-aggregation of carbon and oxygen in Czochralski silicon. *Appl. Phys. Lett.* **47**, 794–796 (1985).
111. H.-D. Chiou, J. Moody, R. Sandfort, and F. Shimura, Effect of oxygen and nitrogen on slip in CZ silicon wafers. *In* "VLSI Science and Technology/1984" (K. E. Bean and G. A. Rozgonyi, eds.), pp. 59–65, Electrochem. Soc., Princeton, New Jersey, 1984).
112. F. Shimura and R. S. Hockett, Nitrogen effect on oxygen precipitation. *Appl. Phys. Lett.* **48**, 224–226 (1986).
113. F. Shimura and H. Tsuya, Multistep repeated annealing for CZ-silicon wafers: Oxygen and induced defect behavior. *J. Electrochem. Soc.* **129**, 2089–2095 (1982).
114. H. Tsuya, F. Shimura, K. Ogawa, and T. Kawamura, A study on intrinsic gettering in CZ silicon crystals: Evaluation, thermal history dependence, and enhancement. *J. Elecrochem. Soc.* **129**, 374–379 (1982).
115. G. Fraundorf, P. Fraundorf, R. A. Craven, R. A. Frederick, J. W. Moody, and R. W. Shaw, The effects of thermal history during crystal growth on O precipitation in Czochralski silicon. *J. Electrochem. Soc.* **132**, 1701–1704 (1985).
116. F. Shimura, H. Tsuya, and T. Kawamura, Thermally induced defect behavior and effective intrinsic gettering sink in silicon wafers. *J. Electrochem. Soc.* **128**, 1579–1583 (1981).
117. H.-D. Chiou and L. W. Shive, Test method for oxygen precipitation in silicon. *In* "VLSI Science and Technology/1985" (W. M. Bullis and S. Broydo, eds.), pp. 429–435. Electrochem. Soc., Princeton, New Jersey, 1985).
118. H.-D. Chiou, Oxygen precipitation behavior and control in silicon crystals. *Solid State Technol.* pp. 77–81 (1987).

119. F. Shimura, Redissolution of precipitated oxygen in Czochralski-grown silicon wafers. *Appl. Phys. Lett.* **39**, 987–989 (1981).

120. F. Shimura, Carbon enhancement effect on oxygen precipitation in Czochralski silicon. *J. Appl. Phys.* **59**, 3251–3254 (1986).

121. F. M. Livingston, S. Messoloras, R. C. Newman, B. C. Pike, R. J. Stewart, M. J. Binns, W. P. Brown, and J. G. Wilkes, An infrared and neutron scattering analysis of the precipitation of oxygen in dislocation-free silicon. *J. Phys. C* **17**, 6253–6276 (1984).

122. C. Y. Kung, Oxygen precipitation behavior in high carbon CZ silicon. *In* "VLSI Science and Technology/1985" (W. M. Bullis and S. Broydo, eds.), pp. 446–455. Electrochem. Soc., Princeton, New Jersey, 1985.

123. F. Shimura, J. P. Baiardo, and P. Fraundorf, Infrared absorption study on carbon and oxygen behavior in Czochralski silicon crystals. *Appl. Phys. Lett.* **46**, 941–943 (1985).

124. M. R. Brozel, R. C. Newman, and D. H. Totterdell, Interstitial defects involving carbon in irradiated silicon. *J. Phys. C* **8**, 243–248 (1975).

125. P. Fraundorf, G. K. Fraundorf, and F. Shimura, Clustering of oxygen atoms around carbon in silicon. *J. Appl. Phys.* **58**, 4049–4055 (1985).

126. J. P. Kalejs, L. A. Ladd, and U. Gösele, Self-interstitial enhanced carbon diffusion in silicon. *Appl. Phys. Lett.* **45**, 268–269 (1984).

127. G. D. Watkins and K. L. Brower, EPR observation of the isolated interstitial carbon atom in silicon. *Phys. Rev. Lett.* **36**, 1329–1332 (1976).

128. R. C. Newman and J. Wakefield, The diffusivity of carbon in silicon. *J. Phys. Chem. Solids* **19**, 230–234 (1961).

129. A. J. R. de Kock and W. M. van de Wijgert, The effect of doping on the formation of swirl defects in dislocation-free Czochralski-grown silicon crystals. *J. Cryst. Growth* **49**, 718–734 (1980).

130. C. W. Pearce and G. A. Rozgonyi, Intrinsic gettering in heavily doped Si substrates for epitaxial devices. *In* "VLSI Science and Technology/1982" (C. J. Dell'Oca and W. M. Bullis, eds.), pp. 53–59. Electrochem. Soc., Princeton, New Jersey, 1982.

131. H. Tsuya, Y. Kondo, and M. Kanamori, Behaviors of thermally induced microdefects in heavily doped silicon wafers. *Jpn. J. Appl. Phys.* **22**, L16-L18 (1983).

132. F. Secco d'Aragona, J. W. Rose, and P. L. Fejes, Outdiffusion, defects and gettering behavior of epitaxial n/n$^+$ and p/p$^+$ wafers used for CMOS technology. *In* "VLSI Science and Technology/1985" (W. M. Bullis and S. Broydo, eds.), pp. 106–117. Electrochem. Soc., Princeton, New Jersey, 1985.

133. S. M. Hu, Effect of ambients on oxygen precipitation in silicon. *Appl. Phys. Lett.* **36**, 561–564 (1980).

134. U. Gösele and T. Tan, The role of vacancies and self-interstitials in diffusion and agglomeration phenomena in silicon. *In* "Aggregation Phenomena of Point Defects in Silicon" (E. Sirtl and J. Goorissen, eds.), pp. 17–36 (1983).

135. H. M. James and R. L. Lark-Horovitz, Localized electronic states in bombarded semiconductors. *Z. Phys. Chem.* **198**, 107–126 (1951).

136. E. J. Blount, Energy levels in irradiated germanium. *J. Appl. Phys.* **30**, 1218–1221 (1959).

137. J. A. Van Vechten and C. D. Thurmond, Entropy of ionization levels of defects in semiconductors. *Phys. Rev. B: Solid State* [3] **14**, 3539–3550 (1976).

138. H. Harada, T. Ito, N. Ozawa, and T. Abe, Incorporation of oxygen impurity into silicon crystals during Czochralski growth. *In* "VLSI Science and Technology/1985" (W. M. Bullis and S. Broydo, eds.), pp. 526–535. Electrochem. Soc., Princeton, New Jersey, 1985.

139. K. G. Barraclough and R. W. Series, Oxygen content of n$^+$ and p$^+$ Czochralski silicon. *In* "Reduced Temperature Processing for VLSI" (R. Reif and G. R. Srinvasan, eds.), pp. 452–463. Electrochem. Soc., Princeton, New Jersey, 1986.

140. T. Y. Tan and W. K. Tice, Oxygen precipitation and the generation of dislocations in silicon. *Philos. Mag.* [8], **34**, 615–631 (1976).

141. D. M. Maher, A. Staudinger, and J. R. Patel, Characterization of structural defects in annealed silicon containing oxygen. *J. Appl. Phys.* **47**, 3813-3825 (1976).

142. F. A. Ponce, T. Yamashita, and S. Hahn, Structure of thermally induced microdefects in Czochralski silicon after high-temperature annealing. *Appl. Phys. Lett.* **43**, 1051-1053 (1983).

143. A. Bourret, J. Thibault-Desseaux, and D. N. Seidman, Early stages of oxygen segregation and precipitation in silicon. *J. Appl. Phys.* **55**, 825-836 (1984).

144. H. Bender, Investigation of the oxygen-related lattice defects in Czochralski silicon by means of electron microscopy techniques. *Phys. Status Solidi* A **86**, 245-261 (1984).

145. Y. Matsushita, Thermally induced microdefects in Czochralski-grown silicon crystals. *J. Cryst. Growth* **56**, 516-525 (1982).

146. N. Yamamoto, P. M. Petroff, and J. R. Patel, Rod-like defects in oxygen-rich Czochralski grown silicon. *J. Appl. Phys.* **54**, 3475-3478 (1983).

147. W. Bergholz, J. L. Hutchison, and P. Pirouz, Precipitation of oxygen at 485°C: Direct evidence for accelerated diffusion of oxygen in silicon? *J. Appl. Phys.* **58**, 3419-3424 (1985).

147a. A. Bourret, Defects induced by oxygen precipitation in silicon: a new hypothesis involving hexagonal silicon, *In* "Microscopy of Semiconducting Materials, 1987" (A. G. Cullis and P. D. Augustus, eds.), pp. 39-48, Inst. of Physics, Bristol and Philadelphia, 1987.

148. K. Tempelhoff and F. Spiegelberg, Precipitation of oxygen in dislocation-free silicon. *In* "Semiconductor Silicon 1981" (H. R. Huff and E. Sirtl, eds.), pp. 585-595. Electrochem. Soc., Princeton, New Jersey, 1981.

149. K. H. Yang, R. Anderson, and H. F. Kappert, Identification of oxide precipitates in annealed silicon crystals. *Appl. Phys. Lett.* **33**, 225-227 (1978).

150. F. Shimura, Octahedral precipitates in high temperature annealed Czochralski-grown silicon. *J. Cryst. Growth* **54**, 588-591 (1981).

151. E. R. Lippincott, A. Van Valkenburg, C. E. Weier, and E. N. Bunting, Infrared studies on polymorphs of silicon dioxide and germanium dioxide. *J. Res. Natl. Bur. Stand.* **61**, 61-70 (1958).

152. W. R. Runyan, "Silicon Semiconductor Technology." McGraw-Hill, New York, 1975.

153. K. Yasutake, M. Umeno, and H. Kawabe, Oxygen precipitation and microdefects in Czochralski-grown silicon crystals. *Phys. Status Solidi* A **83**, 207-217 (1984).

154. G. C. Weatherly, Loss of coherency of growing particles by the prismatic punching of dislocation loops. *Philos. Mag.* [8] **17**, 791-799 (1968).

155. M. F. Ashby and L. Johnson, On the generation of dislocations at misfitting particles in a ductile matrix. *Philos. Mag.* [8], **20**, 1009-1022 (1969).

156. S. Mahajan, G. A. Rozgonyi, and D. Brasen, A model for the formation of stacking faults in silicon. *Appl. Phys. Lett.* **30**, 73-75 (1977).

157. F. Shimura, TEM observations of pyramidal hillocks formed on (001) silicon wafers during chemical etching. *J. Electrochem. Soc.* **127**, 910-913 (1980).

158. H. Shimizu, M. Fujita, T. Aoshima, and Y. Sugino, Dependance of warpage of Czochralski-grown silicon wafers on oxygen concentration and its application to MOS image-sensor device. *Jpn. J. Appl. Phys.* **25**, 68-74 (1986).

159. C.-O. Lee and P. J. Tobin, The effect of CMOS processing on oxygen precipitation, wafer warpage, and flatness. *J. Electrochem. Soc.* **133**, 2147-2151 (1986).

160. J. R. Carruthers, Silicon microdefects and contamination control in MOS device processing. *In* "Defects in Silicon" (W. M. Bullis and L. C. Kimerling, eds.), pp. 375-387. Electrochem. Soc., Princeton, New Jersey, 1983.

161. K. Honda, A. Ohsawa, and N. Toyokura, Breakdown in silicon oxide—Correlation with Cu precipitates. *Appl. Phys. Lett.* **45**, 270-271 (1984).

162. K. Honda, A. Ohsawa, and N. Toyokura, Breakdown in silicon oxide—Correlation with Fe precipitates. *Appl. Phys. Lett.* **46**, 582-584 (1985).

163. W. K. Tice and T. Y. Tan, Nucleation of CuSi precipitate colonies in oxygen-rich silicon. *Appl. Phys. Lett.* **28**, 564–565 (1976).

164. R. B. Marcus, M. Robinson, T. T. Sheng, S. E. Haszko, S. P. Murarka, and L. E. Katz, Electrical activity of epitaxial stacking faults. *J. Electrochem. Soc.* **124**, 425–430 (1977).

165. O. Paz, E. Hearn, and E. Fayo, $POCl_3$ and boron gettering of LSI silicon devices: Similarities and differences. *J. Electrochem. Soc.* **126**, 1754–1761 (1979).

166. C. S. Fuller, J. A. Ditzenberger, N. B. Hannay, and E. Buehler, Resistivity changes in silicon induced by heat treatment. *Phys. Rev.* **96**, 833 (1954).

167. W. Kaiser, Electrical and optical properties of heat-treated silicon. *Phys. Rev.* **105**, 1751–1756 (1957).

168. C. S. Fuller and R. A. Logan, Effect of heat treatment upon the electrical properties of silicon crystals. *J. Appl. Phys.* **28**, 1427–1436 (1957).

169. P. Capper, A. W. Jones, E. J. Wallhouse, and J. G. Wilkes, The effects of heat treatment on dislocation-free oxygen-containing silicon crystals. *J. Appl. Phys.* **48**, 1646–1655 (1977).

170. A. Kanamori and M. Kanamori, Comparison of two kinds of oxygen donors in silicon by resistivity measurements. *J. Appl. Phys.* **50**, 8095–8101 (1979).

171. V. Cazcarra and P. Zunino, Influence of oxygen on silicon resistivity. *J. Appl. Phys.* **51**, 4206–4211 (1980).

172. S. R. Wilson, M. W. Paulson, and R. B. Gregory, Rapid annealing technology for future VLSI. *Solid State Technol.* June, pp. 185–190 (1985).

173. A. R. Bean and R. C. Newman, The effect of carbon on thermal donor formation in heat treated pulled silicon crystals. *J. Phys. Chem. Solids* **33**, 255–268 (1972).

174. J. Leroueille, Influence of carbon on oxygen behavior in silicon. *Phys. Status, Solidi* A **67**, 177–181 (1981).

175. A. Ohsawa, R. Takizawa, K. Honda, A. Shibatomi, and S. Ohkawa, Influence of carbon and oxygen on donor formation at 700°C in Czochralski-grown silicon. *J. Appl. Phys.* **53**, 5733–5737 (1982).

176. A. C. M. Wang and S. Kakihara, Leakage and h_{FE} degradation in microwave bipolar transistors. *IEEE Trans. Electron Devices* **ED-21**, 667–674 (1974).

177. G. H. Plantinga, Influence of dislocations on properties of shallow diffused transistors. *IEEE Trans. Electron Devices* **ED-16**, 394–400 (1969).

178. M. V. Whelan, Leakage currents of n^+p silicon diodes with different amounts of dislocations. *Solid-State Electron.* **12**, 963–968 (1969).

179. J. E. Lawrence, Correlation of silicon material characteristics and device performance. *In* "Semiconductor Silicon 1973" (H. R. Huff and R. R. Burgess, eds.), pp. 17–34. Electrochem. Soc., Princeton, New Jersey, 1973.

180. H. Kressel, A review of the effect of imperfections on the electrical breakdown of p-n junctions. *RCA Rev.,* **28**, 175–207 (1967).

181. W. Shockley and W. T. Read, Statistics of the recombinations of holes and electrons. *Phys. Rev.* **87**, 835–842 (1952).

182. D. V. McCaughan and B. C. Wonsiewicz, Effects of dislocations on the properties of metal SiO_2-silicon capacitors. *J. Appl. Phys.* **45**, 4982–4984 (1974).

183. G. H. Schwuttke, K. Brack, and E. W. Hearn, The influence of stacking faults on leakage currents of FET devices. *Microelectron, Reliab.* **10**, 467–470 (1971).

184. K. V. Ravi, C. J. Varker, and C. E. Volk, *J. Electrochem. Soc.* **120**, 533–541 (1973).

185. S. P. Murarka, T. E. Seidel, J. V. Dalton, J. M. Dishman, and M. H. Read, A study of stacking faults during CMOS processing: Origin, elimination and contribution to leakage. *J. Electrochem. Soc.* **127**, 717–724 (1980).

186. J. M. Dishman, S. E. Haszko, R. B. Marcus, S. P. Murarka, and T. T. Sheng, Electrically active stacking faults in CMOS integrated circuits. *J. Appl. Phys.* **50**, 2689–2696 (1979).

187. H. Strack, K. R. Mayer, and B. O. Kolbesen, The detrimental influence of stacking faults on the refresh time of MOS memories. *Solid-State Electron.* **22**, 135-140 (1979).

188. H. Shiraki, J. Matsui, T. Kawamura, M. Hanaoka, and T. Sakaki, Bright spots in the image of silcon vidicon. *Jpn. J. Appl. Phys.* **10**, 213-217 (1971).

189. Y. Hokari and H. Shiraki, Video defects in charge-coupled image sensors. *Jpn. J. Appl. Phys.* **16**, 213-217 (1977).

190. K. Tanikawa, Y. Ito, and H. Sei, Evaluation of dark-current nonuniformity in a charge-coupled device. *Appl. Phys. Lett.* **28**, 285-287 (1976).

191. S. Prussin, S. P. Li, and R. H. Cockrum, The effect of oxidation-expanded defects upon MOS parameters. *J. Appl. Phys.* **48**, 4613-4617 (1977).

192. P. S. D. Lin, R. B. Marcus, and T. T. Sheng, Leakage and breakdown in thin oxide capacitors—Correlation with decorated stacking faults. *J. Electrochem. Soc.* **130**, 1878-1883 (1983).

193. A. Goetzberger and W. Schockley, Metal precipitates in silicon p-n junctions. *J. Appl. Phys.* **31**, 1821-1824 (1960).

194. F. Shimura and H. R. Huff, VLSI silicon material criteria. *In* "VLSI Handbook" (E. G. Einspruch, ed.), pp. 191-269. Academic Press, New York, 1985).

195. Y. Takano, H. Kozuka, M. Ogirima, and M. Maki, Annealing effect of mechanical damage on minute defects, heavy metals and oxygen atoms in silicon crystal. *In* "Semiconductor Silicon 1981" (H. R. Huff and R. J. Kriegler, eds.), pp. 743-755. Electrochem. Soc., Princeton, New Jersey, 1981.

196. H. Shirai, A. Yamaguchi, and F. Shimura, Effect of backside polysilicon layer on oxygen precipitation in CZ silicon. *Appl. Phys. Lett.* (to be published).

197. R. Reif and G. R. Srinivasan, eds., "Reduced Temperature Processing for VLSI." Electrochem. Soc., Princeton, New Jersey, 1986.

198. J. A. Lauge, Sources of semiconductor wafer contamination. *Semicond. Int.* Apr., pp. 124-128 (1983).

199. J. W. Medernach, V. A. Wells, and L. L. Witherspoon, Study of extrinsic gettering of epitaxial substrates, *Semicond. Int.* Feb., pp. 106-110 (1987).

200. R. A. Craven, Internal gettering in Czochralski silicon. *Semicond. Int.* Sept., pp. 134-139 (1985).

201. L. Baldi, G. Cerofolini, and G. Ferla, Heavy metal gettering in silicon-device processing. *J. Electrochem. Soc.* **127**, 164-169 (1980).

202. M. Ogino, T. Usami, M. Watanabe, H. Sekine, and T. Kawaguchi, Two-step thermal anneal and its application to a CCD sensor and CMOS LSI. *J. Electrochem. Soc.* **130**, 1397-1402 (1983).

203. H. R. Huff, H. F. Schaake, J. T. Robinson, S. C. Baber, and D. Wong, Some observation on oxygen precipitation/gettering in device processed Czochralski silicon. *J. Electrochem. Soc.* **130**, 1551-1554 (1983).

204. G. F. Cerofolini and M. L. Polignano, A comparison of gettering techniques for very large scale integration. *J. Appl. Phys.* **55**, 579-585 (1984).

205. C. N. Anagnostopoulos, E. T. Nelson, J. P. Lavine, K. Y. Wong, and D. N. Nichols, Latch-up and image crosstalk suppression by internal gettering. *IEEE J. Solid-State Circuits* **SC-19**, 91-97 (1984).

206. L. Jastrzebski, R. Soydan, B. Goldsmith, and J. T. McGinn, Internal gettering in bipolar process. *J. Electrochem. Soc.* **131**, 2944-2953 (1984).

207. D. Pomerantz, A cause and cure of stacking faults in silicon epitaxial layers. *J. Appl. Phys.* **38**, 5020-5026 (1967).

208. J. E. Lawrence, Metallographic analysis of gettered silicon. *Trans. Metall. Soc. AIME* **242**, 484-489 (1968).

209. G. H. Schwuttke, K. Yang, and H. Kappert, Lifetime control in silicon through impact sound stressing. *Phys. Status Solidi* A **42**, 553–564 (1977).

210. C. W. Pearce and V. J. Zaleckas, A new approach to lattice damage gettering. *J. Electrochem. Soc.* **126**, 1436–1437 (1979).

211. Y. Hayafuji, T. Yamada, and Y. Aoki, Laser damage gettering and its application to lifetime improvement in silicon. *J. Electrochem. Soc.* **128**, 1975–1980 (1981).

212. D. Elwell and S. Hahn, Effect of laser backside damage upon mechanical properties of silicon. *J. Electrochem. Soc.* **131**, 1395–1400 (1984).

213. P. M. Sandow, VLSI applications of laser annealing. *Solid State Technol.* July, pp. 74–78 (1980).

214. K. Takemura, F. Toyokawa, Y. Ohshita, A. Ishitani, and H. Tsuya, A novel and high throughput extrinsic gettering technique of Si wafer using excimer laser irradiation. *In* "Gettering and Defect Engineering in the Semiconductor Technology" (H. Richter, ed.), pp. 318–322. Acad. Sci. G. D. R., Frankfurt, 1987.

215. C. M. Hsieh, J. R. Mathews, H. D. Seidel, K. A. Pickar, and C. M. Drum, Ion-implantation-damage gettering effect in silicon photodiode array camera target. *Appl. Phys. Lett.* **22**, 238–240 (1973).

216. S. Prussin, Ion implantation gettering: A fundamental approach. *Solid State Technol.* July, pp. 52–54 (1981).

217. T. E. Seidel, R. L. Meek, and A. G. Cullis, Direct comparison of ion-damage gettering and phosphorus diffusion gettering of Au in Si. *J. Appl. Phys.* **46**, 600–609 (1975).

218. H. J. Geipel and W. K. Tice, Critical microstructure for ion-implantation gettering effect in silicon. *Appl. Phys. Lett.* **30**, 325–327 (1977).

219. A. G. Cullis, T. E. Seidel, and R. L. Meek, Comparative study of annealed neon-, argon-, and krypton-ion implantation damage in silicon. *J. Appl. Phys.* **49**, 5188–5198 (1978).

220. K. Murase and H. Harada, Argon implantation gettering for a 'through-oxide' arsenic-implanted layer. *J. Appl. Phys.* **48**, 4404–4406 (1977).

221. K. D. Beyer and T. H. Yeh, Impurity gettering of silicon damage generated by ion implantation through SiO_2 layers. *J. Electrochem. Soc.* **129**, 2527–2530 (1982).

222. R. A. Moline and A. G. Cullis, Residual defects in Si produced by recoil implantation of oxygen. *Appl. Phys. Lett.* **26**, 551–553 (1975).

223. G. A. Rozgonyi, P. M. Petroff, and M. H. Read, Elimination of oxidation-induced stacking faults by preoxidation gettering of silicon wafers. I. Phosphorus diffusion-induced misfit dislocations. *J. Electrochem. Soc.* **122**, 1725–1729 (1975).

224. W. F. Tseng, T. Koji, J. W. Mayer, and T. E. Seidel, Simultaneous gettering of Au in silicon by phosphorus and dislocations. *Appl. Phys. Lett.* **33**, 442–444 (1978).

225. D. Lecrosnier, J. Paugam, F. Richou, G. Pelous, and F. Beniere, Influence of phosphorus-induced point defects on a gold-gettering mechanism in silicon. *J. Appl. Phys.* **51**, 1036–1038 (1980).

226. R. L. Meek and T. E. Seidel, Enhanced solubility and ion pairing of Cu and Au in heavily doped silicon at high temperatures. *J. Phys. Chem. Solids* **36**, 731–740 (1975).

227. M. C. Chen and V. J. Silvestri, Post-epitaxial polysilicon and Si_3N_4 gettering in silicon. *J. Electrochem. Soc.* **129**, 1294–1299 (1982).

228. G. Keefe-Fraundorf, D. E. Hill, and R. A. Craven, Backside gettering of defects in silicon. *Proc. EMTAS '83 Conf.* (*Soc. Manuf. Eng.*), *1983* EE83, pp. 131–143 (1983).

229. W. T. Stacy, M. C. Arst, K. N. Ritz, J. G. de Groot, and M. H. Norcott, The microstructure of polysilicon backsurface gettering. *In* "Defects in Silicon" (W. M. Bullis and L. C. Kimerling, eds.), pp. 423–432. Electrochem. Soc., Princeton, New Jersey, 1983.

230. D. E. Hill, Gettering of gold in silicon wafers using various backside gettering techniques. *In* "Defects in Silicon" (W. M. Bullis and L. C. Kimerling, eds.), pp. 433–441. Electrochem. Soc., Princeton, New Jersey, 1983.

231. P. M. Petroff, G. A. Rozgonyi, and T. T. Sheng, Elimination of process-induced stacking faults by preoxidation gettering of Si wafers. Si_3N_4 process. *J. Electrochem. Soc.* **123**, 565–570 (1976).

232. M. C. Chen and V. J. Silvestri, Pre- and postepitaxial gettering of oxidation and epitaxial stacking faults in silicon. *J. Electrochem. Soc.* **128**, 389–395 (1981).

233. K. Tanno, F. Shimura, and T. Kawamura, Microdefect elimination in reduced pressure epitaxy on silicon wafer by back-damage Si_3N_4 film technique. *J. Electrochem. Soc.* **128**, 395–399 (1981).

234. D. C. Gupta, Effect of enhanced gettering on device performance. *Solid State Technol.* Aug., pp. 149–151 (1983).

235. A. S. M. Salih, H. J. Kim, R. F. Davis, and G. A. Rozgonyi, Extrinsic gettering via the controlled introduction of misfit dislocations. *In* "Semiconductor Processing" (D. C. Gupta, ed.), pp. 272–282. Am. Soc. Test. Mater., Philadelphia, Pennsylvania, 1984).

236. G. A. Rozgonyi, R. P. Deysher, and C. W. Pearce, The identification, annihilation, and suppression of nucleation sites responsible for silicon epitaxial stacking faults. *J. Electrochem. Soc.* **123**, 1910–1915 (1976).

237. T. Y. Tan, E. E. Gardner, and W. K. Tice, Intrinsic gettering by oxide precipitate induced dislocations in Czochralski Si. *Appl. Phys. Lett.* **30**, 175–176 (1977).

238. W. K. Tice and T. Y. Tan, Precipitation of oxygen and intrinsic gettering in silicon. *In* "Defects in Semiconductors" (J. Narayan and T. Y. Tan, eds.), pp. 367–380. North-Holland Publ., Amsterdam, 1981.

239. H. J. Ruiz and G. P. Pollack, High-temperature annealing behavior of oxygen in silicon. *J. Electrochem. Soc.* **125**, 128–130 (1978).

240. U. Gösele and T. Y. Tan, The influence of point defects on diffusion and gettering in silicon. *In* "Impurity Diffusion and Gettering in Silicon" (R. B. Fair, C. W. Pearce, and J. Washburn, eds.), pp. 105–116. Mater. Res. Soc., Pittsburgh, 1985.

241. J. Gass, H. H. Muller, H. Stussi, and S. Schweitzer, Oxygen diffusion in silicon and the influence of different dopants. *J. Appl. Phys.* **51**, 2030–2037 (1980).

242. H. Tsuya, K. Ogawa, and F. Shimura, Improved intrinsic gettering technique for high-temperature-treated CZ silicon wafers. *Jpn. J. Appl. Phys.* **20**, L31–L34 (1981).

243. K. Nagasawa, Y. Matsushita, and S. Kishino, A new intrinsic gettering technique using microdefects in Czochralski silicon crystal: A new double preannealing technique. *Appl. Phys. Lett.* **37**, 622–624 (1980).

244. H. Peibt and H. Raidt, Nucleation of oxygen precipitations and efficiency of internal gettering centers in Czochralski silicon. *Phys. Status. Solidi* A **68**, 253–260 (1981).

245. F. R. N. Nabarro, "Theory of Crystal Dislocations." Oxford Univ. Press, London and New York, 1967.

246. W. T. Read, Jr., Theory of dislocations in germanium, *Philos. Mag.* **45**, 775–796 (1954).

247. F. C. Champion, Some physical consequences of elementary defects in diamonds, *Proc. R. Soc. London* A **234**, 541–556 (1956).

248. E. G. Colas and E. R. Weber, Reduction of iron solubility in silicon with oxygen precipitates. *Appl. Phys. Lett.* **48**, 1371–1373 (1986).

249. N. Nauka, J. Lagowski, H. C. Gatos, and C.-J. Li, Intrinsic gettering in oxygen-free silicon. *Appl. Phys. Lett.* **46**, 673–675 (1985).

250. K. Nauka, J. Lagowski, H. C. Gatos, and O. Ueda, New intrinsic gettering process in silicon based on interaction of silicon interstitials. *J. Appl. Phys.* **60**, 615–621 (1986).

251. O. Ueda, K. Nauka, J. Lagowski, and H. C. Gatos, Identification of intrinsic gettering centers in oxygen-free silicon crystals. *J. Appl. Phys.* **60**, 622–626 (1986).

252. M. Futagami, K. Hoshi, N. Isawa, T. Suzuki, Y. Okubo, Y. Kato, and Y. Okamoto, CMOS static RAM devices fabricated on a high oxygen-content MCZ wafer. *In* "Semiconductor Silicon 1986" (H. R. Huff, T. Abe, and B. Kolbesen, eds.), pp. 939–948. Electrochem. Soc., Princeton, New Jersey, 1986.

Chapter 8

Silicon Wafer Criteria for VLSI/ULSI Technology

Historically, it has been convenient to distinguish the different generations of electronic device technology according to the number of device components per chip. Very-large-scale integration (VLSI) and ultra-large-scale integration (ULSI have been generally defined as those that cover the ranges of 2^{16}–2^{21} (64K–2M) and 2^{21}–2^{26} (2M–64M) components.[1] In the beginning of 1984, several Japanese semiconductor device manufacturers demonstrated the fabrication of 1-Mbit DRAM and 256-kbit SRAM devices.[2] Although the term "VLSI" has been used since 1976,[3] the demonstration just mentioned may suggest that 1984 is really the first year of the VLSI era.[4] Three years later, in 1987, several IC manufacturers started the mass production of 1-Mbit DRAM devices, and announced the beginning of the ULSI era with the development of 4-Mbit and 16-Mbit DRAM.

As repeatedly described in earlier chapters, there are many possible causes for device yield loss; however, the interrelationship of IC design, fabrication processes, and silicon wafer parameters has been more critical with the onset of the VLSI/ULSI era. The standards required for silicon products are very stringent in order to achieve high yield in mass-produced VLSI/ULSI devices. Closer cooperation between manufacturers of silicon wafers and ICs is essential to ensure the availability of superior wafers for consistent circuit performance and fabrication line productivity.[5] Uniting many concepts presented in Refs. 4–7 and especially in Ref. 8, the silicon wafer criteria for VLSI/ULSI technology are discussed in this final chapter of this book.

8.1 High-Technology Silicon Wafer Concept

8.1.1 Demands for Silicon Wafers

The development of electronic devices has been achieved as the result of smaller design rules and increased functionality of integrated circuits. Sub-

Table 8.1 MOS DRAM Circuit Parameters and Technological Features with the Degree of Circuit Integration[a]

Parameter	ULSI	VLSI	LSI	MSI
Components/chip	10^7–10^9	10^5–10^7	10^3–10^5	10^2–10^3
Design rule (μm)	<1	1–3	3–5	5–10
Mask levels	12–18	8–15	6–10	5–6
Chip area (mm^2)	50–100	25–50	10–25	10
Storage cell dielectric thickness (Å)	100–150	150–400	400–900	900–1200
Junction depth (μm)	0.1–0.2	0.2–0.5	0.5–1.2	1.2–2.0

[a] After Huff.[7]

micrometer design rules and increased circuit density, however, have required concurrent improvements in various characteristics of silicon wafers to enhance device performance and production yield. For example, Table 8.1 summarizes some circuit parameters and technological features of MOS DRAM circuitry with the degree of circuit integration.[6] The IC characteristics have been used to guide the design of silicon wafers, which ultimately improve the consistency and performance of IC products.

8.1.2 Multizoned Wafer Structure

A multizoned silicon wafer has been designed for VLSI/ULSI fabrication with specific structural, chemical, and mechanical characteristics to support the IC density and performance goals for leading-edge MOS digital circuit applications such as megabit DRAMS.[4,6,7] A schematic illustration of the multizoned structure for CMOS- and bipolar-IC devices is shown in Fig. 8.1.[4] This structure consists of four principal zones in the silicon wafer used for VLSI/ULSI device fabrication. Device components are physically fabricated in the *device zone*. This zone may be either the polished wafer front surface—that is, a part of the *denuded zone* described below—or an epitaxial layer that has been almost exclusively used for bipolar ICs and has been attractively used for recent CMOS ICs as well. The device zone must be free of unwanted impurities, structural imperfections, and wafer strain. Moreover, this zone is critical to achieve the flatness parameters required to ensure the photolithographic depth of focus needed for submicrometer feature size resolution. Under the device zone, the denuded zone, in which electronic elements function, and an *intrinsic gettering* (IG) *zone* are formed by the combination of heat treatments including a process of oxygen outdiffusion at a high temperature. The IG zone, however, generally exhibits an incubation period until it becomes functionally effective. As was discussed in Section 7.2.4, this incubation period depends on the wafer characteristics and the

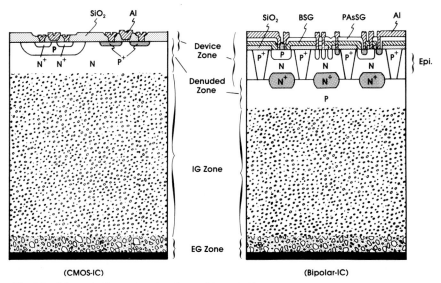

Fig. 8.1   Schematic illustration showing multizoned silicon wafers for MOS and bipolar VLSI/ULSI circuits. (After Shimura and Craven.[4])

detailed IC process conditions that control the extent of oxygen outdiffusion, precipitation, and formation of related IG sinks.

To enhance gettering and to resist warpage during thermal cycles, an *extrinsic gettering* (EG) *zone* with a backside film of polysilicon, possibly in combination with silicon nitride, is recommended. The application of a polysilicon film on the wafer back surface creates a long-lasting EG effect throughout the IC fabrication process. In addition, an oxide seal zone formed with a CVD-grown oxide film may be used to minimize autodoping phenomena.[6]

8.2   VLSI/ULSI Wafer Characteristics

8.2.1   Test-Device Electronic Parameters

General Remarks   The basic microscopic processes of leakage current generation are the same for both electric field–induced and metallurgical junctions. Minority carriers are generated at extended crystallographic imperfections and at point-defect generation–recombination centers within the bulk or near surface regions as well as at the SiO$_2$/Si interface. The magnitude of the leakage current is generally determined by the number of structural imperfections, and by their degree of metallic decoration, intersecting the active device space-charge region. The macroscopic manifestations of the leakage current, however, are significantly dependent on both the function of the electronic device and the circuit architecture. The sensitivity of both MOS and bipolar circuits to leakage currents has continuously increased due to the greater circuit complexity and evermore demanding application requirements. The effect of structural defects on electrical properties has been discussed in Section 7.3; however, it is useful to mention the test-device electronic parameters from the viewpoint of silicon wafer criteria.

Gate Oxide The silicon material, due to its influence on gate oxide quality, impacts two important device parameters: (1) oxide leakage current and (2) oxide breakdown voltage. The gate oxide leakage current should probably be less than 10^{-12} Å at 10 V.[9] Excessive gate oxide leakage current, greater than approximately 10^{-8} Å, and reduced oxide breakdown voltage have been shown to be related to oxidation-induced surface or near-surface defects. It has also recently been suggested that metallic precipitates reduce the oxide breakdown strength by causing a local thinning of the oxide thickness.[10] In any case, extrinsic gettering has been utilized to reduce significantly the oxide leakage current, and this also significantly shifted the oxide breakdown voltage distribution to higher voltages.[11] Polysilicon-gate LOCOS-configured structures simulating IC architectures, rather than metal gates, are preferred for assessing gate oxide quality. A test mask set with a variety of periphery-to-area ratios as well as a structure with a gate oxide area equal to the total thin oxide gate area is recommended to simulate the chip behavior. The improvement of oxide breakdown failure in the range $2-5 \times 10^6$ V/cm where circuits operate has been related to silicon substrate characteristics.[12,13] The yield variability for silicon wafers has been clearly indicated by the following analysis of the breakdown voltage data using log-normal statistics[14]:

$$\ln \ln(1 - P_f)^{-1} = K_1 \xi \tag{8.1}$$

where P_f is the cumulative probability of test-device failure at a given electric field ξ, and K_1 is a constant. It has been shown that EG-treated wafers result in significant reduction in low-electric-field breakdown events compared with nongettered wafers.[13] Epitaxial wafers even without EG have shown the best electric field characteristic in the range of $0-6 \times 10^6$ V/cm.[13] It might be expected, therefore, that the combination of epitaxial wafer and extrinsic gettering yields the best result. Intrinsic gettering has also been reported to be successful in improving the gate oxide breakdown characteristic.[12,15]

Sequential oxidations and oxide strips of CZ silicon have been shown to improve oxide breakdown and approach FZ silicon behavior.[16] Furthermore, it has recently been shown that significantly different starting silicon wafer characteristics dramatically influence the breakdown voltage distribution.[17] The experiments for investigating various influential factors on oxide breakdown characteristics for different silicon materials, including high $[O_i]_0$ CZ, low $[O_i]_0$ CZ, oxygen-lean MCZ, and FZ silicon wafers, however, have shown the difficulty in correlating the oxide integrity with the initial oxygen concentration.[18] That is, lower $[O_i]_0$ silicon material does not necessarily result in higher oxide breakdown voltage.

Time-dependent measurements of oxide breakdown for a fixed electric field, given by the following equation, are also beneficial for assessing oxide integrity[14]:

$$\ln \ln(1 - P_f)^{-1} = K_2 \ln t \tag{8.2}$$

where K_2 is a constant and t is the time.

The devices processed in low-particulate chemicals exhibit a reduction by a factor of approximately two in cumulative failure for times up to 10^5–10^6 sec. The importance of the quality of the chemicals and DI water utilized in wafer processing cannot be overemphasized. Reduced oxide breakdown voltage and spurious leakage currents may also occur when the interface state density is uncontrolled (i.e., greater than approximately 10^{11}–10^{12}/cm^2 eV) due to, for example, residual impurities from chemicals, DI water, and gases. Further research is required to understand oxide breakdown for thin (150–250 Å) oxides and its relationship to the silicon wafer, IC process conditions,[19] and method of measurement.[20,21]

Flat-Band Voltage Shift   The flat-band voltage shift is important in order to assess the electrically active sodium and potassium content in the oxide. These impurities may arise from various sources during wafer preparation and device processing. A measurement temperature of 300°C for about 5 min at a positive gate electric field of approximately 2.5×10^6 V/cm is useful to assess the presence of potassium and sodium ions in the oxide.[22] A 0.20-V shift is a practical upper limit for a metal gate product. Polysilicon gate oxides, especially when grown in a chlorine-containing oxidation ambient, exhibit improved device characteristics generally resulting in essentially zero flat-band voltage shift.[23,24]

Minority-Carrier Generation Lifetime   The minority-carrier generation lifetime optimization for DRAMs requires consideration of two apparently contradictory requirements.[25] In one case, a high lifetime is desired to facilitate the retention of charge for an extended period of time, that is, high circuit refresh time. On the other hand, the high-lifetime material is especially susceptible to discharge of the DRAM stored logic state due to transient electronic phenomena. For example, alpha-particle-induced minority-carrier currents result in the loss of stored information. Intrinsic gettering in conjunction with the formation of a surface denuded zone can reduce the influence of the transient phenomena and increase circuit performance and yield. This is accomplished by fabricating the circuit and its voltage-dependent space-charge region in a sufficiently deep but essentially defect-free surface zone of high-lifetime material.[15] The denuded zone should be sufficiently thin, however, so that the bulk gettering sites are close to the circuit region to ensure effective gettering. It should be noted that the bulk gettering sites can also generate minority carriers during device operation,[26] thereby requiring optimization of the denuded-zone depth for the specific circuit application. This procedure also has beneficial effects on CCD imager performance by collecting minority carriers generated by spurious radiation[27] and on CMOS circuit latch-up control by reducing the recombination lifetime in the substrate.[28]

Since ICs typically operate in the temperature range of 75°C, the contribution of the diffusion current from the bulk becomes an important factor due to the exponential increase in n_i^2 with increasing temperature. A significantly reduced bulk lifetime due to IG, for example, would increase the diffusion current; however, increased substrate doping would substantially decrease the diffusion current. The utilization of a lightly doped epitaxial layer on a heavily doped substrate is particularly useful for reducing the diffusion current. This material configuration is capable of achieving both a high generation lifetime in the vicinity of the surface and a low bulk diffusion current. The thickness of the epitaxial layer is also of critical importance.

A sufficiently high near-surface generation lifetime (300–1000 μsec) at room temperature may be indicative of acceptable material purity at the circuit operating temperature where the diffusion current would be expected to dominate. Although the achievement of "lifetime doping"[29] is still far away, improvements in process and gettering techniques have significantly improved minority-carrier lifetime during the past decade.[30]

Diode Leakage Current   The diode leakage current should probably be less than 10^{-10} Å at 10 V for typical test-device structures of approximately 10^{-3} cm². The leakage current i_L has been observed to be related to the OSF density N_{OSF} (cm²) as follows[31]:

$$i_L = 1.5 \times 10^{-8} N_{OSF}^{0.3} + i_0 \quad (A/cm^2) \tag{8.3}$$

where i_0 is approximately 1×10^{-9} A/cm². In particular, the leakage current has been found to depend on the number of stacking faults (N_I) intersecting the *p-n* junction boundaries[31]:

$$i_L = 1.5 \times 10^{-12} N_I^{2.2} + 2 \times 10^{-10} \quad (A) \tag{8.4}$$

Although this particular study indicated the stacking faults were not decorated by metallic impurities (or more properly, metallic decoration was not resolved by their chemical etching), it is generally believed that metallic decoration may play a key role in excessive leakage currents as was discussed in Section 7.3. In any case, it is extremely difficult to define a unique association between leakage current and the number of defects because both the degree of metallic decoration and defect location relative to the p-n junction appear to be important.

Threshold Voltage   The threshold voltage, transconductance, and body effect are three basic MOS device parameters affecting circuit performance such as access time and power dissipation.[32,33] Shifting of transistor threshold voltage by ion implantation is now extensively utilized throughout the IC industry. Typically, four different transistors may be fabricated: (1) unimplanted, (2) enhancement, boron only, (3) depletion, phosphorus only, and (4) enhancement or depletion depending on the relative dose and energy of both the boron and phosphorus implants. In some cases, the threshold voltage variation of a given transistor type may reflect the wafer resistivity gradient. Assuming no more than 0.03 V difference between two bits is desired, one can perform a model calculation for the difference in substrate doping N_A for a Long-channel transistor to account for the 0.03 V (assuming all other terms remain constant). The threshold voltage V_{th} for an unimplanted n-channel transistor may be written as[35]:

$$V_{th} = -\frac{Q_f}{C_{ox}} + 2\,E_{Fb} + \phi_{gs} + \frac{1}{C_{ox}}(2K_{Si}\varepsilon_0|e|N_A)^{1/2}(2E_{Fb} + |V_{BB}|)^{1/2} \quad (8.5)$$

where Q_f is the fixed interface charge per unit area at the SiO_2/Si interface (taken as $10^{10}/cm^2$); C_{ox} is the gate oxide capacitance per unit area (taken for a gate oxide thickness of 250 Å); E_{Fb} is the bulk Fermi energy relative to the intrinsic Fermi energy (assuming a bulk resistivity of 4 Ω cm); ϕ_{gs} is the work-function difference between the phosphorus-doped polysilicon gate and the p-type silicon substrate (taken as -0.87 V); K_{Si} is the dielectric constant of silicon (11.7); ε_0 is the dielectric permittivity of vacuum; and V_{BB} is the substrate back-gate bias (taken as -3.5 V).

The above parameters yield a threshold voltage of 0.26 V for 4-Ω cm silicon, which is indicative of the usefulness of ion implantation to shift the threshold voltage to higher values. In the present analysis, a positive 0.03-V shift in threshold voltage corresponds to an increase in N_A of 0.4×10^{15} cm^{-3}, or approximately 10%. Considering the potential influence of circuit design and IC fabrication process variations, it appears that in the worst case a 10% variation in dopant concentration can be tolerated. The importance of resistivity gradients for non–ion-implanted epitaxial layers should also be considered. Finally, the importance of uniform resistivity even within a single bit should be taken into account as regards depletion-region uniformity.[36]

8.2.2   Structural Characteristics

Czochralski silicon wafers of 150 mm diameter have become more commonly used in the IC industry. Inspection by X-ray topography should reveal no bulk structural defects in the virgin large-diameter silicon wafers. Although the electrical influence of these and additional surface irregularities[37] is not yet sufficiently understood, such irregularities have historically been reduced in the evolution toward a more perfect silicon wafer.

Dislocation-free silicon wafers are readily available. An OSF density less than $3/cm^2$, which corresponds to a total OSF of 530 for a 150-diameter wafer, is desired.[38] Metallic impurity clusters that originate surface microdefects can be easily eliminated by either EG or IG treatments; however, decoration of extended lattice defects such as dislocations and stacking faults is not as easily gettered.[38,39]

In some cases at the 1-Mbit and 4-Mbit levels, and most probably at the 16-Mbit level, traditional techniques in which NMOS devices are fabricated on silicon substrates will be replaced by either NMOS on epitaxial silicon or, more likely, CMOS on epitaxial silicon. This allows reduction of diffusion current of circuit area, minimization of static power dissipation, simplification of circuit design, reduction of latch-up, and increase of operational speed. Epitaxial silicon wafers should pass the same OSF and surface microdefect criteria currently being developed for substrate materials. In any case, EG is recommended to ensure reduced surface microdefects during subsequent IC processing. Polysilicon gettering is beneficial because it is continuous, clean, and its effectiveness does not anneal out, especially during high-temperature IC processing. Wafer uniformity as regards its structural and chemical characteristics and reproducibility from crystal to crystal is essential for reproducible device characteristics.[40]

8.2.3   Chemical Characteristics

Cleanliness   Wafer cleanliness is perhaps more important than ever before as the feature size approaches typical particle sizes. A front surface particle density less than $0.03/cm^2$ for 1-μm particles[41] and eventually for ≤ 0.5-μm particles appears necessary. In fact, it has recently been proposed that the particle size should be no larger than 10% of the desired linewidth.[42,43] Automated measurement methodologies will be required for more objective quantification, reproducible classification, and statistical understanding. A specification for back-surface particles less than $0.05/cm^2$ for ≤ 1-μm particles also appears important. Back-surface particles can interfere with effective chucking for optimized resolution and lithographic pattern fidelity. The particles can also be transferred to the front surface of other wafers in cleaning solutions and during thermal processing.

A wafer front surface should be haze-free, clean, and covered with a chemical or native oxide in order to avoid additional contamination from the environment. The detailed surface chemistry must be understood to ensure a reproducible hydrophilic surface. In particular, exposure of the native silicon surface to the ambient must be avoided because of increased electrostatic attraction to airborne contaminants. The characteristics of the native oxide and its surface chemistry are important for influencing subsequent thin oxide growth kinetics.[44] The related influence of preoxidation clean-ups on the oxide growth rate has also been reported.[45]

Impurities   Oxygen is a major impurity in a CZ silicon wafer. Carbon and nitrogen are subsidiary impurities influencing the mechanical strength as well as the oxygen precipitation. The interstitial oxygen content should be tailored to the silicon wafer user's specification. The exact value will depend on whether the user desires to utilize IG, which exhibits a critical value of approximately 14 ppma, although this value is dependent on many variables as discussed in Section 7.2.4.[46] A tolerance goal of ± 2 ppma oxygen may eventually be needed to minimize wafer-to-wafer variability in the precipitation of oxygen. Tight control of the oxygen concentration is especially important in the vicinity of the critical value for oxygen precipitation. In other words, the specification regarding $[O_i]_0$ should be especially tight around the critical value. Accordingly, it is also essential to establish the measurement accuracy within the required tolerance. A limit of the oxygen radial gradient to less than 3% is also important to ensure uniform circuit performance.[48] The key point is that the initial oxygen concentration and especially the oxygen precipitation must be controlled and reproducible to a specifically defined thermal test to ensure consistent IG while maintaining sufficient mechanical strength of the wafer. The concentration of carbon, which is another critical impurity, should be lowered at least to levels less than 0.6 ppma,[27] and perhaps as low as 0.3 ppma, in order to minimize its effect on oxygen precipitation.

The control of metallic impurities is also essential. Neutron activation analysis (NAA) measurements have indicated that the best 4-Kbit DRAM hold time was achieved with less than approximately 0.01 ppba total metallic impurities such as copper and gold.[48] Other investigations show that a generation lifetime of approximately 100 μsec is achieved for this gold concentration,[49] and that lifetimes higher than 1000 μsec can be achieved for fully processed and appropriately gettered silicon wafers.[30] It has recently been reported for an MOS device that the generation lifetime, surface generation velocity, and dielectric breakdown strength of SiO_2 are dramatically degraded when the surface iron concentration exceeds 1×10^{12}, 5×10^{12}, and 1×10^{13} atoms/cm^2, respectively.[50] Further work is required to ascertain the influence of critical levels for additional metallic impurities such as chromium, iron, and nickel. Although relating device lifetime to circuit refresh time is not necessarily straightforward, the metallic content during and at the end of the IC process, as well as that in incoming wafers, is critical. The preferential transfer of metallic impurities from heavily doped *p*-type to heavily doped *n*-type regions due to their increased solubility in *n*-type material has also been observed to decrease device and IC performance by the formation of emitter–collector pipes and excessive leakage currents.[51,52] The control of specific metallic impurities to levels less than 0.001 ppba in the bulk is estimated to be required for future ULSI circuits.[53]

8.2.4 Mechanical Characteristics

Wafer Dimensions In order to attain more chips, silicon wafers of 150 mm in diameter are becoming dominant in the silicon-based IC industry, and interest in 200-mm-diameter wafers has been accelerating. A diameter tolerance $\leq \pm 0.2$ mm will be required to ensure accurate placement of large diameter wafers onto stepper lithographic equipment and to minimize wafer breakage during robotic transfer operations. Improved length tolerance, $\leq \pm 1.5$ mm, in the orientation flat will also be required to facilitate automated step-and-repeat alignment. A thickness of 625 and 675 μm and a tolerance of $\leq \pm 10 \mu$m have become standard for 150-mm-diameter wafers. Larger-diameter wafers are thicker to ensure mechanical stability. Many possible scenarios must be considered when larger-diameter wafers are inserted into a previously established fabrication line. It was reported, for example, that thicker wafers (508 μm thick) exhibited less effective EG, for the identical IC process, compared with thinner wafers (355 μm thick).[54] It is therefore indicated that the effect of going to 200-mm-diameter wafers from 150-mm-diameter ones will be of extreme importance with respect to the effectiveness of EG. In addition, the combined influence of increased wafer thickness and lower-temperature IC processes on gettering effectiveness requires careful consideration as discussed in Section 7.4.

Flatness, Warpage, and Edge Contour The utilization of stepper lithographic technology has reemphasized the importance of the wafer mechanical properties, which ultimately affect the ability to resolve and reproduce 1-μm or submicrometer linewidths and spaces in varying topologies.[55] The group of properties generically referred to as *flatness* will become more important and perhaps more controversial than ever in order to ensure 1-μm or submicrometer feature size resolution. Wafer parallelism (i.e., taper) will be required to be less than 10 μm, especially when stepper processing employs back-surface referencing. The line resolution R_L and depth of focus δ_f are expressed by the following equations:

$$R_L = K\lambda/A_L \tag{8.6}$$

$$\delta_f = \pm \lambda/2A_L \tag{8.7}$$

where K is a constant, λ the exposure wavelength, and A_L the lens numerical aperture.

A global flatness of approximately 3–4 μm may be required for small feature size (~ 2 μm) projection printer applications. The local site flatness of approximately ± 1 μm over a 20×20 mm^2 field of view will ensure that each step and repeat field is sufficiently flat to enable lens focusing down to 1 μm linewidth resolution.[56] Optimizing the optical lithography and related etching processes is also required to ensure that the benefits of a specific wafer flatness are realized.[57] Of course, control of plastic deformation of the wafers during IC processing is essential. Increased wafer diameters require thicker wafers to ensure the control of plastic deformation and wafer warpage.[58,59]

The ability to control the polished surface curvature per customers' specifications may be necessary.[60] Bow and warp values less than 10 μm have been suggested to facilitate the desired patterning and registration in point-contact vacuum chuck step-and-repeat lithography. It should be noted, however, that differential front- and back-surface film stresses developed during IC processing as well as wafer warpage due to nonoptimized oxygen precipitation[61] can modify the wafer curvature. It has been reported that vacuum chucks are effective for a polished wafer with convex backside, rather than concave backside.[58,60] On the other hand, significantly reduced slip dislocations have been found in a wafer with concave backside, rather than convex backside.[59,62] These observations might require further clarification for the effect of wafer curvature on device performance.

The edge contour and shaping of the wafer are also important. Crack-free wafer edges minimize potential stress-raisers at edge cracks, which can increase the critical activation energy for slip and wafer breakage. They are also helpful in reducing a source of silicon dust generation and ensuring high throughput of IC automatic process handling equipment.[63,64]

8.3   Concluding Remarks

The wafer parameters are classified as electrical, structural, chemical, and mechanical material characteristics. The trends and target values for silicon wafers for VLSI/ULSI technology are summarized in Table 8.2.[65] It is important to understand the interrelationships among silicon material characteristics, IC fabrication, and circuit design in order to ensure the successful

Table 8.2   Selected Values and Trends of VSLI/ULSI Silicon Wafers

Material property	Value, trend
Electrical	
Oxide breakdown voltage	$\leq 1\%$ Failure for electric fields $\leq 5 \times 10^6$ V/cm
Flat-band voltage shift	≤ 0.2 V (Specifically defined metal gate test)
Generation lifetime	$\sim 300–1000$ μsec
Resistivity variation	$\leq 10\%$
Chemical	
Cleanliness	$\leq 0.03/cm^2$ Particles (≤ 0.5 μm) on wafer front surface; $\leq 0.05/cm^2$ (≤ 1 μm) on wafer back surface: no wafer back surface stain
Oxygen concentration	Customer-specified; ± 2 ppma
Oxygen radial gradient	$\leq 3\%$
Carbon concentration	≤ 0.3 ppma
Metals	
Bulk	≤ 0.001 ppba
Surface	$\leq 10^{11}/cm^2$ for specific metals
Structural	
Grown-in dislocation	$0/cm^2$
OSF	$\leq 3/cm^2$ (Specifically defined oxidation test)
Mechanical	
Diameter	≥ 150 mm
Tolerance	≤ 0.2 mm
Thickness	625, 675 μm
Tolerance	≤ 10 μm
Orientation flat tolerance	≤ 1.5 mm
Total thickness variation	≤ 10 μm
Global flatness	≤ 3 μm
Bow	≤ 10 μm
Warp	≤ 10 μm
Local site flatness	≤ 1.0 $\mu m/20 \times 20$ mm^2 field
Wafer curvature	Convex or concave specified by customer
Edge contour	Chip-free

fabrication of VLSI/ULSI circuits. This approach is believed to lead to an effective correlation of the starting-material characteristics with IC circuit performance and to the development of appropriate starting-material specifications with the ultimate goal of improving the consistency of IC performance and yield. The wafer characteristics for future circuits are expected to become even more stringent with the increasing pervasiveness of VLSI/ULSI technology.

The substitution of CMOS for NMOS designs to reduce static power dissipation and to simplify circuit design is becoming more commonplace for large and small system applications. CMOS will indeed become the mainstream technology for future VLSI/ULSI circuits. CMOS circuits, however, are susceptible to failure from latch-up at small p^+ to n^+ design rules. A variety of latch-up prevention methods have been investigated.[66,67] However, it appears that the utilization of epitaxial silicon wafers, perhaps in conjunction with additional IC process schemes such as trench isolation to raise the margin of latch-up trigger current (i.e., increase in the latch-up voltage), will provide a universal solution to the latch-up phenomenon. This approach will become more dominant as commercialization of the 4-Mbit and 16-Mbit DRAM is achieved. Design-rule limitations and related circuit performance considerations for even higher density bit DRAMs have been outlined recently.[1]

Ultimately, the growth of more perfect silicon crystals will be critical to obtain uniform, reproducible wafer characteristics in order to ensure consistent IC performance. Advanced crystal growth methodologies such as continuous pulling, the use of magnetic fields, and automated crystal growth algorithms may present useful improvements. The further understanding and correlation of IC malfunctions with test-device electronic parameters, and ultimately with silicon wafer parameters, is essential to effectively design the silicon wafer characteristics for future circuits. Closer working relationships will be required between the silicon wafer manufacturer and the IC manufacturer in order to effectively develop and fabricate advanced IC products.

Finally, although it may not be as straightforward, it should be noted that the above goals strongly depends on the development of advanced diagnostic techniques to effectively assess the device impact of silicon wafer characteristics as well as the individual circuit design and fabrication process technologies. Present state-of-the-art diagnostics, such as the detection of surface particles, evaluation of the native oxide, and the related analysis of impurities, may not be sufficient for VLSI/ULSI requirements. The development of improved diagnostic techniques will indeed be a major driving force for establishing VLSI and, ultimately, ULSI quality silicon.

References

1. S. Broydo and C. M. Osburn, eds., "ULSI Science and Technology/1987." Electrochem. Soc., Princeton, New Jersey, 1987.
2. *IEEE Int. Solid-State Circuit Conf., 1984* (1984).
3. S. M. Sze, ed., "VLSI Technology." McGraw-Hill, New York, 1983.
4. F. Shimura and R. A. Craven, Process-induced microdefects in VLSI silicon wafers. *In* "The Physics of VLSI" (J. C. Knights, ed.), pp. 205–219. Am. Inst. Phys., New York, 1984.

5. H. R. Huff and R. F. Holt, Computer automated IC manufacturing demands ULSI wafers. *Solid State Technol.* Sept., pp. 193–194 (1985).
6. H. R. Huff, Silicon wafers engineered for ULSI circuits. *Semicond. Int.* July, pp. 82–85 (1985).
7. J. E. Lawrence and H. R. Huff, Silicon material properties for VLSI circuitry. *In* "VLSI Electronics Microstructure Science" (N. G. Einspruch, ed.), Vol. 5, pp. 51–102. Academic Press, New York, 1982.
8. H. R. Huff and F. Shimura, Silicon material criteria for VLSI electronics. *Solid State Technol.* Mar., pp. 103–118 (1985).
9. P. S. D. Lin, R. B. Marcus, and T. T. Sheng, Leakage and breakdown in thin oxide capacitors—Correlation with decorated stacking faults. *J. Electrochem. Soc.* **130**, 1878–1883 (1983).
10. K. Honda, A. Ohsawa, and N. Toyokura, Breakdown in silicon oxide—Correlation with Cu precipitates. *Appl. Phys. Lett.* **45**, 270–271 (1984).
11. B. H. Yun, Improved oxide of metal-oxide-silicon capacitors resulting from backside argon implantation. *Appl. Phys. Lett.* **39**, 330–332 (1981).
12. K. Yamabe, K. Taniguchi, and Y. Matsushita, Thickness dependence of dielectric breakdown failure of thermal SiO_2 films. *In* "Defects in Silicon" (W. M. Bullis and L. C. Kimerling, eds.), pp. 629–638. Electrochem. Soc., Princeton, New Jersey, 1983.
13. H. R. Huff, J. Glick, and P. Trammel, Breakdown voltage and lifetime studies on LOCOS configured polysilicon gate oxides. *J. Electrochem. Soc.* **130**, 240C–242C (1983).
14. D. Wolters, T. Hoogestyn, and H. Kraaij, MOS wearout and breakdown statistics. *In* "The Physics of MOS Insulators" (G. Lucovsky, S. T. Pantelides, and F. L. Galeener, eds.), pp. 349–352. Pergamon, Oxford, 1980.
15. H. Ohtsuka, M. Nakamura, and M. Watanabe, Intrinsic gettering technique for MOS VLSI fabrication. *Nikkei Electron.*, Aug. 31, pp. 138–154 (1981) (in Japanese).
16. M. Itsumi and F. Kiyosumi, Origin and elimination of defects in SiO_2 thermally grown on Czochralski silicon substrates. *Appl. Phys. Lett.* **40**, 496–498 (1982).
17. H. Abe, F. Kiyosumi, K. Yoshioka, and M. Ino, Analysis of defects in thin SiO_2 thermally grown on Si substrate. *Proc. IEDM* pp. 372–375 (1985).
18. F. Shimura, M. P. Guse, and R. J. Crepin, unpublished.
19. D. A. Baglee and P. L. Shah, Ultra thin-gate dielectric processes for VLSI applications. *In* "VLSI Electronics Microstructure Science" (N. G. Einspruch, ed.), Vol. 7, Academic Press, New York, 1983.
20. P. Heiman, An operational definition for breakdown of thin thermal oxides. *IEEE Trans. Electron Devices* **ED-30**, 1366–1368 (1983).
21. E. Harari, Conduction of trapping of electrons in highly stressed ultrathin films of thermal SiO_2. *Appl. Phys. Lett.* **30**, 601–603 (1977).
22. J. P. Stagg, Drift mobilities of Na^+ and K^+ ions in SiO_2 films. *Appl. Phys. Lett.* **31**, 532–533 (1977).
23. C. M. Osburn and E. Bassous, Improved dielectric reliability of SiO_2 films with polycrystalline silicon electrodes. *J. Electrochem. Soc.* **122**, 89–92 (1975).
24. D. W. Ormond, Dielectric breakdown of silicon dioxide thin film capacitors using polycrystalline silicon and aluminum electrodes. *J. Electrochem. Soc.* **126**, 162–164 (1979).
25. H. Otsuka, K. Watanabe, H. Nishimura, H. Iwai, and H. Nihira, The effect of substrate materials on holding time degradation in MOS dynamic RAMs. *IEEE Trans. Electron Device Lett.* **EDL-3**, 182–184 (1982).
26. S. N. Chakravarti, P. L. Garbarino, and K. Murty, Oxygen precipitation effects on Si n^+p junction leakage behavior. *Appl. Phys. Lett.* **40**, 581–583 (1982).
27. M. Ogino, T. Usami, M. Watanabe, H. Sekine, and T. Kawaguchi, Two-step thermal anneal and its application to a CCD sensor and CMOS LSI. *J. Electrochem. Soc.* **130**, 1397–1402 (1983).

28. C. N. Anagnostopoulos, E. T. Nelson, J. P. Lavine, K. Y. Wang, and D. D. Nichols, Latch-up and image crosstalk suppression by internal gettering. *IEEE Trans. Electron Devices* **ED-31**, 225–231 (1984).

29. E. Spenke, History and future needs in silicon technology. *In* "Semiconductor Silicon 1969" (R. R. Haberecht and E. L. Kern, eds.), pp. 1–35. Electrochem. Soc., Princeton, New Jersey, 1969.

30. P. K. Chatterjee, G. W. Taylor, A. F. Tasch, Jr., and H.-S. Fu, Leakage studies in high-density dynamic MOS memory devices. *IEEE J. Solid-State Circuits* **SC-14**, 486–498 (1979).

31. S. P. Murarka, T. E. Seidel, J. V. Dalton, J. M. Dishman, and M. H. Read, A study of stacking faults during CMOS processing: Origin, elimination and contribution to leakage. *J. Electrochem. Soc.* **127**, 716–724 (1980).

32. C. H. Stapper and P. B. Hwang, Simulation of FET device parameters for LSI manufacturing. *In* "Semiconductor Silicon 1977" (H. R. Huff and E. Sirtl, eds.), pp. 955–967. Electrochem. Soc., Princeton, New Jersey, 1977.

33. J. T. Clemens and E. F. Labuda, A statistical analysis of the threshold voltage of a double insulator (Al_2O_3/SiO_2) IGFET integrated circuit technology. *In* "Semiconductor Silicon 1973" (H. R. Huff and R. R. Burgess, eds.), pp. 779–790. Electrochem. Soc., Princeton, New Jersey, 1973.

34. Deleted in proof.

35. P. Richman, "MOS Field-Effect Transistors and Integrated Circuits." Wiley (Interscience), New York, 1973.

36. J. R. Brews, W. Fichtner, E. H. Nicollian, and S. M. Sze, A generalized guide for MOSFET miniaturization. *IEEE Electron Device Lett.* **EDL-1**, 2–4 (1980).

37. K. Kugimiya, Characterization of microdeformation and crystal defects in silicon wafer surfaces. *J. Electrochem. Soc.* **130**, 2123–2125 (1983).

38. E. J. Janssens and G. L. Declerck, The mechanisms of lifetime improvement by HCl oxidation. *In* "Semiconductor Characterization Techniques" (P. A. Barnes and G. A. Rozgonyi, eds.), pp. 376–385. Electrochem. Soc., Princton, New Jersey, 1978.

39. C. W. Pearce, L. E. Katz, and T. E. Seidel, Considerations regarding gettering in integrated circuits. *In* "Semiconductor Silicon 1981" (H. R. Huff, R. J. Kriegler, and Y. Takeishi, eds.), pp. 705–723. Electrochem. Princeton, New Jersey, 1981.

40. C. J. Varker, Electrical microscopy of test parameter inhomogeneities resulting from microdefects in processed silicon wafers. *IEEE Trans. Electron Devices* **ED-27**, 2205–2212 (1980).

41. A. P. Lane, Particle sources and control methods in components and materials. *Standard Electron. Lab. Presentation* Aug. 22 (1984).

42. M. Parikh and U. Kaempf, SMIF: A technology for wafer cassette transfer in VLSI manufacturing. *Solid State Technol.* July, pp. 111–115 (1984).

43. M. Tosa, What are the problems in semiconductor materials? *Denshi Zairyo* Aug., pp. 22–25 (1984) (in Japanese).

44. K. K. Ng, W. J. Polito, and J. R. Ligenza, Growth kinetics of thin silicon dioxide in a controlled ambient oxidation system. *Appl. Phys. Lett.* **44**, 626–628 (1984).

45. F. N. Schwettmann, K. L. Chiang, and W. A. Brown, Variation of silicon dioxide growth rate with pre-oxidation clean. *Ext. Abstr., Electrochem. Soc. Meet.* Vol. 78-1, pp. 688–689 (1978).

46. F. Shimura and H. Tsuya, Oxygen precipitation factors in silicon. *J. Electrochem. Soc.* **129**, 1062–1066 (1982).

47. H. R. Huff, H. F. Shaake, J. T. Robinson, S. C. Baber, and D. Wong, Some observations on oxygen precipitation/gettering in device processed Czochralski silicon. *J. Electrochem. Soc.* **130**, 1551–1555 (1983).

48. L. E. Katz, P. F. Schmidt, and C. W. Pearce, Neutron activation study of a gettering treatment for Czochralski silicon substrates. *J. Electrochem. Soc.* **128**, 620–624 (1981).
49. F. Richou, G. Pelous, and D. Lecrosnier, Thermal generation of carriers in gold-doped silicon. *J. Appl. Phys.* **51**, 6252–6257 (1980).
50. R. Takizawa, T. Nakanishi, and A. Ohsawa, Degradation of metal-oxide-semiconductor devices caused by iron impurities on the silicon wafer surface. *J. Appl. Phys.* **62**, 4933–4935 (1987).
51. J. A. Amick, Copper degradation of silicon devices and the behavior of copper in silicon. *Ext. Abstr. Electrochem. Soc. Meet.*, Vol. 76-2, pp. 874–877 (1976).
52. P. J. Ward, A survey of ion contamination in silicon substrates and its impact on circuit yield. *J. Electrochem. Soc.* **129**, 2573–2576 (1982).
53. J. A. Keenan and G. B. Larrabee, Characterization of silicon materials of VLSI. *In* "VLSI Electronics Microstructure Science" (N. G. Einspruch and G. B. Larrabee, eds.), Vol. 6, pp. 1–72. Academic Press, New York, 1983.
54. J. M. Dishman, S. E. Haszko, R. B. Marcus, S. P. Muraka, and T. T. Sheng, Electrically active stacking faults in CMOS integrated circuits. *J. App. Phys.* **50**, 2689–2696 (1979).
55. L. Denes, The effect of wafer flatness on yield by off-line computer simulation of the microphotolithographic process. *In* "Semiconductor Processing" (D. C. Gupta, ed.), pp. 143–159. Am. Soc. Test. Mater., Philadelphia, Pennsylvania, 1984.
56. S. Sasayama, Automation and high resolution lithography. *SEMI West Tech. Proc.* pp. 5–10 (1984).
57. H. R. Rottmann, Stepper pattern size errors at one micron. *Kodak Microelectron. Semin.-—Interface '83* (1983).
58. S. Takasu, Silicon wafer for VLSI. *In* "VLSI Science and Technology/1984" (K. E. Bean and G. A. Rozgonyi, eds.), pp. 490–499. Electrochem. Soc., Princeton, New Jersey, 1984.
59. B. Leroy and C. Plougonvan, Warpage of silicon wafers, *J. Electrochem. Soc.* **127**, 961–970 (1980).
60. S. Takasu, Wafer design. *In* "Semiconductor Technologies" (J. Nishizawa, ed.), pp. 1–19. North-Holland Publ., Amsterdam, 1984.
61. K. G. Moerschel, C. W. Pearce, and R. E. Reusser, A study on the effects of oxygen content, initial bow, and furnace processing on warpage of three-inch diameter silicon wafers. *In* "Semiconductor Silicon 1977" (H. R. Huff and E. Sirtl, eds.), pp. 170–181. Electrochem. Soc., Princeton, New Jersey, 1977.
62. L. D. Dyer, H. R. Huff, and W. W. Boyd, Plastic deformation in central regions of epitaxial silicon slices. *J. Appl. Phys.* **42**, 5680–5688 (1971).
63. W. R. Bottom and J. S. Wenstrand, Trends in wafer fab and their driving economic forces. *Solid State Technol.* Aug., pp. 173–180 (1983).
64. T. Makimoto and H. Nagatomo, Automation in semiconductor manufacturing. *International Electron Device Meeting, 1982*, pp. 11–15 (1982).
65. F. Shimura and H. R. Huff, VLSI silicon material criteria. *In* "VLSI Handbook" (N. G. Einspruch, ed.), pp. 191–269. Academic Press, New York, 1985.
66. A. G. Lewis, Latch up suppression in fine-dimension shallow p-well CMOS circuits. *IEEE Trans. Electron Devices* **ED-31**, 1472–1481 (1984).
67. T. Yamaguchi, S. Morimoto, G. H. Kawamoto, and J. C. DeLacy, Process and device performance of 1 μm-channel n-well CMOS technology. *IEEE Trans. Electron Devices* **ED-31**, 205–214 (1984).

Appendixes

A	Activity of nuclide
A_L	Lens numerical aperture
B	Magnetic field
B	Magnetic flux density
b	Burgers vector
c	Light velocity
C	Capacitance
D	Diffusion coefficient
d	Interplanar distance
E	Young's modulus, electric field strength, or energy
E_c	Energy of conduction band bottom
E_F	Fermi energy level
E_g	Energy bandgap
E_n	Energy level
E_v	Energy of valence band top
e	Absolute value of electron charge
F	Lorentz force
F	Force
F_d	Defect factor
f_c	Conversion factor
$f(E)$	Fermi–Dirac distribution function
G	Gibbs free energy
G_g	Microscopic crystal growth rate
G_m	Melt level drop rate
G_p	Crystal pulling rate
G_r	Grashof number
G_s	Macroscopic solidification rate
g	Fraction solidified
g	Gravity acceleration
$\mathbf{g}_{hkl}$	Diffraction vector
H	Enthalpy
Ha	Hartman number
h	Planck's constant
I	Intensity

(continues)

Appendix I   (*Continued*)

I_0	Initial intensity
i	Electric current
J	Impurity flux
J_e	Electric current density
K	Thermal conductivity
K_{ox}	Dielectric constant of SiO_2
K_{Si}	Dielectric constant of Si
k	Boltzmann's constant
k_{eff}	Effective segregation coefficient
k_0	Equilibrium segregation coefficient
l	Azimuthal quantum number
M	Nucleus mass
M_a	Angular momentum
M_p	Particle momentum
m_e	Electron mass
m_h	Hole mass
m^*	Effective mass of carrier
m_1	Magnetic quantum number
m_0	Electron rest mass
N	Coordination number
$[N]$	Doping concentration
N_A^-	Ionized acceptor concentration
N_c	Effective state density in conduction band
N_D^+	Ionized donor concentration
$N(E)$	State density
N_v	Effective state density in valence band
n	Principle quantum number
$\bar{n}$	Reflective index
n_e	Electron density
n_h	Hole density
n_i	Intrinsic Carrier density
P_f	Cumulative probability of test device failure
Pr	Prandtl number
Q	Activation energy
Q_f	Fixed oxide charge
Q_{it}	Interface trapped charge
Q_m	Mobil ionic charge
Q_{ot}	Oxide trapped charge
R_c	Circuit resistance
R_g	Generation rate
R_L	Line resolution
R_p	Total parallel resistance
R_R	Recombination rate
R_s	Sample resistance
R_{sp}	Spreading resistance
Ra	Rayleigh number
Re	Reynolds number
r	Radius

S	Entropy
s	Spin quantum number
$\mathbf{s}$	Surface recombination velocity
T	Temperature
t_r	Relaxation time
V	Voltage or volume
V_0	Potential energy or excitation voltage
v	Velocity
W	Pattern width
w_n	Weight fraction
Z	Atomic number
z	Valence
α	Absorption coefficient
β	Thermal expansion coefficient
δ	Thickness of diffusion boundary layer
δ_f	Depth of focus
ε_0	Dielectric permittivity of vacuum
ε_{ox}	Dielectric permittivity of silicon dioxide
ε_{Si}	Dielectric permittivity of silicon
θ_B	Bragg angle
λ	Wavelength
λ_i	Decay constant
μ	Particle mass
μ_a	Linear absorption coefficient
μ_D	Drift mobility
μ_e	Electron mobility
μ_n	Hole mobility
ν	Frequency
ν_k	Kinematical viscosity
ξ	Electric field
ρ	Resistivity
ρ_c	Crystal density
ρ_m	Melt density
ρ_s	Density of substance
σ	Conductivity
σ_i	Cross section
σ_T	Thermal stress
τ	Lifetime
τ_g	Generation lifetime
τ_r	Recombination lifetime
τ_{LY}	Lower yield stress
τ_{UY}	Upper yield stress
Φ	Irradiation particle flux
ϕ	Work function
ϕ_{bi}	Built-in potential
ψ	Wave function
ω	Rotation rate

Appendix II List of Abbreviations

AEM	Analytical electron microscopy (or microscope)
AES	Auger electron spectroscopy
bcc	Body-centered cubic
CCD	Charge-coupled device
CCZ	Continuous-charging Czochralski (crystal growth method)
CG	Chemical gettering
CMOS	Complementary metal oxide semiconductor (transistor)
CPAA	Charged-particle activation analysis
CTEM	Conventional transmission electron microscopy (or microscope)
CVD	Chemical vapor deposition
CZ	Czochralksi (crystal growth method)
DI	Deionized (water)
DRAM	Dynamic random-access memory
EBIC	Electron beam induced current
EDX	Energy-dispersive X-ray spectrometer
EELS	Electron energy loss spectroscopy
EG	Extrinsic (or external) getting
EMF	Electromotive force
ESCA	Electron spectroscopy for chemical analysis
fcc	Face-centered cubic
FPD	Focal-plane deviation
FTIR	Fourier-transform infrared (spectrometer)
FZ	Float-zone (crystal growth method)
HMCZ	Horizontal magnetic-field-applied Czochralski (crystal growth method)
HRTEM	High-resolution transmission electron microscopy (or microscope)
HVEM	High-voltage electron microscopy (or microscope)
IC	Integrated circuit
IG	Intrinsic (or internal) gettering
IGFET	Insulated-gate field-effect transistor
IR	Infrared (absorption)
IMA or IMMA	Ion microprobe mass analysis (IMA or IMMA $\equiv$ SIMS)
LPCVD	Low-pressure chemical vapor deposition
LPE	Liquid-phase epitaxy
LSI	Large-scale integration
MBE	Molecular-beam epitaxy
MCCG	Magnetic-field-applied continuous-charging Czochralski (crystal growth method)
MCZ	Magnetic-field-applied Czochralski (crystal growth method)
MG-Si	Metallurgical-grade silicon
MIS	Metal insulator semiconductor (transistor)
MISFET	Metal insulator semiconductor field-effect transistor
MOS	Metal oxide semiconductor (transistor)
MOSFET	Metal oxide semiconductor field-effect transistor
MOST	Metal oxide semiconductor transistor
MSI	Medium-scale integration
NAA	Neutron activation analysis
NMOS	n-Channel metal oxide semiconductor (transistor)
OISF or OSF	Oxidation-induced stacking fault
PDC	Precipitate dislocation complex

PEM	Photoelectromagnetic (effect)
PMOS	*p*-Channel metal oxide semiconductor (transistor)
POGO	Preoxidation gettering at other side
RAE	Resistive anode encoder
RAM	Random-access memory
RIE	Reactive-ion etching
RTA	Rapid thermal annealing
RTP	Rapid thermal processing
SEM	Scanning electron microscopy (or microscope)
SG-Si	Semiconductor-grade silicon
SIMS	Secondary ion mass spectrometry ($\equiv$ IMA or IMMA)
SN	Signal-to-noise (ratio)
SOI	Silicon epitaxy on insulator
SOT	Scanning oscillator technique
SPE	Solid-phase epitaxy
SRAM	Static random-access memory
SRH	Schockley–Read–Hall
SSI	Small-scale integration
STEM	Scanning transmission electron microscopy (or microscope)
TCE	Trichloroethylen
TEM	Transmission electron microscopy (or microscope)
TIR	Total indicating reading
TTV	Total thickness variation
ULSI	Ultra-large-scale integration
VLSI	Very-large-scale integration
VMCZ	Vertical magnetic-field-applied Czochralski (crystal growth method)
VPE	Vapor phase epitaxy
XRC	X-ray rocking curve
XRT	X-ray topography

Appendix III   Basic Physical Constants

Absolute elementary charge (e)	1.602×10^{-20} emu
Boltzmann's constant (k)	0.863×10^{-4} eV K^{-1}
Electron rest mass (m_0)	9.109×10^{-28} g
Permittivity of vacuum (ε_0)	8.854×10^{-14} F cm^{-1}
Permittivity of Si (ε_{Si})	$1.04 \ \times 10^{-12}$ F cm^{-1}
Permittivity of SiO$_2$ (ε_{ox})	$3.4 \ \times 10^{-13}$ F cm^{-1}
Planck's constant (h)	6.626×10^{-27} erg sec
Speed of light (c)	2.998×10^{10} cm sec^{-1}

Appendix IV Single Bond Energy of Relevant Elements[a]

Bond	eV/bond	Bond	eV/bond
H—H	4.51	Si—H	3.05
C—C	3.60	Si—C	3.01
O—O	1.44	Si—O	3.83
Cl—Cl	2.52	Si—Cl	3.72
Ge—Ge	1.63	Si—S	2.35
N—N	1.67	Si—F	5.61
P—P	2.23	Si—I	2.21
As—As	1.39	Si—Br	3.00
Si—Si	1.83		

[a] From L. Pauling, "The Nature of the Chemical Bond." Cornell Univ. Press, Ithaca, New York, 1960.

Appendix V Thermo-mechanical Properties of Crystalline Silicon[a]

Hardness (orientation dependent)	7 Moh, 1000 Vickers, 950–1150 Knoop
Elastic constants	C_{11}:1.6740 × 10^{12} dyn/cm^2 C_{12}:0.6523 × 10^{12} dyn/cm^2 C_{44}:0.7959 × 10^{12} dyn/cm^2
Temperature coefficients of elastic constants	KC_{11}: $-75 \times 10^{-6}/°C$ KC_{12}: $-24.5 \times 10^{-6}/°C$ KC_{44}: $-55.5 \times 10^{-6}/°C$
Young's modulus	1.9 × 10^{12} dyn/cm^2, [111] direction
Modulus of rupture (in bending)	700–3500 kg/cm^2
Breaking strength (in compression)	4900–5600 kg/cm^2
Linear thermal coefficient of expansion	2.33 × 10$^{-6}/°C$
Surface tension	720 dyn/cm (freezing point)

[a] From W. R. Runyan, "Silicon Semiconductor Technology." McGraw-Hill, New York, 1965.

Appendix VI Properties of Deposited Silicon Dioxide[a]

	Deposition			
	Plasma	$SiH_4 + O_2$	$Si(OC_2H_5)_4$	$SiCl_2H_2 + N_2O$
Temperature (°C)	200	450	700	900
Composition	$SiO_{1.9}(H)$	$SiO_2(H)$	SiO_2	$SiO_2(Cl)$
Density (g/cm³)	2.3	2.1	2.2	2.2
Refractive index	1.47	1.44	1.46	1.46
Dielectric strength (10^6 V/cm)	3–6	8	10	10
Etch rate (100:1 = H_2O:HF) (Å/min)	400	60	30	30

[a] From A. C. Adams, Dielectric and polysilicon film deposition, in "VLSI Technology" (S. M. Sze, ed.), pp. 93–129 McGraw-Hill, New York, 1983.

Appendix VII   Diffusivity of Elements in SiO_2[a]

Element	Diffusivity at 1100°C (cm^2/sec)	Diffusivity at 1100°C (cm^2/sec)
B	3×10^{-17} to 2×10^{-14}	2×10^{-16} to 5×10^{-14}
Ga	5.3×10^{-11}	5×10^{-8}
P	2.9×10^{-16} to 2×10^{-13}	2×10^{-15} to 7.6×10^{-13}
Sb	9.9×10^{-17}	1.5×10^{-14}

[a] From M. Ghezzo and D. M. Brown, Diffusivity of B, Ga, P, As, and Sb in SiO_2, *J. Electrochem. Soc.* **120**, 146–148 (1973).

Appendix VIII   Color Chart for Thermally Grown SiO_2 Films Observed Perpendicularly under Daylight Fluorescent Lighting

Film thickness (μm)	Color and comment	Film thickness (μm)	Color and comment
0.05_0	Tan	0.72	Blue-green to green (quite broad)
0.07_5	Brown		
0.10_0	Dark violet to red-violet	0.77	"Yellowish"
0.12_5	Royal blue	0.80	Orange (rather broad for orange)
0.15_0	Light blue to metallic blue		
0.17_5	Metallic to very light yellow-green	0.82	Salmon
0.20_0	Light gold or yellow (slightly metallic)	0.85	Dull, light red-violet
		0.86	Violet
0.22_5	Gold with slight yellow-orange	0.87	Blue-violet
0.25_0	Orange to melon	0.89	Blue
0.27_5	Red-violet	0.92	Blue-green
0.30_0	Blue to violet-blue	0.95	Dull yellow-green
0.31_0	Blue	0.97	Yellow to "yellowish"
0.32_5	Blue to blue-green	0.99	Orange
0.34_5	Light green	1.00	Carnation pink
0.35_0	Green to yellow-green	1.02	Violet-red
0.36_5	Yellow-green	1.05	Red-violet
0.39	Yellow	1.06	Violet
0.41_2	Light orange	1.07	Blue-violet
0.42_6	Carnation pink	1.10	Green
0.44_3	Violet-red	1.11	Yellow-green
0.46_5	Red-violet	1.12	Green
0.47_6	Violet	1.18	Violet
0.48_0	Blue-violet	1.19	Red-violet
0.49_3	Blue	1.21	Violet-red
0.50_2	Blue-green	1.24	Carnation pink to salmon
0.54_0	Yellow-green	1.25	Orange
0.56	Green-yellow	1.28	"Yellowish"
0.57_4	Yellow to "yellowish"	1.32	Sky blue to green-blue
0.58_5	Light orange or yellow to pink borderline	1.40	Orange
		1.45	Violet
0.60_0	Carnation pink	1.46	Blue-violet
0.63_0	Violet-red	1.50	Blue
0.68	"Bluish"	1.54	Dull yellow-green

[a] From W. A. Pliskin and E. E. Conrad, Nondestructive determination of thickness and refractive index of transparent films, *IBM J. Res. Dev.* **8**, 43–51 (1964).

Appendix IX Typical Reaction for Depositing Dielectric and Polysilicon Films[a]

Product	Reactants	Deposition temperature (°C)
Silicon dioxide	$SiH_4 + CO_2 + H_2$	850–950
	$SiCl_2H_2 + N_2O$	850–900
	$SiH_4 + N_2O$	750–850
	$SiH_4 + NO$	650–750
	$Si(OC_2H_5)_4$	650–750
	$SiH_4 + O_2$	400–450
Silicon nitride	$SiH_4 + NH_3$	700–900
	$SiCl_2H_2 + NH_3$	650–750
Plasma silicon nitride	$SiH_4 + NH_3$	200–350
	$SiH_4 + N_2$	200–350
Plasma silicon dioxide	$SiH_4 + N_2O$	200–350
Polysilicon	SiH_4	600–650

[a] From A. C. Adams, Dielectric and polysilicon film deposition, *in* "VLSI Technology" (S. M. Sze, ed.), p. 93–129 McGraw-Hill, New York, 1983.

Appendix X Properties of Silicon Nitride[a]

	LPCVD	Plasma
Temperature (°C)	700–800	250–350
Composition	$Si_3N_4(H)$	SiN_xH_y
Si/N ratio	0.75	0.8–1.2
Atomic % of H	4–8	20–25
Refractive index	2.01	1.8–2.5
Density (g/cm^3)	2.9–3.1	2.4–2.8
Dielectric constant	6–7	6–9
Resistivity (Ω cm)	10^{16}	10^6–10^{15}
Dielectric strength (10^6 V/cm)	10	5
Energygap (eV)	5	4–5

[a] From A. C. Adams, Dielectric and polysilicon film deposition, *in* "VLSI Technology" (S. M. Sze, ed.), pp. 93–129 McGraw-Hill, New York, 1983.

Appendix XI   Lowest Binary Eutectic Temperatures and Resistivities of Several Silicides[a]

Silicide	Lowest binary eutectic temperature ($^\circ$C)	Specific resistivity ($10^{-6}\ \Omega$ cm)
WSi_2	1440	70
$MoSi_2$	1410	100
$TaSi_2$	1385	35–45
VSi_2	1385	50–55
$ZrSi_2$	1355	35–40
$TiSi_2$	1330	13–25
$HfSi_2$	1300	45–50
$NbSi_2$	1295	50
$CoSi_2$	1195	18–25
$NiSi_2$	966	50–60
$PtSi$	830	28–35
Pd_2Si_2	720	30–35

[a] From S. P. Muraka, Refractory silicides for integrated circuits, *J. Vac. Sci. Technol.* **17**, 775 (1980).

Appendix XII   Hydrogen Peroxide-Based Immersion Cleaning Procedure for Silicon Product[a]

A. Preliminary cleaning (if necessary)
  1. Remove bulk of photoresist film (if present) by plasma oxidation stripping, or immersion in organic photoresist stripper, or with a hot 1:2 v/v H_2O_2–H_2SO_4 mixture if adequate safety precautions are exercised.
  2. Rinse with water (see note on water purity for entire processing).
  3. Transfer the wafers to a clean Teflon holder. Pick up wafers with Teflon or polypropylene plastic tweezers.
B. Removal of residual organic contaminants and certain metals
  1. Prepare a fresh mixture of H_2O:NH_4OH:H_2O_1 = 5:1:1 (*solution 1*) by measuring the following reagents into a beaker of fused silica (opaque silica ware is acceptable):
     a. volumes of water
     b. 1 volume of ammonium hydroxide (29%, electronic grade, w/w based on NH_3)
     c. 1 volume of ammonium hydroxide (30%, unstabilized electronic grade, w/w)
  2. Stir the solution with a clean rod of fused quartz.
  3. Submerge holder with wafers in the cold solution and place the beaker on a hot plate.
  4. Heat to 75 to 80°C. Then reduce heating to maintain the solution at 80°C for an additional 10 min. (Vigorous bubbling is due to oxygen evolution. Make sure not to boil the solution so as to prevent rapid decomposition of the H_2O_2 and volatilization of the ammonia.)
  5. Overflow-quench the solution by placing the beaker under running water for about 1 min.
  6. Remove holder with wafers and immediately place it in a cascade water rinse tank for 5 min.

(continues)

Appendix XII   (*Continued*)

C.   Stripping of thin hydrous oxide film ($HF:H_2O = 1:50$)
1.   Submerge wafer assembly from step B-6 directly in an agitated mixture of 1 volume hydrofluoric acid (49%, electronic grade) and 50 volumes of water.
2.   Allow to remain in the solution for only 15 sec. Exposed silicon should repel the HF solution. Use a polypropylene beaker for this step.
3.   Transfer the wafer assembly to a water tank, but rinse for only 20 to 30 sec with agitation to remove the HF solution (this minimizes regrowth of a hydrous oxide film).
4.   Transfer the wafer assembly immediately, without drying, into the hot *solution 2* of step D.

D.   Desorption of remaining atomic and ionic contaminants
1.   Prepare a fresh mixture of $H_2O:HCl:H_2O_2 = 6:1:1$ (*solution 2*) by measuring the following reagents into a beaker of fused quartz:
     a.   6 volumes of water
     b.   1 volume of hydrochloric acid (37%, electronic grade)
     c.   1 volume of hydrogen peroxide (30%, unstabilized, electronic grade)
2.   Place the beaker on a hot plate and heat to 75 to 80°C.
3.   Submerge the still-wet wafers in the holder (after step B-6 or C-3) in the hot solution.
4.   Maintain the solution at 80°C for 10 to 15 min.
5.   Overflow-quench as in step B-5.
6.   Continue the rinsing at this stage for a total of 20 min in a cascade rinsing system.

E.   Drying of the wafers
1.   Transfer the holder with the wet wafers into a wafer centrifuge.
2.   Apply a final water rinse during spinning.
3.   Allow to dry while gradually increasing the spinning speed (to avoid aerosol formation from the water droplets).
4.   Remove the wafers by dump transfer for high-temperature processing. If single-wafer handling must be used, handle the wafers only at the edge with plastic tweezers.

F.   Storage
1.   Avoid storage of cleaned wafers, preferably by immediate continuation of processing. If storage is unavoidable, store the wafers in closed glass containers cleaned with hot *solution 1*, followed by water rinsing and oven-drying.

Note concerning processing water and reagents: All water used for preparing the reagent mixtures or for rinsing should be thoroughly deionized and ultrafiltered, with a resistivity in the 10 to 20 MΩ range at 18 to 23°C. All reagents should be electronic grade, preferably ultrafiltered for particulate impurities.

---

[a] From W. Kern and D. A. Puotinen, Cleaning solutions based on hydrogen peroxide for use in silicon semiconductor technology, *RCA Rev.* **31**, 187–206 (1970); W. Kern, Hydrogen peroxide solutions for silicon wafer cleaning, *RCA Engineer* **28-4**, 99–105 (1983); W. Kern, Purifying Si and SiO$_2$ surfaces with hydrogen peroxide, *Semiconductor International*, April 1984, pp. 94–99.

Index